T0185877

Graduate Texts in Mathematics

Series Editors:

Sheldon Axler
San Francisco State University, San Francisco, CA, USA

Kenneth Ribet
University of California, Berkeley, CA, USA

Advisory Board:

Alejandro Adem, *University of British Columbia*
David Eisenbud, *University of California, Berkeley & MSRI*
Irene M. Gamba, *The University of Texas at Austin*
J.F. Jardine, *University of Western Ontario*
Jeffrey C. Lagarias, *University of Michigan*
Ken Ono, *Emory University*
Jeremy Quastel, *University of Toronto*
Fadil Santosa, *University of Minnesota*
Barry Simon, *California Institute of Technology*

Graduate Texts in Mathematics bridge the gap between passive study and creative understanding, offering graduate-level introductions to advanced topics in mathematics. The volumes are carefully written as teaching aids and highlight characteristic features of the theory. Although these books are frequently used as textbooks in graduate courses, they are also suitable for individual study.

More information about this series at http://www.springer.com/series/136

Jean-François Le Gall

Brownian Motion, Martingales, and Stochastic Calculus

 Springer

Jean-François Le Gall
Département de Mathématiques
Université Paris-Sud
Orsay Cedex, France

Translated from the French language edition:
'Mouvement brownien, martingales et calcul stochastique' by Jean-François Le Gall
Copyright © Springer-Verlag Berlin Heidelberg 2013
Springer International Publishing is part of Springer Science+Business Media
All Rights Reserved.

ISSN 0072-5285   ISSN 2197-5612   (electronic)
Graduate Texts in Mathematics
ISBN 978-3-319-80961-8   ISBN 978-3-319-31089-3   (eBook)
DOI 10.1007/978-3-319-31089-3

Mathematics Subject Classification (2010): 60H05, 60G44, 60J65, 60H10, 60J55, 60J25

© Springer International Publishing Switzerland 2016
Softcover reprint of the hardcover 1st edition 2016
This work is subject to copyright. All rights are reserved by the Publisher, whether the whole or part of
the material is concerned, specifically the rights of translation, reprinting, reuse of illustrations, recitation,
broadcasting, reproduction on microfilms or in any other physical way, and transmission or information
storage and retrieval, electronic adaptation, computer software, or by similar or dissimilar methodology
now known or hereafter developed.
The use of general descriptive names, registered names, trademarks, service marks, etc. in this publication
does not imply, even in the absence of a specific statement, that such names are exempt from the relevant
protective laws and regulations and therefore free for general use.
The publisher, the authors and the editors are safe to assume that the advice and information in this book
are believed to be true and accurate at the date of publication. Neither the publisher nor the authors or
the editors give a warranty, express or implied, with respect to the material contained herein or for any
errors or omissions that may have been made.

Printed on acid-free paper

This Springer imprint is published by Springer Nature
The registered company is Springer International Publishing AG Switzerland

Preface

This book originates from lecture notes for an introductory course on stochastic calculus taught as part of the master's program in probability and statistics at Université Pierre et Marie Curie and then at Université Paris-Sud. The aim of this course was to provide a concise but rigorous introduction to the theory of stochastic calculus for continuous semimartingales, putting a special emphasis on Brownian motion. This book is intended for students who already have a good knowledge of advanced probability theory, including tools of measure theory and the basic properties of conditional expectation. We also assume some familiarity with the notion of uniform integrability (see, for instance, Chapter VII in Grimmett and Stirzaker [30]). For the reader's convenience, we record in Appendix A2 those results concerning discrete time martingales that we use in our study of continuous time martingales.

The first chapter is a brief presentation of Gaussian vectors and processes. The main goal is to arrive at the notion of a Gaussian white noise, which allows us to give a simple construction of Brownian motion in Chap. 2. In this chapter, we discuss the basic properties of sample paths of Brownian motion and the strong Markov property with its classical application to the reflection principle. Chapter 2 also gives us the opportunity to introduce, in the relatively simple setting of Brownian motion, the important notions of filtrations and stopping times, which are studied in a more systematic and abstract way in Chap. 3. The latter chapter discusses continuous time martingales and supermartingales and investigates the regularity properties of their sample paths. Special attention is given to the optional stopping theorem, which in connection with stochastic calculus yields a powerful tool for lots of explicit calculations. Chapter 4 introduces continuous semimartingales, starting with a detailed discussion of finite variation functions and processes. We then discuss local martingales, but as in most of the remaining part of the course, we restrict our attention to the case of continuous sample paths. We provide a detailed proof of the key theorem on the existence of the quadratic variation of a local martingale. Chapter 5 is at the core of this book, with the construction of the stochastic integral with respect to a continuous semimartingale, the proof in this setting of the celebrated Itô formula, and several important applications (Lévy's characterization theorem for Brownian motion, the Dambis–Dubins–Schwarz representation of a continuous martingale as a time-changed Brownian motion, the Burkholder–Davis–Gundy inequalities, the representation of Brownian martingales as stochastic integrals, Girsanov's theorem and the Cameron–Martin formula, etc.). Chapter 6, which presents the fundamental ideas of the theory of Markov processes with emphasis on the case of Feller semigroups, may appear as a digression to our main topic. The results of this chapter, however, play an important role in Chap. 7, where we combine tools of the theory of Markov processes with techniques of stochastic calculus to investigate connections of Brownian motion with partial differential equations, including the probabilistic solution of the classical Dirichlet problem. Chapter 7 also derives the conformal invariance of planar Brownian motion and applies this property to the skew-product decomposition, which in turn leads to asymptotic laws such as the celebrated Spitzer theorem on Brownian windings. Stochastic differential equations, which are another very important application of stochastic calculus and in fact motivated Itô's invention of this theory, are studied in detail in Chap. 8, in the case of Lipschitz continuous coefficients. Here again the general theory developed in Chap. 6 is used in our study of the Markovian properties of solutions of stochastic differential equations, which play a crucial role in many applications. Finally, Chap. 9 is devoted to local times of continuous semimartingales. The construction of local times in this setting, the study of their regularity properties, and the proof of the density of occupation formula are very convincing illustrations of the power of stochastic calculus techniques. We conclude Chap. 9 with a brief discussion of Brownian local times, including Trotter's theorem and the famous Lévy theorem identifying the law of the local time process at level 0.

A number of exercises are listed at the end of every chapter, and we strongly advise the reader to try them. These exercises are especially numerous at the end of Chap. 5, because stochastic calculus is primarily a technique, which can only be mastered by treating a sufficient number of explicit calculations. Most of the exercises come from exam problems for courses taught at Université Pierre et Marie Curie and at Université Paris-Sud or from exercise sessions accompanying these courses.

Although we say almost nothing about applications of stochastic calculus in other fields, such as mathematical finance, we hope that this book will provide a strong theoretical background to the reader interested in such applications. While presenting all tools of stochastic calculus in the general setting of continuous semimartingales, together with some of the most important developments of the theory, we have tried to keep this text to a reasonable size, without making any concession to mathematical rigor. The reader who wishes to go further in the theory and applications of stochastic calculus may consult the classical books of Karatzas and Shreve [49], Revuz and Yor [70], or Rogers and Williams [72]. For a historical perspective on the development of the theory, we recommend Itô's original papers [41] and McKean's book [57], which greatly helped to popularize Itô's work. More information about stochastic differential equations can be found in the books by Stroock and Varadhan [77], Ikeda and Watanabe [35], and Øksendal [66]. Stochastic calculus for semimartingales with jumps, which we do not present in this book, is treated in Jacod and Shiryaev [44] or Protter [63] and in the classical treatise of Dellacherie and Meyer [13, 14]. Many other references for further reading appear in the notes and comments at the end of every chapter.

I wish to thank all those who attended my stochastic calculus lectures in the last 20 years and who contributed to this book through their questions and comments. I am especially indebted to Marc Yor, who left us too soon. Marc taught me most of what I know about stochastic calculus, and his numerous remarks helped me to improve this text.

Orsay, France Jean-François Le Gall
January 2016

Interconnections between chapters

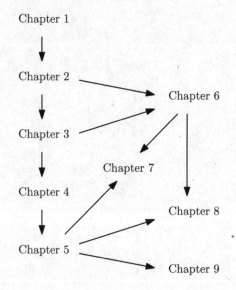

The original version of this book was revised. An erratum can be found at DOI
10.1007/978-3-319-31089-3_10

Contents

Contents   xiii

Chapter 1
Gaussian Variables and Gaussian Processes

Gaussian random processes play an important role both in theoretical probability and in various applied models. We start by recalling basic facts about Gaussian random variables and Gaussian vectors. We then discuss Gaussian spaces and Gaussian processes, and we establish the fundamental properties concerning independence and conditioning in the Gaussian setting. We finally introduce the notion of a Gaussian white noise, which will be used to give a simple construction of Brownian motion in the next chapter.

1.1 Gaussian Random Variables

Throughout this chapter, we deal with random variables defined on a probability space $(\Omega, \mathscr{F}, P)$. For some of the existence statements that follow, this probability space should be chosen in an appropriate way. For every real $p \geq 1$, $L^p(\Omega, \mathscr{F}, P)$, or simply L^p if there is no ambiguity, denotes the space of all real random variables X such that $|X|^p$ is integrable, with the usual convention that two random variables that are a.s. equal are identified. The space L^p is equipped with the usual norm.

A real random variable X is said to be a *standard Gaussian* (or *normal*) variable if its law has density

$$p_X(x) = \frac{1}{\sqrt{2\pi}} \exp(-\frac{x^2}{2})$$

with respect to Lebesgue measure on $\mathbb{R}$. The complex Laplace transform of X is then given by

$$E[e^{zX}] = e^{z^2/2}, \qquad \forall z \in \mathbb{C}.$$

© Springer International Publishing Switzerland 2016
J.-F. Le Gall, *Brownian Motion, Martingales, and Stochastic Calculus*,
Graduate Texts in Mathematics 274, DOI 10.1007/978-3-319-31089-3_1

To get this formula (and also to verify that the complex Laplace transform is well defined), consider first the case when $z = \lambda \in \mathbb{R}$:

$$E[e^{\lambda X}] = \frac{1}{\sqrt{2\pi}} \int_{\mathbb{R}} e^{\lambda x} e^{-x^2/2} \, dx = e^{\lambda^2/2} \frac{1}{\sqrt{2\pi}} \int_{\mathbb{R}} e^{-(x-\lambda)^2/2} \, dx = e^{\lambda^2/2}.$$

This calculation ensures that $E[e^{zX}]$ is well-defined for every $z \in \mathbb{C}$, and defines a holomorphic function on $\mathbb{C}$. By analytic continuation, the identity $E[e^{zX}] = e^{z^2/2}$, which is true for every $z \in \mathbb{R}$, must also be true for every $z \in \mathbb{C}$.

By taking $z = i\xi$, $\xi \in \mathbb{R}$, we get the characteristic function of X:

$$E[e^{i\xi X}] = e^{-\xi^2/2}.$$

From the expansion

$$E[e^{i\xi X}] = 1 + i\xi E[X] + \cdots + \frac{(i\xi)^n}{n!} E[X^n] + O(|\xi|^{n+1}),$$

as $\xi \to 0$ (this expansion holds for every $n \geq 1$ when X belongs to all spaces L^p, $1 \leq p < \infty$, which is the case here), we get

$$E[X] = 0, \quad E[X^2] = 1$$

and more generally, for every integer $n \geq 0$,

$$E[X^{2n}] = \frac{(2n)!}{2^n n!}, \quad E[X^{2n+1}] = 0.$$

If $\sigma > 0$ and $m \in \mathbb{R}$, we say that a real random variable Y is *Gaussian* with $\mathcal{N}(m, \sigma^2)$-distribution if Y satisfies any of the three equivalent properties:

(i) $Y = \sigma X + m$, where X is a standard Gaussian variable (i.e. X follows the $\mathcal{N}(0, 1)$-distribution);
(ii) the law of Y has density

$$p_Y(y) = \frac{1}{\sigma\sqrt{2\pi}} \exp{-\frac{(y-m)^2}{2\sigma^2}};$$

(iii) the characteristic function of Y is

$$E[e^{i\xi Y}] = \exp(im\xi - \frac{\sigma^2}{2}\xi^2).$$

We have then

$$E[Y] = m, \quad \text{var}(Y) = \sigma^2.$$

By extension, we say that Y is Gaussian with $\mathcal{N}(m, 0)$-distribution if $Y = m$ a.s. (property (iii) still holds in that case).

Sums of independent Gaussian variables Suppose that Y follows the $\mathcal{N}(m, \sigma^2)$-distribution, Y' follows the $\mathcal{N}(m', \sigma'^2)$-distribution, and Y and Y' are independent. Then $Y + Y'$ follows the $\mathcal{N}(m + m', \sigma^2 + \sigma'^2)$-distribution. This is an immediate consequence of (iii).

Proposition 1.1 *Let $(X_n)_{n \geq 1}$ be a sequence of real random variables such that, for every $n \geq 1$, X_n follows the $\mathcal{N}(m_n, \sigma_n^2)$-distribution. Suppose that X_n converges in L^2 to X. Then:*

(i) *The random variable X follows the $\mathcal{N}(m, \sigma^2)$-distribution, where $m = \lim m_n$ and $\sigma = \lim \sigma_n$.*
(ii) *The convergence also holds in all L^p spaces, $1 \leq p < \infty$.*

Remark The assumption that X_n converges in L^2 to X can be weakened to convergence in probability (and in fact the convergence in distribution of the sequence $(X_n)_{n \geq 1}$ suffices to get part (i)). We leave this as an exercise for the reader.

Proof

(i) The convergence in L^2 implies that $m_n = E[X_n]$ converges to $E[X]$ and $\sigma_n^2 = \mathrm{var}(X_n)$ converges to $\mathrm{var}(X)$ as $n \to \infty$. Then, setting $m = E[X]$ and $\sigma^2 = \mathrm{var}(X)$, we have for every $\xi \in \mathbb{R}$,

$$E[e^{i\xi X}] = \lim_{n \to \infty} E[e^{i\xi X_n}] = \lim_{n \to \infty} \exp\left(im_n\xi - \frac{\sigma_n^2}{2}\xi^2\right) = \exp\left(im\xi - \frac{\sigma^2}{2}\xi^2\right),$$

showing that X follows the $\mathcal{N}(m, \sigma^2)$-distribution.
(ii) Since X_n has the same distribution as $\sigma_n N + m_n$, where N is a standard Gaussian variable, and since the sequences (m_n) and (σ_n) are bounded, we immediately see that

$$\sup_n E[|X_n|^q] < \infty, \qquad \forall q \geq 1.$$

It follows that

$$\sup_n E[|X_n - X|^q] < \infty, \qquad \forall q \geq 1.$$

Let $p \geq 1$. The sequence $Y_n = |X_n - X|^p$ converges in probability to 0 and is uniformly integrable because it is bounded in L^2 (by the preceding bound with $q = 2p$). It follows that this sequence converges to 0 in L^1, which was the desired result.

$\square$

1.2   Gaussian Vectors

Let E be a d-dimensional Euclidean space (E is isomorphic to $\mathbb{R}^d$ and we may take $E = \mathbb{R}^d$, with the usual inner product, but it will be more convenient to work with an abstract space). We write $\langle u, v \rangle$ for the inner product in E. A random variable X with values in E is called a *Gaussian vector* if, for every $u \in E$, $\langle u, X \rangle$ is a (real) Gaussian variable. (For instance, if $E = \mathbb{R}^d$, and if $X_1, \ldots, X_d$ are are independent Gaussian variables, the property of sums of independent Gaussian variables shows that the random vector $X = (X_1, \ldots, X_d)$ is a Gaussian vector.)

Let X be a Gaussian vector with values in E. Then there exist $m_X \in E$ and a nonnegative quadratic form q_X on E such that, for every $u \in E$,

$$E[\langle u, X \rangle] = \langle u, m_X \rangle,$$
$$\mathrm{var}(\langle u, X \rangle) = q_X(u).$$

Indeed, let $(e_1, \ldots, e_d)$ be an orthonormal basis on E, and write $X = \sum_{i=1}^d X_j e_j$ in this basis. Notice that the random variables $X_j = \langle e_j, X \rangle$ are Gaussian. It is then immediate that the preceding formulas hold with $m_X = \sum_{j=1}^d E[X_j] e_j \overset{\text{(not.)}}{=} E[X]$, and, if $u = \sum_{j=1}^d u_j e_j$,

$$q_X(u) = \sum_{j,k=1}^d u_j u_k \, \mathrm{cov}(X_j, X_k).$$

Since $\langle u, X \rangle$ follows the $\mathcal{N}(\langle u, m_X \rangle, q_X(u))$-distribution, we get the characteristic function of the random vector X,

$$E[\exp(\mathrm{i}\langle u, X \rangle)] = \exp(\mathrm{i}\langle u, m_X \rangle - \frac{1}{2} q_X(u)). \qquad (1.1)$$

Proposition 1.2 *Under the preceding assumptions, the random variables $X_1, \ldots, X_d$ are independent if and only if the covariance matrix $(\mathrm{cov}(X_j, X_k))_{1 \le j,k \le d}$ is diagonal or equivalently if and only if q_X is of diagonal form in the basis $(e_1, \ldots, e_d)$.*

Proof If the random variables $X_1, \ldots, X_d$ are independent, the covariance matrix $(\mathrm{cov}(X_j, X_k))_{j,k=1,\ldots d}$ is diagonal. Conversely, if this matrix is diagonal, we have for every $u = \sum_{j=1}^d u_j e_j \in E$,

$$q_X(u) = \sum_{j=1}^d \lambda_j u_j^2 \,,$$

where $\lambda_j = \text{var}(X_j)$. Consequently, using (1.1),

$$E\Big[\exp\Big(i\sum_{j=1}^{d} u_j X_j\Big)\Big] = \prod_{j=1}^{d} \exp(iu_j E[X_j] - \frac{1}{2}\lambda_j u_j^2) = \prod_{j=1}^{d} E[\exp(iu_j X_j)],$$

which implies that $X_1, \ldots, X_d$ are independent. $\qquad\square$

With the quadratic form q_X, we associate the unique symmetric endomorphism γ_X of E such that

$$q_X(u) = \langle u, \gamma_X(u)\rangle$$

(the matrix of γ_X in the basis $(e_1, \ldots, e_d)$ is $(\text{cov}(X_j, X_k))_{1 \le j,k \le d}$ but of course the definition of γ_X does not depend on the choice of a basis). Note that γ_X is nonnegative in the sense that its eigenvalues are all nonnegative.

From now on, to simplify the statements, we restrict our attention to centered Gaussian vectors, i.e. such that $m_X = 0$, but the following results are easily adapted to the non-centered case.

Theorem 1.3

(i) *Let γ be a nonnegative symmetric endomorphism of E. Then there exists a Gaussian vector X such that $\gamma_X = \gamma$.*

(ii) *Let X be a centered Gaussian vector. Let $(\varepsilon_1, \ldots, \varepsilon_d)$ be a basis of E in which γ_X is diagonal, $\gamma_X \varepsilon_j = \lambda_j \varepsilon_j$ for every $1 \le j \le d$, where*

$$\lambda_1 \ge \lambda_2 \ge \cdots \ge \lambda_r > 0 = \lambda_{r+1} = \cdots = \lambda_d$$

so that r is the rank of γ_X. Then,

$$X = \sum_{j=1}^{r} Y_j \varepsilon_j,$$

where Y_j, $1 \le j \le r$, are independent (centered) Gaussian variables and the variance of Y_j is λ_j. Consequently, if P_X denotes the distribution of X, the topological support of P_X is the vector space spanned by $\varepsilon_1, \ldots, \varepsilon_r$. Furthermore, P_X is absolutely continuous with respect to Lebesgue measure on E if and only if $r = d$, and in that case the density of X is

$$p_X(x) = \frac{1}{(2\pi)^{d/2}\sqrt{\det \gamma_X}} \exp -\frac{1}{2}\langle x, \gamma_X^{-1}(x)\rangle.$$

Proof

(i) Let $(\varepsilon_1, \ldots, \varepsilon_d)$ be an orthonormal basis of E in which γ is diagonal, $\gamma(\varepsilon_j) = \lambda_j \varepsilon_j$ for $1 \leq j \leq d$, and let $Y_1, \ldots, Y_d$ be independent centered Gaussian variables with $\mathrm{var}(Y_j) = \lambda_j$, $1 \leq j \leq d$. We set

$$X = \sum_{j=1}^{d} Y_j \varepsilon_j.$$

Then, if $u = \sum_{j=1}^{d} u_j \varepsilon_j$,

$$q_X(u) = E\left[\left(\sum_{j=1}^{d} u_j Y_j \right)^2 \right] = \sum_{j=1}^{d} \lambda_j u_j^2 = \langle u, \gamma(u) \rangle.$$

(ii) Let $Y_1, \ldots, Y_d$ be the coordinates of X in the basis $(\varepsilon_1, \ldots, \varepsilon_d)$. Then the matrix of γ_X in this basis is the covariance matrix of $Y_1, \ldots, Y_d$. The latter covariance matrix is diagonal and, by Proposition 1.2, the variables $Y_1, \ldots, Y_d$ are independent. Furthermore, for $j \in \{r+1, \ldots, d\}$, we have $E[Y_j^2] = 0$ hence $Y_j = 0$ a.s.

Then, since $X = \sum_{j=1}^{r} Y_j \varepsilon_j$ a.s., it is clear that $\mathrm{supp} \, P_X$ is contained in the subspace spanned by $\varepsilon_1, \ldots, \varepsilon_r$. Conversely, if O is a rectangle of the form

$$O = \{u = \sum_{j=1}^{r} \alpha_j \varepsilon_j : a_j < \alpha_j < b_j, \ \forall 1 \leq j \leq r\},$$

we have $P[X \in O] = \prod_{j=1}^{r} P[a_j < Y_j < b_j] > 0$. This is enough to get that $\mathrm{supp} \, P_X$ is the subspace spanned by $\varepsilon_1, \ldots, \varepsilon_r$.

If $r < d$, since the vector space spanned by $\varepsilon_1, \ldots, \varepsilon_r$ has zero Lebesgue measure, the distribution of X is singular with respect to Lebesgue measure on E. Suppose that $r = d$, and write Y for the random vector in $\mathbb{R}^d$ defined by $Y = (Y_1, \ldots, Y_d)$. Note that the bijection $\varphi(y_1, \ldots, y_d) = \sum y_j \varepsilon_j$ maps Y to X. Then, writing $y = (y_1, \ldots, y_d)$, we have

$$E[g(X)] = E[g(\varphi(Y))]$$

$$= \frac{1}{(2\pi)^{d/2}} \int_{\mathbb{R}^d} g(\varphi(y)) \exp\left(-\frac{1}{2} \sum_{j=1}^{d} \frac{y_j^2}{\lambda_j} \right) \frac{dy_1 \ldots dy_d}{\sqrt{\lambda_1 \ldots \lambda_d}}$$

$$= \frac{1}{(2\pi)^{d/2} \sqrt{\det \gamma_X}} \int_{\mathbb{R}^d} g(\varphi(y)) \exp\left(-\frac{1}{2} \langle \varphi(y), \gamma_X^{-1}(\varphi(y)) \rangle \right) dy_1 \ldots dy_d$$

$$= \frac{1}{(2\pi)^{d/2} \sqrt{\det \gamma_X}} \int_E g(x) \exp\left(-\frac{1}{2} \langle x, \gamma_X^{-1}(x) \rangle \right) dx,$$

since Lebesgue measure on E is by definition the image of Lebesgue measure on $\mathbb{R}^d$ under φ (or under any other vector isometry from $\mathbb{R}^d$ onto E). In the second equality, we used the fact that $Y_1, \ldots, Y_d$ are independent Gaussian variables, and in the third equality we observed that

$$\langle \varphi(y), \gamma_X^{-1}(\varphi(y)) \rangle = \langle \sum_{j=1}^d y_j \varepsilon_j, \sum_{j=1}^d \frac{y_j}{\lambda_j} \varepsilon_j \rangle = \sum_{j=1}^d \frac{y_j^2}{\lambda_j}.$$

$\square$

1.3   Gaussian Processes and Gaussian Spaces

From now on until the end of this chapter, we consider only centered Gaussian variables, and we frequently omit the word "centered".

Definition 1.4 A (centered) *Gaussian space* is a closed linear subspace of $L^2(\Omega, \mathscr{F}, P)$ which contains only centered Gaussian variables.

For instance, if $X = (X_1, \ldots, X_d)$ is a centered Gaussian vector in $\mathbb{R}^d$, the vector space spanned by $\{X_1, \ldots, X_d\}$ is a Gaussian space.

Definition 1.5 Let $(E, \mathscr{E})$ be a measurable space, and let T be an arbitrary index set. A *random process* (indexed by T) with values in E is a collection $(X_t)_{t \in T}$ of random variables with values in E. If the measurable space $(E, \mathscr{E})$ is not specified, we will implicitly assume that $E = \mathbb{R}$ and $\mathscr{E} = \mathscr{B}(\mathbb{R})$ is the Borel σ-field on $\mathbb{R}$.

Here and throughout this book, we use the notation $\mathscr{B}(F)$ for the Borel σ-field on a topological space F. Most of the time, the index set T will be $\mathbb{R}_+$ or another interval of the real line.

Definition 1.6 A (real-valued) random process $(X_t)_{t \in T}$ is called a (centered) *Gaussian process* if any finite linear combination of the variables X_t, $t \in T$ is centered Gaussian.

Proposition 1.7 *If $(X_t)_{t \in T}$ is a Gaussian process, the closed linear subspace of L^2 spanned by the variables X_t, $t \in T$, is a Gaussian space, which is called the Gaussian space generated by the process X.*

Proof It suffices to observe that an L^2-limit of centered Gaussian variables is still centered Gaussian, by Proposition 1.1. $\square$

We now turn to independence properties in a Gaussian space. We need the following definition.

Definition 1.8 Let H be a collection of random variables defined on $(\Omega, \mathscr{F}, P)$. The σ-field *generated by H*, denoted by $\sigma(H)$, is the smallest σ-field on Ω such that all variables $\xi \in H$ are measurable for this σ-field. If $\mathscr{C}$ is a collection of subsets of Ω, we also write $\sigma(\mathscr{C})$ for the smallest σ-field on Ω that contains all elements of $\mathscr{C}$.

The next theorem shows that, in some sense, independence is equivalent to orthogonality in a Gaussian space. This is a very particular property of the Gaussian distribution.

Theorem 1.9 *Let H be a centered Gaussian space and let $(H_i)_{i \in I}$ be a collection of linear subspaces of H. Then the subspaces H_i, $i \in I$, are (pairwise) orthogonal in L^2 if and only the σ-fields $\sigma(H_i), i \in I$, are independent.*

Remark It is crucial that the vector spaces H_i are subspaces of a common Gaussian space H. Consider for instance a random variable X distributed according to $\mathscr{N}(0, 1)$ and another random variable ε independent of X and such that $P[\varepsilon = 1] = P[\varepsilon = -1] = 1/2$. Then $X_1 = X$ and $X_2 = \varepsilon X$ are both distributed according to $\mathscr{N}(0, 1)$. Moreover, $E[X_1 X_2] = E[\varepsilon]E[X^2] = 0$. Nonetheless X_1 and X_2 are obviously not independent (because $|X_1| = |X_2|$). In this example, (X_1, X_2) is not a Gaussian vector in $\mathbb{R}^2$ despite the fact that both coordinates are Gaussian variables.

Proof Suppose that the σ-fields $\sigma(H_i)$ are independent. Then, if $i \neq j$, if $X \in H_i$ and $Y \in H_j$,

$$E[XY] = E[X]E[Y] = 0,$$

so that the linear spaces H_i are pairwise orthogonal.

Conversely, suppose that the linear spaces H_i are pairwise orthogonal. From the definition of the independence of an infinite collection of σ-fields, it is enough to prove that, if $i_1, \ldots, i_p \in I$ are distinct, the σ-fields $\sigma(H_{i_1}), \ldots, \sigma(H_{i_p})$ are independent. To this end, it is enough to verify that, if $\xi_1^1, \ldots, \xi_{n_1}^1 \in H_{i_1}, \ldots, \xi_1^p, \ldots, \xi_{n_p}^p \in H_{i_p}$ are fixed, the vectors $(\xi_1^1, \ldots, \xi_{n_1}^1), \ldots, (\xi_1^p, \ldots, \xi_{n_p}^p)$ are independent (indeed, for every $j \in \{1, \ldots, p\}$, the events of the form $\{\xi_1^j \in A_1, \ldots, \xi_{n_j}^j \in A_{n_j}\}$ give a class stable under finite intersections that generates the σ-field $\sigma(H_{i_j})$, and the desired result follows by a standard monotone class argument, see Appendix A1). However, for every $j \in \{1, \ldots, p\}$ we can find an orthonormal basis $(\eta_1^j, \ldots, \eta_{m_j}^j)$ of the linear subspace of L^2 spanned by $\{\xi_1^j, \ldots, \xi_{n_j}^j\}$. The covariance matrix of the vector

$$(\eta_1^1, \ldots, \eta_{m_1}^1, \eta_1^2, \ldots, \eta_{m_2}^2, \ldots, \eta_1^p, \ldots, \eta_{m_p}^p)$$

is then the identity matrix (for $i \neq j$, $E[\eta_l^i \eta_{l'}^j] = 0$ because H_i and H_j are orthogonal). Moreover, this vector is Gaussian because its components belong to H. By Proposition 1.2, the components of the latter vector are independent random variables. This implies in turn that the vectors $(\eta_1^1, \ldots, \eta_{m_1}^1), \ldots, (\eta_1^p, \ldots, \eta_{m_p}^p)$ are independent. Equivalently, the vectors $(\xi_1^1, \ldots, \xi_{n_1}^1), \ldots, (\xi_1^p, \ldots, \xi_{n_p}^p)$ are independent, which was the desired result. $\qquad\square$

As an application of the previous theorem, we now discuss conditional expectations in the Gaussian framework. Again, the fact that these conditional expectations can be computed as orthogonal projections (as shown in the next corollary) is very particular to the Gaussian setting.

Corollary 1.10 *Let H be a (centered) Gaussian space and let K be a closed linear subspace of H. Let p_K denote the orthogonal projection onto K in the Hilbert space L^2, and let $X \in H$.*

(i) *We have*

$$E[X \mid \sigma(K)] = p_K(X).$$

(ii) *Let $\sigma^2 = E[(X - p_K(X))^2]$. Then, for every Borel subset A of $\mathbb{R}$, the random variable $P[X \in A \mid \sigma(K)]$ is given by*

$$P[X \in A \mid \sigma(K)](\omega) = Q(\omega, A),$$

where $Q(\omega, \cdot)$ denotes the $\mathcal{N}(p_K(X)(\omega), \sigma^2)$-distribution:

$$Q(\omega, A) = \frac{1}{\sigma\sqrt{2\pi}} \int_A dy \, \exp\left(-\frac{(y - p_K(X)(\omega))^2}{2\sigma^2}\right)$$

(and by convention $Q(\omega, A) = \mathbf{1}_A(p_K(X))$ if $\sigma = 0$).

Remarks

(a) Part (ii) of the statement can be interpreted by saying that the conditional distribution of X knowing $\sigma(K)$ is $\mathcal{N}(p_K(X), \sigma^2)$.

(b) For a general random variable X in L^2, one has

$$E[X \mid \sigma(K)] = p_{L^2(\Omega, \sigma(K), P)}(X).$$

Assertion (i) shows that, in our Gaussian framework, this orthogonal projection coincides with the orthogonal projection onto the space K, which is "much smaller" than $L^2(\Omega, \sigma(K), P)$.

(c) Assertion (i) also gives the principle of linear regression. For instance, if (X_1, X_2, X_3) is a (centered) Gaussian vector in $\mathbb{R}^3$, the best approximation in L^2 of X_3 as a (not necessarily linear) function of X_1 and X_2 can be written $\lambda_1 X_1 + \lambda_2 X_2$ where λ_1 and λ_2 are computed by saying that $X_3 - (\lambda_1 X_1 + \lambda_2 X_2)$ is orthogonal to the vector space spanned by X_1 and X_2.

Proof

(i) Let $Y = X - p_K(X)$. Then Y is orthogonal to K and, by Theorem 1.9, Y is independent of $\sigma(K)$. Then,

$$E[X \mid \sigma(K)] = E[p_K(X) \mid \sigma(K)] + E[Y \mid \sigma(K)] = p_K(X) + E[Y] = p_K(X).$$

(ii) For every nonnegative measurable function f on $\mathbb{R}_+$,

$$E[f(X) \mid \sigma(K)] = E[f(p_K(X) + Y) \mid \sigma(K)] = \int P_Y(dy) f(p_K(X) + y),$$

where P_Y is the law of Y, which is $\mathcal{N}(0, \sigma^2)$ since Y is centered Gaussian with variance σ^2. In the second equality, we also use the following general fact: if Z is a $\mathcal{G}$-measurable random variable and if Y is independent of $\mathcal{G}$ then, for every nonnegative measurable function g, $E[g(Y, Z) \mid \mathcal{G}] = \int g(y, Z) P_Y(dy)$. Property (ii) immediately follows.

$\square$

Let $(X_t)_{t \in T}$ be a (centered) Gaussian process. The covariance function of X is the function $\Gamma : T \times T \longrightarrow \mathbb{R}$ defined by $\Gamma(s, t) = \mathrm{cov}(X_s, X_t) = E[X_s X_t]$. This function characterizes the collection of *finite-dimensional marginal distributions* of the process X, that is, the collection consisting for every choice of the distinct indices $t_1, \ldots, t_p$ in T of the law of the vector $(X_{t_1}, \ldots, X_{t_p})$. Indeed this vector is a centered Gaussian vector in $\mathbb{R}^p$ with covariance matrix $(\Gamma(t_i, t_j))_{1 \leq i, j \leq p}$.

Remark One can define in an obvious way the notion of a non-centered Gaussian process. The collection of finite-dimensional marginal distributions is then characterized by the covariance function and the mean function $t \mapsto m(t) = E[X_t]$.

Given a function Γ on $T \times T$, one may ask whether there exists a Gaussian process X whose Γ is the covariance function. The function Γ must be symmetric ($\Gamma(s, t) = \Gamma(t, s)$) and of positive type in the following sense: if c is a real function on T with finite support, then

$$\sum_{T \times T} c(s) c(t)\, \Gamma(s, t) \geq 0.$$

Indeed, if Γ is the covariance function of the process X, we have immediately

$$\sum_{T \times T} c(s) c(t)\, \Gamma(s, t) = \mathrm{var}\left(\sum_T c(s) X_s \right) \geq 0.$$

Note that when T is finite, the problem of the existence of X is solved under the preceding assumptions on Γ by Theorem 1.3.

The next theorem solves the existence problem in the general case. This theorem is a direct consequence of the Kolmogorov extension theorem, which in the particular case $T = \mathbb{R}_+$ is stated as Theorem 6.3 in Chap. 6 below (see e.g. Neveu [64, Chapter III], or Kallenberg [47, Chapter VI] for the general case). We omit the proof as this result will not be used in the sequel.

Theorem 1.11 *Let Γ be a symmetric function of positive type on $T \times T$. There exists, on an appropriate probability space $(\Omega, \mathscr{F}, P)$, a centered Gaussian process whose covariance function is Γ.*

Example Consider the case $T = \mathbb{R}$ and let μ be a finite measure on $\mathbb{R}$, which is also symmetric (i.e. $\mu(-A) = \mu(A)$). Then set, for every $s, t \in \mathbb{R}$,

$$\Gamma(s, t) = \int e^{i\xi(t-s)} \mu(d\xi).$$

It is easy to verify that Γ has the required properties. In particular, if c is a real function on $\mathbb{R}$ with finite support,

$$\sum_{\mathbb{R} \times \mathbb{R}} c(s)c(t)\, \Gamma(s, t) = \int_{\mathbb{R}} |\sum_{\mathbb{R}} c(s)e^{i\xi s}|^2 \mu(d\xi) \geq 0.$$

The process Γ enjoys the additional property that $\Gamma(s, t)$ only depends on $t - s$. It immediately follows that any (centered) Gaussian process $(X_t)_{t \in \mathbb{R}}$ with covariance function Γ is stationary (in a strong sense), meaning that

$$(X_{t_1+t}, X_{t_2+t}, \ldots, X_{t_n+t}) \stackrel{(d)}{=} (X_{t_1}, X_{t_2}, \ldots, X_{t_n})$$

for any choice of $t_1, \ldots, t_n, t \in \mathbb{R}$. Conversely, any stationary Gaussian process X indexed by $\mathbb{R}$ has a covariance of the preceding type (this is Bochner's theorem, which we will not use in this book), and the measure μ is called the spectral measure of X.

1.4 Gaussian White Noise

Definition 1.12 Let $(E, \mathscr{E})$ be a measurable space, and let μ be a σ-finite measure on $(E, \mathscr{E})$. A *Gaussian white noise* with intensity μ is an isometry G from $L^2(E, \mathscr{E}, \mu)$ into a (centered) Gaussian space.

Hence, if $f \in L^2(E, \mathscr{E}, \mu)$, $G(f)$ is centered Gaussian with variance

$$E[G(f)^2] = \|G(f)\|^2_{L^2(\Omega, \mathscr{F}, P)} = \|f\|^2_{L^2(E, \mathscr{E}, \mu)} = \int f^2 \, d\mu.$$

If $f, g \in L^2(E, \mathscr{E}, \mu)$, the covariance of $G(f)$ and $G(g)$ is

$$E[G(f)G(g)] = \langle f, g \rangle_{L^2(E, \mathscr{E}, \mu)} = \int fg \, d\mu.$$

In particular, if $f = \mathbf{1}_A$ with $\mu(A) < \infty$, $G(\mathbf{1}_A)$ is $\mathscr{N}(0, \mu(A))$-distributed. To simplify notation, we will write $G(A) = G(\mathbf{1}_A)$.

Let $A_1, \ldots, A_n \in \mathscr{E}$ be disjoint and such that $\mu(A_j) < \infty$ for every j. Then the vector

$$(G(A_1), \ldots, G(A_n))$$

is a Gaussian vector in $\mathbb{R}^n$ and its covariance matrix is diagonal since, if $i \neq j$,

$$E[G(A_i)G(A_j)] = \langle \mathbf{1}_{A_i}, \mathbf{1}_{A_j} \rangle_{L^2(E, \mathscr{E}, \mu)} = 0.$$

From Proposition 1.2, we get that the variables $G(A_1), \ldots, G(A_n)$ are independent.

Suppose that $A \in \mathscr{E}$ is such that $\mu(A) < \infty$ and that A is the disjoint union of a countable collection $A_1, A_2, \ldots$ of measurable subsets of E. Then, $\mathbf{1}_A = \sum_{j=1}^{\infty} \mathbf{1}_{A_j}$ where the series converges in $L^2(E, \mathscr{E}, \mu)$, and by the isometry property this implies that

$$G(A) = \sum_{j=1}^{\infty} G(A_j)$$

where the series converges in $L^2(\Omega, \mathscr{F}, P)$ (since the random variables $G(A_j)$ are independent, an easy application of the convergence theorem for discrete martingales also shows that the series converges a.s.).

Properties of the mapping $A \mapsto G(A)$ are therefore very similar to those of a measure depending on the parameter $\omega \in \Omega$. However, one can show that, if ω is fixed, the mapping $A \mapsto G(A)(\omega)$ does not (in general) define a measure. We will come back to this point later.

Proposition 1.13 *Let $(E, \mathscr{E})$ be a measurable space, and let μ be a σ-finite measure on $(E, \mathscr{E})$. There exists, on an appropriate probability space $(\Omega, \mathscr{F}, P)$, a Gaussian white noise with intensity μ.*

Proof We rely on elementary Hilbert space theory. Let $(f_i, i \in I)$ be a total orthonormal system in the Hilbert space $L^2(E, \mathscr{E}, \mu)$. For every $f \in L^2(E, \mathscr{E}, \mu)$,

$$f = \sum_{i \in I} \alpha_i f_i$$

where the coefficients $\alpha_i = \langle f, f_i \rangle$ are such that

$$\sum_{i \in I} \alpha_i^2 = \|f\|^2 < \infty.$$

On an appropriate probability space $(\Omega, \mathscr{F}, P)$ we can construct a collection $(X_i)_{i \in I}$, indexed by the same index set I, of independent $\mathscr{N}(0, 1)$ random variables (see [64, Chapter III] for the existence of such a collection – in the sequel we will only need the case when I is countable, and then an elementary construction using only the existence of Lebesgue measure on $[0, 1]$ is possible), and we set

$$G(f) = \sum_{i \in I} \alpha_i X_i.$$

The series converges in L^2 since the X_i, $i \in I$, form an orthonormal system in L^2. Then clearly G takes values in the Gaussian space generated by X_i, $i \in I$. Furthermore, G is an isometry since it maps the orthonormal basis $(f_i, i \in I)$ to an orthonormal system. □

We could also have deduced the previous result from Theorem 1.11 applied with $T = L^2(E, \mathscr{E}, \mu)$ and $\Gamma(f, g) = \langle f, g \rangle_{L^2(E, \mathscr{E}, \mu)}$. In this way we get a Gaussian process $(X_f, f \in L^2(E, \mathscr{E}, \mu))$ and we just have to take $G(f) = X_f$.

Remark In what follows, we will only consider the case when $L^2(E, \mathscr{E}, \mu)$ is separable. For instance, if $(E, \mathscr{E}) = (\mathbb{R}_+, \mathscr{B}(\mathbb{R}_+))$ and μ is Lebesgue measure, the construction of G only requires a sequence $(\xi_n)_{n \geq 0}$ of independent $\mathscr{N}(0, 1)$ random variables, and the choice of an orthonormal basis $(\varphi_n)_{n \geq 0}$ of $L^2(\mathbb{R}_+, \mathscr{B}(\mathbb{R}_+), dt)$: We get G by setting

$$G(f) = \sum_{n \geq 0} \langle f, \varphi_n \rangle \xi_n.$$

See Exercise 1.18 for an explicit choice of $(\varphi_n)_{n \geq 0}$ when $E = [0, 1]$.

Our last proposition gives a way of recovering the intensity $\mu(A)$ of a measurable set A from the values of G on the atoms of finer and finer partitions of A.

Proposition 1.14 *Let G be a Gaussian white noise on $(E, \mathscr{E})$ with intensity μ. Let $A \in \mathscr{E}$ be such that $\mu(A) < \infty$. Assume that there exists a sequence of partitions of A,*

$$A = A_1^n \cup \ldots \cup A_{k_n}^n$$

whose "mesh" tends to 0, in the sense that

$$\lim_{n\to\infty}\left(\sup_{1\leq j\leq k_n}\mu(A_j^n)\right)=0.$$

Then,

$$\lim_{n\to\infty}\sum_{j=1}^{k_n}G(A_j^n)^2=\mu(A)$$

in L^2.

Proof For every fixed n, the variables $G(A_1^n),\ldots,G(A_{k_n}^n)$ are independent. Furthermore, $E[G(A_j^n)^2]=\mu(A_j^n)$. We then compute

$$E\left[\left(\sum_{j=1}^{k_n}G(A_j^n)^2-\mu(A)\right)^2\right]=\sum_{j=1}^{k_n}\mathrm{var}(G(A_j^n)^2)=2\sum_{j=1}^{k_n}\mu(A_j^n)^2,$$

because, if X is $\mathcal{N}(0,\sigma^2)$, $\mathrm{var}(X^2)=E(X^4)-\sigma^4=3\sigma^4-\sigma^4=2\sigma^4$. Then,

$$\sum_{j=1}^{k_n}\mu(A_j^n)^2\leq\left(\sup_{1\leq j\leq k_n}\mu(A_j^n)\right)\mu(A)$$

tends to 0 as $n\to\infty$ by assumption.   □

Exercises

Exercise 1.15 Let $(X_t)_{t\in[0,1]}$ be a centered Gaussian process. We assume that the mapping $(t,\omega)\mapsto X_t(\omega)$ from $[0,1]\times\Omega$ into $\mathbb{R}$ is measurable. We denote the covariance function of X by K.

1. Show that the mapping $t\mapsto X_t$ from $[0,1]$ into $L^2(\Omega)$ is continuous if and only if K is continuous on $[0,1]^2$. In what follows, we assume that this condition holds.
2. Let $h:[0,1]\to\mathbb{R}$ be a measurable function such that $\int_0^1|h(t)|\sqrt{K(t,t)}\,dt<\infty$. Show that, for a.e. ω, the integral $\int_0^1 h(t)X_t(\omega)dt$ is absolutely convergent. We set $Z=\int_0^1 h(t)X_t dt$.
3. We now make the stronger assumption $\int_0^1|h(t)|dt<\infty$. Show that Z is the L^2-limit of the variables $Z_n=\sum_{i=1}^n X_{\frac{i}{n}}\int_{\frac{i-1}{n}}^{\frac{i}{n}}h(t)dt$ when $n\to\infty$ and infer that Z is a Gaussian random variable.

4. We assume that K is twice continuously differentiable. Show that, for every $t \in [0, 1]$, the limit

$$\dot{X}_t := \lim_{s \to t} \frac{X_s - X_t}{s - t}$$

exists in $L^2(\Omega)$. Verify that $(\dot{X}_t)_{t \in [0,1]}$ is a centered Gaussian process and compute its covariance function.

Exercise 1.16 (*Kalman filtering*) Let $(\epsilon_n)_{n \geq 0}$ and $(\eta_n)_{n \geq 0}$ be two independent sequences of independent Gaussian random variables such that, for every n, ϵ_n is distributed according to $\mathcal{N}(0, \sigma^2)$ and η_n is distributed according to $\mathcal{N}(0, \delta^2)$, where $\sigma > 0$ and $\delta > 0$. We consider two other sequences $(X_n)_{n \geq 0}$ and $(Y_n)_{n \geq 0}$ defined by the properties $X_0 = 0$, and, for every $n \geq 0$, $X_{n+1} = a_n X_n + \epsilon_{n+1}$ and $Y_n = c X_n + \eta_n$, where c and a_n are positive constants. We set

$$\hat{X}_{n/n} = E[X_n \mid Y_0, Y_1, \ldots, Y_n],$$

$$\hat{X}_{n+1/n} = E[X_{n+1} \mid Y_0, Y_1, \ldots, Y_n].$$

The goal of the exercise is to find a recursive formula allowing one to compute these conditional expectations.

1. Verify that $\hat{X}_{n+1/n} = a_n \hat{X}_{n/n}$, for every $n \geq 0$.

2. Show that, for every $n \geq 1$,

$$\hat{X}_{n/n} = \hat{X}_{n/n-1} + \frac{E[X_n Z_n]}{E[Z_n^2]} Z_n,$$

where $Z_n := Y_n - c\hat{X}_{n/n-1}$.

3. Evaluate $E[X_n Z_n]$ and $E[Z_n^2]$ in terms of $P_n := E[(X_n - \hat{X}_{n/n-1})^2]$ and infer that, for every $n \geq 1$,

$$\hat{X}_{n+1/n} = a_n \left(\hat{X}_{n/n-1} + \frac{cP_n}{c^2 P_n + \delta^2} Z_n \right).$$

4. Verify that $P_1 = \sigma^2$ and that, for every $n \geq 1$, the following induction formula holds:

$$P_{n+1} = \sigma^2 + a_n^2 \frac{\delta^2 P_n}{c^2 P_n + \delta^2}.$$

Exercise 1.17 Let H be a (centered) Gaussian space and let H_1 and H_2 be linear subspaces of H. Let K be a closed linear subspace of H. We write p_K for the orthogonal projection onto K. Show that the condition

$$\forall X_1 \in H_1, \forall X_2 \in H_2, \quad E[X_1 X_2] = E[p_K(X_1) p_K(X_2)]$$

implies that the σ-fields $\sigma(H_1)$ and $\sigma(H_2)$ are conditionally independent given $\sigma(K)$. (This means that, for every nonnegative $\sigma(H_1)$-measurable random variable X_1, and for every nonnegative $\sigma(H_2)$-measurable random variable X_2, one has $E[X_1 X_2 | \sigma(K)] = E[X_1 | \sigma(K)] E[X_2 | \sigma(K)]$.) *Hint*: Via monotone class arguments explained in Appendix A1, it is enough to consider the case where X_1, resp. X_2, is the indicator function of an event depending only on finitely many variables in H_1, resp. in H_2.

Exercise 1.18 (*Lévy's construction of Brownian motion*) For every $t \in [0, 1]$, we set $h_0(t) = 1$, and then, for every integer $n \geq 0$ and every $k \in \{0, 1, \ldots, 2^n - 1\}$,

$$h_k^n(t) = 2^{n/2} \mathbf{1}_{[(2k)2^{-n-1},(2k+1)2^{-n-1})}(t) - 2^{n/2} \mathbf{1}_{[(2k+1)2^{-n-1},(2k+2)2^{-n-1})}(t).$$

1. Verify that the functions $h_0, (h_k^n)_{n \geq 1, 0 \leq k \leq 2^n - 1}$ form an orthonormal basis of $L^2([0, 1], \mathcal{B}([0, 1]), dt)$. (*Hint*: Observe that, for every fixed $n \geq 0$, any function $f : [0, 1) \to \mathbb{R}$ that is constant on every interval of the form $[(j - 1)2^{-n}, j2^{-n})$, for $1 \leq j \leq 2^n$, is a linear combination of the functions $h_0, (h_k^m)_{0 \leq m < n, 0 \leq k \leq 2^m - 1}$.)
2. Suppose that $N_0, (N_k^n)_{n \geq 1, 0 \leq k \leq 2^n - 1}$ are independent $\mathcal{N}(0, 1)$ random variables. Justify the existence of the (unique) Gaussian white noise G on $[0, 1]$, with intensity dt, such that $G(h_0) = N_0$ and $G(h_k^n) = N_k^n$ for every $n \geq 0$ and $0 \leq k \leq 2^n - 1$.
3. For every $t \in [0, 1]$, set $B_t = G([0, t])$. Verify that

$$B_t = t N_0 + \sum_{n=0}^{\infty} \left(\sum_{k=0}^{2^n - 1} g_k^n(t) N_k^n \right),$$

where the series converges in L^2, and the functions $g_k^n : [0, 1] \to [0, \infty)$ are given by

$$g_k^n(t) = \int_0^t h_k^n(s) \, ds.$$

Note that the functions g_k^n are continuous and satisfy the following property: For every fixed $n \geq 0$, the functions $g_k^n, 0 \leq k \leq 2^n - 1$, have disjoint supports and are bounded above by $2^{-n/2}$.

4. For every integer $m \geq 0$ and every $t \in [0, 1]$ set

$$B_t^{(m)} = t N_0 + \sum_{n=0}^{m-1} \left(\sum_{k=0}^{2^n - 1} g_k^n(t) N_k^n \right).$$

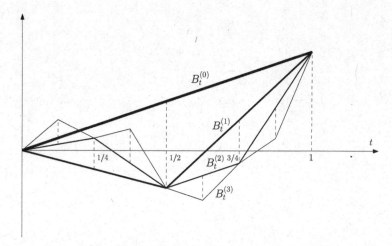

Fig. 1.1 Illustration of the construction of $B_t^{(m)}$ in Exercise 1.18, for $m = 0, 1, 2, 3$. For the clarity of the figure, lines become thinner when m increases. The lengths of the dashed segments are determined by the values of N_0 and N_k^m for $m = 0, 1, 2$

See Fig. 1.1 for an illustration. Verify that the continuous functions $t \mapsto B_t^{(m)}(\omega)$ converge uniformly on $[0, 1]$ as $m \to \infty$, for a.a. ω. (*Hint:* If N is $\mathcal{N}(0, 1)$-distributed, prove the bound $P(|N| \geq a) \leq e^{-a^2/2}$ for $a \geq 1$, and use this estimate to bound the probability of the event $\{\sup\{|N_k^n| : 0 \leq k \leq 2^n - 1\} > 2^{n/4}\}$, for every fixed $n \geq 0$.)

5. Conclude that we can, for every $t \geq 0$, select a random variable B_t' which is a.s. equal to B_t, in such a way that the mapping $t \mapsto B_t'(\omega)$ is continuous for every $\omega \in \Omega$.

Notes and Comments

The material in this chapter is standard. We refer to Adler [1] and Lifshits [55] for more information about Gaussian processes. The more recent book [56] by Marcus and Rosen develops striking applications of the known results about Gaussian processes to Markov processes and their local times. Exercise 1.16 involves a simple particular case of the famous Kalman filter, which has numerous applications in technology. See [49] or [62] for the details of the construction in Exercise 1.18.

Chapter 2
Brownian Motion

In this chapter, we construct Brownian motion and investigate some of its properties. We start by introducing the "pre-Brownian motion" (this is not a canonical terminology), which is easily defined from a Gaussian white noise on $\mathbb{R}_+$ whose intensity is Lebesgue measure. Going from pre-Brownian motion to Brownian motion requires the additional property of continuity of sample paths, which is derived here via the classical Kolmogorov lemma. The end of the chapter discusses several properties of Brownian sample paths, and establishes the strong Markov property, with its classical application to the reflection principle.

2.1 Pre-Brownian Motion

Throughout this chapter, we argue on a probability space $(\Omega, \mathscr{F}, P)$. Most of the time, but not always, random processes will be indexed by $T = \mathbb{R}_+$ and take values in $\mathbb{R}$.

Definition 2.1 Let G be a Gaussian white noise on $\mathbb{R}_+$ whose intensity is Lebesgue measure. The random process $(B_t)_{t \in \mathbb{R}_+}$ defined by

$$B_t = G(\mathbf{1}_{[0,t]})$$

is called *pre-Brownian motion*.

Proposition 2.2 *Pre-Brownian motion is a centered Gaussian process with covariance*

$$K(s,t) = \min\{s,t\} \overset{\text{(not.)}}{=} s \wedge t.$$

© Springer International Publishing Switzerland 2016
J.-F. Le Gall, *Brownian Motion, Martingales, and Stochastic Calculus*,
Graduate Texts in Mathematics 274, DOI 10.1007/978-3-319-31089-3_2

Proof By the definition of a Gaussian white noise, the variables B_t belong to a common Gaussian space, and therefore $(B_t)_{t\geq0}$ is a Gaussian process. Moreover, for every $s, t \geq 0$,

$$E[B_s B_t] = E[G([0,s])G([0,t])] = \int_0^\infty dr\, \mathbf{1}_{[0,s]}(r)\mathbf{1}_{[0,t]}(r) = s \wedge t.$$

$\square$

The next proposition gives different ways of characterizing pre-Brownian motion.

Proposition 2.3 *Let $(X_t)_{t\geq0}$ be a (real-valued) random process. The following properties are equivalent:*

(i) *$(X_t)_{t\geq0}$ is a pre-Brownian motion;*
(ii) *$(X_t)_{t\geq0}$ is a centered Gaussian process with covariance $K(s,t) = s \wedge t$;*
(iii) *$X_0 = 0$ a.s., and, for every $0 \leq s < t$, the random variable $X_t - X_s$ is independent of $\sigma(X_r, r \leq s)$ and distributed according to $\mathcal{N}(0, t-s)$;*
(iv) *$X_0 = 0$ a.s., and, for every choice of $0 = t_0 < t_1 < \cdots < t_p$, the variables $X_{t_i} - X_{t_{i-1}}, 1 \leq i \leq p$ are independent, and, for every $1 \leq i \leq p$, the variable $X_{t_i} - X_{t_{i-1}}$ is distributed according to $\mathcal{N}(0, t_i - t_{i-1})$.*

Proof The fact that (i)$\Rightarrow$(ii) is Proposition 2.2. Let us show that (ii)$\Rightarrow$(iii). We assume that $(X_t)_{t\geq0}$ is a centered Gaussian process with covariance $K(s,t) = s \wedge t$, and we let H be the Gaussian space generated by $(X_t)_{t\geq0}$. Then X_0 is distributed according to $\mathcal{N}(0,0)$ and therefore $X_0 = 0$ a.s. Then, fix $s > 0$ and write H_s for the vector space spanned by $(X_r, 0 \leq r \leq s)$, and $\tilde{H}_s$ for the vector space spanned by $(X_{s+u} - X_s, u \geq 0)$. Then H_s and $\tilde{H}_s$ are orthogonal since, for $r \in [0, s]$ and $u \geq 0$,

$$E[X_r(X_{s+u} - X_s)] = r \wedge (s+u) - r \wedge s = r - r = 0.$$

Noting that H_s and $\tilde{H}_s$ are subspaces of H, we deduce from Theorem 1.9 that $\sigma(H_s)$ and $\sigma(\tilde{H}_s)$ are independent. In particular, if we fix $t > s$, the variable $X_t - X_s$ is independent of $\sigma(H_s) = \sigma(X_r, r \leq s)$. Finally, using the form of the covariance function, we immediately get that $X_t - X_s$ is distributed according to $\mathcal{N}(0, t-s)$.

The implication (iii)$\Rightarrow$(iv) is straightforward. Taking $s = t_{p-1}$ and $t = t_p$ we obtain that $X_{t_p} - X_{t_{p-1}}$ is independent of $(X_{t_1}, \ldots, X_{t_{p-1}})$. Similarly, $X_{t_{p-1}} - X_{t_{p-2}}$ is independent of $(X_{t_1}, \ldots, X_{t_{p-2}})$, and so on. This implies that the variables $X_{t_i} - X_{t_{i-1}}$, $1 \leq i \leq p$, are independent.

Let us show that (iv)$\Rightarrow$(i). It easily follows from (iv) that X is a centered Gaussian process. Then, if f is a step function on $\mathbb{R}_+$ of the form $f = \sum_{i=1}^n \lambda_i \mathbf{1}_{(t_{i-1}, t_i]}$, where $0 = t_0 < t_1 < t_2 < \cdots < t_p$, we set

$$G(f) = \sum_{i=1}^n \lambda_i (X_{t_i} - X_{t_{i-1}})$$

(note that this definition of $G(f)$ depends only on f and not on the particular way we have written $f = \sum_{i=1}^{n} \lambda_i \mathbf{1}_{(t_{i-1}, t_i]}$). Suppose then that f and g are two step functions. We can write $f = \sum_{i=1}^{n} \lambda_i \mathbf{1}_{(t_{i-1}, t_i]}$ and $g = \sum_{i=1}^{n} \mu_i \mathbf{1}_{(t_{i-1}, t_i]}$ with the *same* subdivision $0 = t_0 < t_1 < t_2 < \cdots < t_p$ for f and for g (just take the union of the subdivisions arising in the expressions of f and g). It then follows from a simple calculation that

$$E[G(f)G(g)] = \int_{\mathbb{R}_+} f(t)g(t)\,dt,$$

so that G is an isometry from the vector space of step functions on $\mathbb{R}_+$ into the Gaussian space H generated by X. Using the fact that step functions are dense in $L^2(\mathbb{R}_+, \mathscr{B}(\mathbb{R}_+), dt)$, we get that the mapping $f \mapsto G(f)$ can be extended to an isometry from $L^2(\mathbb{R}_+, \mathscr{B}(\mathbb{R}_+), dt)$ into the Gaussian space H. Finally, we have $G([0,t]) = X_t - X_0 = X_t$ by construction. $\qquad\square$

Remark The variant of (iii) where the law of $X_t - X_s$ is not specified but required to only depend on $t - s$ is called the property of stationarity (or homogeneity) and independence of increments. Pre-Brownian motion is thus a special case of the class of processes with stationary independent increments (under an additional regularity assumption, these processes are also called Lévy processes, see Sect. 6.5.2).

Corollary 2.4 *Let $(B_t)_{t\geq 0}$ be a pre-Brownian motion. Then, for every choice of $0 = t_0 < t_1 < \cdots < t_n$, the law of the vector $(B_{t_1}, B_{t_2}, \ldots, B_{t_n})$ has density*

$$p(x_1, \ldots, x_n) = \frac{1}{(2\pi)^{n/2}\sqrt{t_1(t_2 - t_1)\ldots(t_n - t_{n-1})}} \exp\left(-\sum_{i=1}^{n} \frac{(x_i - x_{i-1})^2}{2(t_i - t_{i-1})}\right),$$

where by convention $x_0 = 0$.

Proof The random variables $B_{t_1}, B_{t_2} - B_{t_1}, \ldots, B_{t_n} - B_{t_{n-1}}$ are independent with respective distributions $\mathscr{N}(0, t_1), \mathscr{N}(0, t_2 - t_1), \ldots, \mathscr{N}(0, t_n - t_{n-1})$. Hence the vector $(B_{t_1}, B_{t_2} - B_{t_1}, \ldots, B_{t_n} - B_{t_{n-1}})$ has density

$$q(y_1, \ldots, y_n) = \frac{1}{(2\pi)^{n/2}\sqrt{t_1(t_2 - t_1)\ldots(t_n - t_{n-1})}} \exp\left(-\sum_{i=1}^{n} \frac{y_i^2}{2(t_i - t_{i-1})}\right),$$

and the change of variables $x_i = y_1 + \cdots + y_i$ for $i \in \{1, \ldots, n\}$ completes the argument. Alternatively we could have used Theorem 1.3 (ii). $\qquad\square$

Remark Corollary 2.4, together with the property $B_0 = 0$, determines the collection of *finite-dimensional marginal distributions* of pre-Brownian motion. Property (iv) of Proposition 2.3 shows that a process having the same finite-dimensional marginal distributions as pre-Brownian motion must also be a pre-Brownian motion.

Proposition 2.5 *Let B be a pre-Brownian motion. Then,*

   (i) *$-B$ is also a pre-Brownian motion (symmetry property);*

  (ii) *for every $\lambda > 0$, the process $B_t^\lambda = \frac{1}{\lambda} B_{\lambda^2 t}$ is also a pre-Brownian motion (invariance under scaling);*

 (iii) *for every $s \geq 0$, the process $B_t^{(s)} = B_{s+t} - B_s$ is a pre-Brownian motion and is independent of $\sigma(B_r, r \leq s)$ (simple Markov property).*

Proof (i) and (ii) are very easy. Let us prove (iii). With the notation of the proof of Proposition 2.3, the σ-field generated by $B^{(s)}$ is $\sigma(\tilde{H}_s)$, which is independent of $\sigma(H_s) = \sigma(B_r, r \leq s)$. To see that $B^{(s)}$ is a pre-Brownian motion, it is enough to verify property (iv) of Proposition 2.3, which is immediate since $B_{t_i}^{(s)} - B_{t_{i-1}}^{(s)} = B_{s+t_i} - B_{s+t_{i-1}}$. □

Let B be a pre-Brownian motion and let G be the associated Gaussian white noise. Note that G is determined by B: If f is a step function there is an explicit formula for $G(f)$ in terms of B, and one then uses a density argument. One often writes for $f \in L^2(\mathbb{R}_+, \mathcal{B}(\mathbb{R}_+), dt)$,

$$G(f) = \int_0^\infty f(s)\, dB_s$$

and similarly

$$G(f\mathbf{1}_{[0,t]}) = \int_0^t f(s)\, dB_s \quad , \quad G(f\mathbf{1}_{(s,t]}) = \int_s^t f(r)\, dB_r \,.$$

This notation is justified by the fact that, if $u < v$,

$$\int_u^v dB_s = G((u,v]) = G([0,v]) - G([0,u]) = B_v - B_u.$$

The mapping $f \mapsto \int_0^\infty f(s)\, dB_s$ (that is, the Gaussian white noise G) is then called the *Wiener integral* with respect to B. Recall that $\int_0^\infty f(s)dB_s$ is distributed according to $\mathcal{N}(0, \int_0^\infty f(s)^2 ds)$.

Since a Gaussian white noise is not a "real" measure depending on ω, $\int_0^\infty f(s)dB_s$ is not a "real" integral depending on ω. Much of what follows in this book is devoted to extending the definition of $\int_0^\infty f(s)dB_s$ to functions f that may depend on ω.

2.2 The Continuity of Sample Paths

We start with a general definition. Let E be a metric space equipped with its Borel σ-field.

Definition 2.6 Let $(X_t)_{t \in T}$ be a random process with values in E. The *sample paths* of X are the mappings $T \ni t \mapsto X_t(\omega)$ obtained when fixing $\omega \in \Omega$. The sample paths of X thus form a collection of mappings from T into E indexed by $\omega \in \Omega$.

Let $B = (B_t)_{t \geq 0}$ be a pre-Brownian motion. At the present stage, we have no information about the sample paths of B. We cannot even assert that these sample paths are measurable functions. In this section, we will show that, at the cost of "slightly" modifying B, we can ensure that sample paths are continuous.

Definition 2.7 Let $(X_t)_{t \in T}$ and $(\tilde{X}_t)_{t \in T}$ be two random processes indexed by the same index set T and with values in the same metric space E. We say that $\tilde{X}$ is a *modification* of X if

$$\forall t \in T, \qquad P(\tilde{X}_t = X_t) = 1.$$

This implies in particular that $\tilde{X}$ has the same finite-dimensional marginals as X. Thus, if X is a pre-Brownian motion, $\tilde{X}$ is also a pre-Brownian motion. On the other hand, sample paths of $\tilde{X}$ may have very different properties from those of X. For instance, considering the case where $T = \mathbb{R}_+$ and $E = \mathbb{R}$, it is easy to construct examples where all sample paths of $\tilde{X}$ are continuous whereas all sample paths of X are discontinuous.

Definition 2.8 The process $\tilde{X}$ is said to be *indistinguishable* from X if there exists a negligible subset N of Ω such that

$$\forall \omega \in \Omega \backslash N, \ \forall t \in T, \quad \tilde{X}_t(\omega) = X_t(\omega).$$

Put in a different way, $\tilde{X}$ is indistinguishable from X if

$$P(\forall t \in T, \ X_t = \tilde{X}_t) = 1.$$

(This formulation is slightly incorrect since the set $\{\forall t \in T, \ X_t = \tilde{X}_t\}$ need not be measurable.)

If $\tilde{X}$ is indistinguishable from X then $\tilde{X}$ is a modification of X. The notion of indistinguishability is however much stronger: Two indistinguishable processes have a.s. the same sample paths. In what follows, we will always identify two indistinguishable processes. An assertion such as "there exists a unique process such that ..." should always be understood "up to indistinguishability", even if this is not stated explicitly.

The following observation will play an important role. Suppose that $T = I$ is an interval of $\mathbb{R}$. If the sample paths of both X and $\tilde{X}$ are continuous (except possibly on a negligible subset of Ω), then $\tilde{X}$ is a modification of X if and only if $\tilde{X}$ is indistinguishable from X. Indeed, if $\tilde{X}$ is a modification of X we have a.s. $X_t = \tilde{X}_t$ for every $t \in I \cap \mathbb{Q}$ (we throw out a countable union of negligible sets) hence a.s. $X_t = \tilde{X}_t$ for every $t \in I$, by a continuity argument. We get the same result if we only assume that the sample paths are right-continuous, or left-continuous.

Theorem 2.9 (Kolmogorov's lemma)   *Let* $X = (X_t)_{t \in I}$ *be a random process indexed by a bounded interval I of $\mathbb{R}$, and taking values in a complete metric space (E, d). Assume that there exist three reals q, ε, $C > 0$ such that, for every $s, t \in I$,*

$$E[d(X_s, X_t)^q] \leq C |t - s|^{1+\varepsilon}.$$

Then, there is a modification $\tilde{X}$ of X whose sample paths are Hölder continuous with exponent α for every $\alpha \in (0, \frac{\varepsilon}{q})$: This means that, for every $\omega \in \Omega$ and every $\alpha \in (0, \frac{\varepsilon}{q})$, there exists a finite constant $C_\alpha(\omega)$ such that, for every $s, t \in I$,

$$d(\tilde{X}_s(\omega), \tilde{X}_t(\omega)) \leq C_\alpha(\omega) |t - s|^\alpha.$$

In particular, $\tilde{X}$ is a modification of X with continuous sample paths (by the preceding observations such a modification is unique up to indistinguishability).

Remarks

(i) If I is unbounded, for instance if $I = \mathbb{R}_+$, we may still apply Theorem 2.9 successively with $I = [0, 1], [1, 2], [2, 3]$, etc. and we get that X has a modification whose sample paths are *locally* Hölder with exponent α for every $\alpha \in (0, \varepsilon/q)$.

(ii) It is enough to prove that, for every fixed $\alpha \in (0, \varepsilon/q)$, X has a modification whose sample paths are Hölder with exponent α. Indeed, we can then apply this result to every choice of α in a sequence $\alpha_k \uparrow \varepsilon/q$, noting that the resulting modifications are indistinguishable, by the observations preceding the theorem.

Proof   To simplify the presentation, we take $I = [0, 1]$, but the proof would be the same for any bounded interval (closed or not). We fix $\alpha \in (0, \frac{\varepsilon}{q})$.

The assumption of the theorem implies that, for $a > 0$ and $s, t \in I$,

$$P(d(X_s, X_t) \geq a) \leq a^{-q} E[d(X_s, X_t)^q] \leq C a^{-q} |t - s|^{1+\varepsilon}.$$

We apply this inequality to $s = (i - 1)2^{-n}$, $t = i2^{-n}$ (for $i \in \{1, \ldots, 2^n\}$) and $a = 2^{-n\alpha}$:

$$P\left(d(X_{(i-1)2^{-n}}, X_{i2^{-n}}) \geq 2^{-n\alpha}\right) \leq C 2^{nq\alpha} 2^{-(1+\varepsilon)n}.$$

By summing over i we get

$$P\left(\bigcup_{i=1}^{2^n} \{d(X_{(i-1)2^{-n}}, X_{i2^{-n}}) \geq 2^{-n\alpha}\}\right) \leq 2^n \cdot C 2^{nq\alpha - (1+\varepsilon)n} = C 2^{-n(\varepsilon - q\alpha)}.$$

By assumption, $\varepsilon - q\alpha > 0$. Summing now over n, we obtain

$$\sum_{n=1}^{\infty} P\left(\bigcup_{i=1}^{2^n} \{d(X_{(i-1)2^{-n}}, X_{i2^{-n}}) \geq 2^{-n\alpha}\}\right) < \infty,$$

and the Borel–Cantelli lemma implies that, with probability one, we can find a finite integer $n_0(\omega)$ such that

$$\forall n \geq n_0(\omega), \ \forall i \in \{1, \ldots, 2^n\}, \quad d(X_{(i-1)2^{-n}}, X_{i2^{-n}}) \leq 2^{-n\alpha}.$$

Consequently the constant $K_\alpha(\omega)$ defined by

$$K_\alpha(\omega) = \sup_{n \geq 1} \left(\sup_{1 \leq i \leq 2^n} \frac{d(X_{(i-1)2^{-n}}, X_{i2^{-n}})}{2^{-n\alpha}} \right)$$

is finite a.s. (If $n \geq n_0(\omega)$, the supremum inside the parentheses is bounded above by 1, and, on the other hand, there are only finitely many terms before $n_0(\omega)$.)

At this point, we use an elementary analytic lemma, whose proof is postponed until after the end of the proof of Theorem 2.9. We write D for the set of all reals of $[0, 1)$ of the form $i2^{-n}$ for some integer $n \geq 1$ and some $i \in \{0, 1, \ldots, 2^n - 1\}$.

Lemma 2.10 *Let f be a mapping defined on D and with values in the metric space (E, d). Assume that there exists a real $\alpha > 0$ and a constant $K < \infty$ such that, for every integer $n \geq 1$ and every $i \in \{1, 2, \ldots, 2^n - 1\}$,*

$$d(f((i-1)2^{-n}), f(i2^{-n})) \leq K 2^{-n\alpha}.$$

Then we have, for every $s, t \in D$,

$$d(f(s), f(t)) \leq \frac{2K}{1 - 2^{-\alpha}} |t - s|^\alpha.$$

We immediately get from the lemma and the definition of $K_\alpha(\omega)$ that, on the event $\{K_\alpha(\omega) < \infty\}$ (which has probability 1), we have, for every $s, t \in D$,

$$d(X_s, X_t) \leq C_\alpha(\omega) |t - s|^\alpha,$$

where $C_\alpha(\omega) = 2(1 - 2^{-\alpha})^{-1} K_\alpha(\omega)$. Consequently, on the event $\{K_\alpha(\omega) < \infty\}$, the mapping $t \mapsto X_t(\omega)$ is Hölder continuous on D, hence uniformly continuous on D. Since (E, d) is complete, this mapping has a unique continuous extension to $I = [0, 1]$, which is also Hölder with exponent α. We can thus set, for every $t \in [0, 1]$

$$\tilde{X}_t(\omega) = \begin{cases} \lim\limits_{s \to t, s \in D} X_s(\omega) & \text{if } K_\alpha(\omega) < \infty, \\ x_0 & \text{if } K_\alpha(\omega) = \infty, \end{cases}$$

where x_0 is a point of E which can be fixed arbitrarily. Clearly $\tilde{X}_t$ is a random variable.

By the previous remarks, the sample paths of the process $\tilde{X}$ are Hölder with exponent α on $[0, 1]$. We still need to verify that $\tilde{X}$ is a modification of X. To this end, fix $t \in [0, 1]$. The assumption of the theorem implies that

$$\lim_{s \to t} X_s = X_t$$

in probability. Since by definition $\tilde{X}_t$ is the almost sure limit of X_s when $s \to t$, $s \in D$, we conclude that $X_t = \tilde{X}_t$ a.s. $\square$

Proof of Lemma 2.10 Fix $s, t \in D$ with $s < t$. Let $p \geq 1$ be the smallest integer such that $2^{-p} \leq t-s$, and let $k \geq 0$ be the smallest integer such that $k2^{-p} \geq s$. Then, we may write

$$s = k2^{-p} - \varepsilon_1 2^{-p-1} - \ldots - \varepsilon_l 2^{-p-l}$$
$$t = k2^{-p} + \varepsilon_0' 2^{-p} + \varepsilon_1' 2^{-p-1} + \ldots + \varepsilon_m' 2^{-p-m},$$

where l, m are nonnegative integers and $\varepsilon_i, \varepsilon_j' = 0$ or 1 for every $1 \leq i \leq l$ and $0 \leq j \leq m$. Set

$$s_i = k2^{-p} - \varepsilon_1 2^{-p-1} - \ldots - \varepsilon_i 2^{-p-i} \qquad \text{for every } 0 \leq i \leq l,$$
$$t_j = k2^{-p} + \varepsilon_0' 2^{-p} + \varepsilon_1' 2^{-p-1} + \ldots + \varepsilon_j' 2^{-p-j} \text{ for every } 0 \leq j \leq m.$$

Then, noting that $s = s_l, t = t_m$ and that we can apply the assumption of the lemma to each of the pairs (s_0, t_0), (s_{i-1}, s_i) (for $1 \leq i \leq l$) and (t_{j-1}, t_j) (for $1 \leq j \leq m$), we get

$$d(f(s), f(t)) = d(f(s_l), f(t_m))$$

$$\leq d(f(s_0), f(t_0)) + \sum_{i=1}^{l} d(f(s_{i-1}), f(s_i)) + \sum_{j=1}^{m} d(f(t_{j-1}), f(t_j))$$

$$\leq K 2^{-p\alpha} + \sum_{i=1}^{l} K 2^{-(p+i)\alpha} + \sum_{j=1}^{m} K 2^{-(p+j)\alpha}$$

$$\leq 2K (1 - 2^{-\alpha})^{-1} 2^{-p\alpha}$$

$$\leq 2K (1 - 2^{-\alpha})^{-1} (t - s)^{\alpha}$$

since $2^{-p} \leq t - s$. This completes the proof of Lemma 2.10. $\square$

We now apply Theorem 2.9 to pre-Brownian motion.

Corollary 2.11 *Let* $B = (B_t)_{t\geq 0}$ *be a pre-Brownian motion. The process* B *has a modification whose sample paths are continuous, and even locally Hölder continuous with exponent* $\frac{1}{2} - \delta$ *for every* $\delta \in (0, \frac{1}{2})$.

Proof If $s < t$, the random variable $B_t - B_s$ is distributed according to $\mathcal{N}(0, t - s)$, and thus $B_t - B_s$ has the same law as $\sqrt{t - s}\, U$, where U is $\mathcal{N}(0, 1)$. Consequently, for every $q > 0$,

$$E[|B_t - B_s|^q] = (t - s)^{q/2} E[|U|^q] = C_q\,(t - s)^{q/2}$$

where $C_q = E[|U|^q] < \infty$. Taking $q > 2$, we can apply Theorem 2.9 with $\varepsilon = \frac{q}{2} - 1$. It follows that B has a modification whose sample paths are locally Hölder continuous with exponent α for every $\alpha < (q - 2)/(2q)$. If q is large we can take α arbitrarily close to $\frac{1}{2}$. $\qquad\square$

Definition 2.12 A process $(B_t)_{t\geq 0}$ is a *Brownian motion* if:

(i) $(B_t)_{t\geq 0}$ is a pre-Brownian motion.
(ii) All sample paths of B are continuous.

This is in fact the definition of a *real* (or *linear*) Brownian motion *started from* 0. Extensions to arbitrary starting points and to higher dimensions will be discussed later.

The existence of Brownian motion in the sense of the preceding definition follows from Corollary 2.11. Indeed, starting from a pre-Brownian motion, this corollary provides a modification with continuous sample paths, which is still a pre-Brownian motion. In what follows we no longer consider pre-Brownian motion, as we will be interested only in Brownian motion.

It is important to note that the statement of Proposition 2.5 holds without change if pre-Brownian motion is replaced everywhere by Brownian motion. Indeed, with the notation of this proposition, it is immediate to verify that $-B, B^\lambda, B^{(s)}$ have continuous sample paths if B does.

The Wiener measure. Let $C(\mathbb{R}_+, \mathbb{R})$ be the space of all continuous functions from $\mathbb{R}_+$ into $\mathbb{R}$. We equip $C(\mathbb{R}_+, \mathbb{R})$ with the σ-field $\mathscr{C}$ defined as the smallest σ-field on $C(\mathbb{R}_+, \mathbb{R})$ for which the coordinate mappings $\mathrm{w} \mapsto \mathrm{w}(t)$ are measurable for every $t \geq 0$ (alternatively, one checks that $\mathscr{C}$ coincides with the Borel σ-field on $C(\mathbb{R}_+, \mathbb{R})$ associated with the topology of uniform convergence on every compact set). Given a Brownian motion B, we can consider the mapping

$$\Omega \longrightarrow C(\mathbb{R}_+, \mathbb{R})$$

$$\omega \mapsto (t \mapsto B_t(\omega))$$

and one verifies that this mapping is measurable (if we take its composition with a coordinate map $\mathrm{w} \mapsto \mathrm{w}(t)$ we get the random variable B_t, and a simple argument shows that this suffices for the desired measurability).

The **Wiener measure** (or law of Brownian motion) is by definition the image of the probability measure $P(d\omega)$ under this mapping. The Wiener measure, which we denote by $W(dw)$, is thus a probability measure on $C(\mathbb{R}_+, \mathbb{R})$, and, for every measurable subset A of $C(\mathbb{R}_+, \mathbb{R})$, we have

$$W(A) = P(B. \in A),$$

where in the right-hand side $B.$ stands for the random continuous function $t \mapsto B_t(\omega)$.

We can specialize the last equality to a "cylinder set" of the form

$$A = \{w \in C(\mathbb{R}_+, \mathbb{R}) : w(t_0) \in A_0, w(t_1) \in A_1, \ldots, w(t_n) \in A_n\},$$

where $0 = t_0 < t_1 < \cdots < t_n$, and $A_0, A_1, \ldots, A_n \in \mathscr{B}(\mathbb{R})$ (recall that $\mathscr{B}(\mathbb{R})$ stands for the Borel σ-field on $\mathbb{R}$). Corollary 2.4 then gives

$$W(\{w; w(t_0) \in A_0, w(t_1) \in A_1, \ldots, w(t_n) \in A_n\})$$

$$= P(B_{t_0} \in A_0, B_{t_1} \in A_1, \ldots, B_{t_n} \in A_n)$$

$$= \mathbf{1}_{A_0}(0) \int_{A_1 \times \cdots \times A_n} \frac{dx_1 \ldots dx_n}{(2\pi)^{n/2} \sqrt{t_1(t_2 - t_1) \ldots (t_n - t_{n-1})}} \exp\left(-\sum_{i=1}^{n} \frac{(x_i - x_{i-1})^2}{2(t_i - t_{i-1})}\right),$$

where $x_0 = 0$ by convention.

This formula for the W-measure of cylinder sets characterizes the probability measure W. Indeed, the class of cylinder sets is stable under finite intersections and generates the σ-field $\mathscr{C}$, which by a standard monotone class argument (see Appendix A1) implies that a probability measure on $\mathscr{C}$ is characterized by its values on this class. A consequence of the preceding formula for the W-measure of cylinder sets is the (fortunate) fact that the definition of the Wiener measure does not depend on the choice of the Brownian motion B: The law of Brownian motion is uniquely defined!

Suppose that B' is another Brownian motion. Then, for every $A \in \mathscr{C}$,

$$P(B'. \in A) = W(A) = P(B. \in A).$$

This means that the probability that a given property (corresponding to a measurable subset A of $C(\mathbb{R}_+, \mathbb{R})$) holds is the same for the sample paths of B and for the sample paths of B'. We will use this observation many times in what follows (see in particular the second part of the proof of Proposition 2.14 below).

Consider now the special choice of a probability space,

$$\Omega = C(\mathbb{R}_+, \mathbb{R}), \quad \mathscr{F} = \mathscr{C}, \quad P(dw) = W(dw).$$

Then on this probability space, the so-called *canonical process*

$$X_t(\mathrm{w}) = \mathrm{w}(t)$$

is a Brownian motion (the continuity of sample paths is obvious, and the fact that X has the right finite-dimensional marginals follows from the preceding formula for the W-measure of cylinder sets). This is the *canonical construction* of Brownian motion.

2.3   Properties of Brownian Sample Paths

In this section, we investigate properties of sample paths of Brownian motion (Fig. 2.1). We fix a Brownian motion $(B_t)_{t \geq 0}$. For every $t \geq 0$, we set

$$\mathscr{F}_t = \sigma(B_s, s \leq t).$$

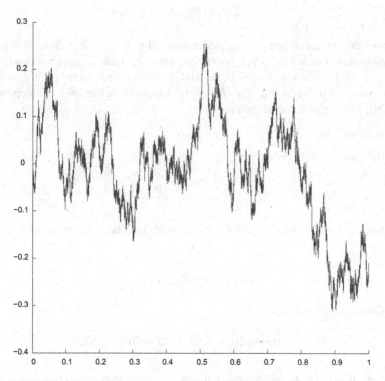

Fig. 2.1 Simulation of a Brownian sample path on the time interval $[0, 1]$

Note that $\mathscr{F}_s \subset \mathscr{F}_t$ if $s \leq t$. We also set

$$\mathscr{F}_{0+} = \bigcap_{s>0} \mathscr{F}_s.$$

We start by stating a useful $0 - 1$ law.

Theorem 2.13 (Blumenthal's zero-one law) *The σ-field $\mathscr{F}_{0+}$ is trivial, in the sense that $P(A) = 0$ or 1 for every $A \in \mathscr{F}_{0+}$.*

Proof Let $0 < t_1 < t_2 < \cdots < t_k$ and let $g : \mathbb{R}^k \longrightarrow \mathbb{R}$ be a bounded continuous function. Also fix $A \in \mathscr{F}_{0+}$. Then, by a continuity argument,

$$E[\mathbf{1}_A\, g(B_{t_1}, \ldots, B_{t_k})] = \lim_{\varepsilon \downarrow 0} E[\mathbf{1}_A\, g(B_{t_1} - B_\varepsilon, \ldots, B_{t_k} - B_\varepsilon)].$$

If $0 < \varepsilon < t_1$, the variables $B_{t_1} - B_\varepsilon, \ldots, B_{t_k} - B_\varepsilon$ are independent of $\mathscr{F}_\varepsilon$ (by the simple Markov property of Proposition 2.5) and thus also of $\mathscr{F}_{0+}$. It follows that

$$E[\mathbf{1}_A\, g(B_{t_1}, \ldots, B_{t_k})] = \lim_{\varepsilon \downarrow 0} P(A)\, E[g(B_{t_1} - B_\varepsilon, \ldots, B_{t_k} - B_\varepsilon)]$$

$$= P(A)\, E[g(B_{t_1}, \ldots, B_{t_k})].$$

We have thus obtained that $\mathscr{F}_{0+}$ is independent of $\sigma(B_{t_1}, \ldots, B_{t_k})$. Since this holds for any finite collection $\{t_1, \ldots, t_k\}$ of (strictly) positive reals, $\mathscr{F}_{0+}$ is independent of $\sigma(B_t, t > 0)$. However, $\sigma(B_t, t > 0) = \sigma(B_t, t \geq 0)$ since B_0 is the pointwise limit of B_t when $t \to 0$. Since $\mathscr{F}_{0+} \subset \sigma(B_t, t \geq 0)$, we conclude that $\mathscr{F}_{0+}$ is independent of itself, which yields the desired result. $\square$

Proposition 2.14

(i) *We have, a.s. for every $\varepsilon > 0$,*

$$\sup_{0 \leq s \leq \varepsilon} B_s > 0, \qquad \inf_{0 \leq s \leq \varepsilon} B_s < 0.$$

(ii) *For every $a \in \mathbb{R}$, let $T_a = \inf\{t \geq 0 : B_t = a\}$ (with the convention $\inf \varnothing = \infty$). Then,*

$$a.s., \quad \forall a \in \mathbb{R}, \quad T_a < \infty.$$

Consequently, we have a.s.

$$\limsup_{t \to \infty} B_t = +\infty, \qquad \liminf_{t \to \infty} B_t = -\infty.$$

Remark It is not a priori obvious that $\sup_{0 \leq s \leq \varepsilon} B_s$ is measurable, since this is an uncountable supremum of random variables. However, we can take advantage of the continuity of sample paths to restrict the supremum to *rational* values of $s \in [0, \varepsilon]$, so that we have a supremum of a countable collection of random variables. We will implicitly use such remarks in what follows.

Proof

(i) Let (ε_p) be a sequence of positive reals strictly decreasing to 0, and let

$$A = \bigcap_p \left\{ \sup_{0 \le s \le \varepsilon_p} B_s > 0 \right\}.$$

Since this is a monotone decreasing intersection, it easily follows that A is $\mathscr{F}_{0+}$-measurable (we can restrict the intersection to $p \ge p_0$, for any choice of $p_0 \ge 1$). On the other hand,

$$P(A) = \lim_{p \to \infty} \downarrow P\left(\sup_{0 \le s \le \varepsilon_p} B_s > 0 \right),$$

and

$$P\left(\sup_{0 \le s \le \varepsilon_p} B_s > 0 \right) \ge P(B_{\varepsilon_p} > 0) = \frac{1}{2},$$

which shows that $P(A) \ge 1/2$. By Theorem 2.13 we have $P(A) = 1$, hence

$$\text{a.s. } \forall \varepsilon > 0, \qquad \sup_{0 \le s \le \varepsilon} B_s > 0.$$

The assertion about $\inf_{0 \le s \le \varepsilon} B_s$ is obtained by replacing B by $-B$.

(ii) We write

$$1 = P\left(\sup_{0 \le s \le 1} B_s > 0 \right) = \lim_{\delta \downarrow 0} \uparrow P\left(\sup_{0 \le s \le 1} B_s > \delta \right),$$

and we use the scale invariance property (see Proposition 2.5 (ii) and the notation of this proposition) with $\lambda = \delta$ to see that, for every $\delta > 0$,

$$P\left(\sup_{0 \le s \le 1} B_s > \delta \right) = P\left(\sup_{0 \le s \le 1/\delta^2} B_s^{\delta} > 1 \right) = P\left(\sup_{0 \le s \le 1/\delta^2} B_s > 1 \right).$$

In the second equality, we use the remarks following the definition of the Wiener measure to observe that the probability of the event $\{\sup_{0 \le s \le 1/\delta^2} B_s > 1\}$ is the same for any Brownian motion B. If we let δ go to 0, we get

$$P\left(\sup_{s \ge 0} B_s > 1 \right) = \lim_{\delta \downarrow 0} \uparrow P\left(\sup_{0 \le s \le 1/\delta^2} B_s > 1 \right) = 1.$$

Then another scaling argument shows that, for every $M > 0$,

$$P\left(\sup_{s \geq 0} B_s > M\right) = 1$$

and replacing B by $-B$ we have also

$$P\left(\inf_{s \geq 0} B_s < -M\right) = 1.$$

The assertions in (ii) now follow easily. For the last one we observe that a continuous function $f : \mathbb{R}_+ \longrightarrow \mathbb{R}$ can visit all reals only if $\lim\sup_{t \to +\infty} f(t) = +\infty$ and $\lim\inf_{t \to +\infty} f(t) = -\infty$.

$\square$

Corollary 2.15 *Almost surely, the function $t \mapsto B_t$ is not monotone on any non-trivial interval.*

Proof Using assertion (i) of Proposition 2.14 and the simple Markov property, we immediately get that a.s. for every rational $q \in \mathbb{Q}_+$, for every $\varepsilon > 0$,

$$\sup_{q \leq t \leq q+\varepsilon} B_t > B_q, \qquad \inf_{q \leq t \leq q+\varepsilon} B_t < B_q.$$

The desired result follows. Notice that we restricted ourselves to rational values of q in order to throw out a *countable* union of negligible sets (and by the way the result would fail if we had considered all real values of q). $\square$

Proposition 2.16 *Let $0 = t_0^n < t_1^n < \cdots < t_{p_n}^n = t$ be a sequence of subdivisions of $[0, t]$ whose mesh tends to 0 (i.e. $\sup_{1 \leq i \leq p_n}(t_i^n - t_{i-1}^n) \to 0$ as $n \to \infty$). Then,*

$$\lim_{n \to \infty} \sum_{i=1}^{p_n}(B_{t_i^n} - B_{t_{i-1}^n})^2 = t,$$

in L^2.

Proof This is an immediate consequence of Proposition 1.14, writing $B_{t_i^n} - B_{t_{i-1}^n} = G((t_{i-1}^n, t_i^n])$, where G is the Gaussian white noise associated with B. $\square$

If $a < b$ and f is a real function defined on $[a, b]$, the function f is said to have infinite variation if the supremum of the quantities $\sum_{i=1}^{p} |f(t_i) - f(t_{i-1})|$, over all subdivisions $a = t_0 < t_1 < \cdots < t_p = b$, is infinite.

Corollary 2.17 *Almost surely, the function $t \mapsto B_t$ has infinite variation on any non-trivial interval.*

Proof From the simple Markov property, it suffices to consider the interval $[0, t]$ for some fixed $t > 0$. We use Proposition 2.16, and note that by extracting a subsequence we may assume that the convergence in this proposition holds a.s. We then observe that

$$\sum_{i=1}^{p_n}(B_{t_i^n} - B_{t_{i-1}^n})^2 \leq \left(\sup_{1\leq i\leq p_n} |B_{t_i^n} - B_{t_{i-1}^n}| \right) \times \sum_{i=1}^{p_n} |B_{t_i^n} - B_{t_{i-1}^n}|.$$

The supremum inside parentheses tends to 0 by the continuity of sample paths, whereas the left-hand side tends to t a.s. It follows that $\sum_{i=1}^{p_n} |B_{t_i^n} - B_{t_{i-1}^n}|$ tends to infinity a.s., giving the desired result. $\qquad\qquad\qquad\qquad\qquad\qquad\qquad\qquad\square$

The previous corollary shows that it is not possible to define the integral $\int_0^t f(s)dB_s$ as a special case of the usual (Stieltjes) integral with respect to functions of finite variation (see Sect. 4.1.1 for a brief presentation of the integral with respect to functions of finite variation, and also the comments at the end of Sect. 2.1).

2.4 The Strong Markov Property of Brownian Motion

Our goal is to extend the simple Markov property (Proposition 2.5 (iii)) to the case where the deterministic time s is replaced by a random time T. We first need to specify the class of admissible random times.

As in the previous section, we fix a Brownian motion $(B_t)_{t\geq 0}$. We keep the notation $\mathscr{F}_t$ introduced before Theorem 2.13 and we also set $\mathscr{F}_\infty = \sigma(B_s, s \geq 0)$.

Definition 2.18 A random variable T with values in $[0, \infty]$ is a *stopping time* if, for every $t \geq 0$, $\{T \leq t\} \in \mathscr{F}_t$.

It is important to note that the value ∞ is allowed. If T is a stopping time, we also have, for every $t > 0$,

$$\{T < t\} = \bigcup_{q\in[0,t)\cap\mathbb{Q}} \{T \leq q\} \in \mathscr{F}_t.$$

Examples The random variables $T = t$ (constant stopping time) and $T = T_a$ are stopping times (notice that $\{T_a \leq t\} = \{\inf_{0\leq s\leq t} |B_s - a| = 0\}$). On the other hand, $T = \sup\{s \leq 1 : B_s = 0\}$ is not a stopping time (arguing by contradiction, this will follow from the strong Markov property stated below and Proposition 2.14). If T is a stopping time, then, for every $t \geq 0$, $T + t$ is also a stopping time.

Definition 2.19 Let T be a stopping time. The *σ-field of the past before T* is

$$\mathscr{F}_T = \{A \in \mathscr{F}_\infty : \forall t \geq 0, A \cap \{T \leq t\} \in \mathscr{F}_t\}.$$

It is easy to verify that $\mathscr{F}_T$ is indeed a σ-field, and that the random variable T is $\mathscr{F}_T$-measurable. Moreover, if we define the real random variable $\mathbf{1}_{\{T<\infty\}}B_T$ by setting

$$1_{\{T<\infty\}}B_T(\omega) = \begin{cases} B_{T(\omega)}(\omega) & \text{if } T(\omega) < \infty, \\ 0 & \text{if } T(\omega) = \infty, \end{cases}$$

then $1_{\{T<\infty\}}B_T$ is also $\mathscr{F}_T$-measurable. To see this, we first observe that

$$1_{\{T<\infty\}}B_T = \lim_{n\to\infty} \sum_{i=0}^{\infty} 1_{\{i2^{-n}\le T<(i+1)2^{-n}\}}B_{i2^{-n}} = \lim_{n\to\infty} \sum_{i=0}^{\infty} 1_{\{T<(i+1)2^{-n}\}}1_{\{i2^{-n}\le T\}}B_{i2^{-n}}.$$

We then note that, for any $s \ge 0$, $B_s 1_{\{s\le T\}}$ is $\mathscr{F}_T$-measurable, because if A is a Borel subset of $\mathbb{R}$ not containing 0 (the case where $0 \in A$ is treated by considering the complementary event) we have

$$\{B_s 1_{\{s\le T\}} \in A\} \cap \{T \le t\} = \begin{cases} \varnothing & \text{if } t < s \\ \{B_s \in A\} \cap \{s \le T \le t\} & \text{if } t \ge s \end{cases}$$

which is $\mathscr{F}_t$-measurable in both cases (write $\{s \le T \le t\} = \{T \le t\} \cap \{T < s\}^c$).

Theorem 2.20 (strong Markov property) *Let T be a stopping time. We assume that $P(T < \infty) > 0$ and we set, for every $t \ge 0$,*

$$B_t^{(T)} = 1_{\{T<\infty\}}(B_{T+t} - B_T).$$

Then under the probability measure $P(\cdot \mid T < \infty)$, the process $(B_t^{(T)})_{t\ge 0}$ is a Brownian motion independent of $\mathscr{F}_T$.

Proof We first consider the case where $T < \infty$ a.s. We fix $A \in \mathscr{F}_T$ and $0 \le t_1 < \cdots < t_p$, and we let F be a bounded continuous function from $\mathbb{R}^p$ into $\mathbb{R}_+$. We will verify that

$$E[1_A F(B_{t_1}^{(T)}, \ldots, B_{t_p}^{(T)})] = P(A) E[F(B_{t_1}, \ldots, B_{t_p})]. \tag{2.1}$$

The different assertions of the theorem then follow. First the case $A = \Omega$ shows that the process $(B_t^{(T)})_{t\ge 0}$ has the same finite-dimensional marginal distributions as B and is thus a Brownian motion (notice that the sample paths of $B^{(T)}$ are continuous). Then (2.1) implies that, for every choice of $0 \le t_1 < \cdots < t_p$, the vector $(B_{t_1}^{(T)}, \ldots, B_{t_p}^{(T)})$ is independent of $\mathscr{F}_T$ and it follows that $B^{(T)}$ is independent of $\mathscr{F}_T$.

Let us prove (2.1). For every integer $n \ge 1$, and for every real $t \ge 0$, we write $[t]_n$ for the smallest real of the form $k2^{-n}$, with $k \in \mathbb{Z}_+$, belonging to the interval $[t, \infty)$. We also set $[\infty]_n = \infty$ by convention. In order to prove (2.1), we observe that we have a.s.

$$F(B_{t_1}^{(T)}, \ldots, B_{t_p}^{(T)}) = \lim_{n\to\infty} F(B_{t_1}^{([T]_n)}, \ldots, B_{t_p}^{([T]_n)}),$$

hence by dominated convergence

$$E[\mathbf{1}_A F(B_{t_1}^{(T)}, \ldots, B_{t_p}^{(T)})]$$

$$= \lim_{n \to \infty} E[\mathbf{1}_A F(B_{t_1}^{([T]_n)}, \ldots, B_{t_p}^{([T]_n)})]$$

$$= \lim_{n \to \infty} \sum_{k=0}^{\infty} E[\mathbf{1}_A \mathbf{1}_{\{(k-1)2^{-n} < T \leq k2^{-n}\}} F(B_{k2^{-n}+t_1} - B_{k2^{-n}}, \ldots, B_{k2^{-n}+t_p} - B_{k2^{-n}})],$$

where to get the last equality we have decomposed the expectation according to the possible values of $[T]_n$. The point now is the fact that, since $A \in \mathscr{F}_T$, the event

$$A \cap \{(k-1)2^{-n} < T \leq k2^{-n}\} = (A \cap \{T \leq k2^{-n}\}) \cap \{T \leq (k-1)2^{-n}\}^c$$

is $\mathscr{F}_{k2^{-n}}$-measurable. By the simple Markov property (Proposition 2.5 (iii)), we have thus

$$E[\mathbf{1}_{A \cap \{(k-1)2^{-n} < T \leq k2^{-n}\}} F(B_{k2^{-n}+t_1} - B_{k2^{-n}}, \ldots, B_{k2^{-n}+t_p} - B_{k2^{-n}})]$$

$$= P(A \cap \{(k-1)2^{-n} < T \leq k2^{-n}\}) E[F(B_{t_1}, \ldots, B_{t_p})],$$

and we just have to sum over k to get (2.1).

Finally, when $P(T = \infty) > 0$, the same arguments give, instead of (2.1),

$$E[\mathbf{1}_{A \cap \{T < \infty\}} F(B_{t_1}^{(T)}, \ldots, B_{t_p}^{(T)})] = P(A \cap \{T < \infty\}) E[F(B_{t_1}, \ldots, B_{t_p})]$$

and the desired result again follows in a straightforward way.   □

An important application of the strong Markov property is the "reflection principle" that leads to the following theorem.

Theorem 2.21 *For every* $t > 0$*, set* $S_t = \sup_{s \leq t} B_s$*. Then, if* $a \geq 0$ *and* $b \in (-\infty, a]$*, we have*

$$P(S_t \geq a, B_t \leq b) = P(B_t \geq 2a - b).$$

In particular, S_t *has the same distribution as* $|B_t|$*.*

Proof We apply the strong Markov property at the stopping time

$$T_a = \inf\{t \geq 0 : B_t = a\}.$$

We already saw (Proposition 2.14) that $T_a < \infty$ a.s. Then, using the notation of Theorem 2.20, we have

$$P(S_t \geq a, B_t \leq b) = P(T_a \leq t, B_t \leq b) = P(T_a \leq t, B_{t-T_a}^{(T_a)} \leq b-a),$$

since $B_{t-T_a}^{(T_a)} = B_t - B_{T_a} = B_t - a$ on the event $\{T_a \leq t\}$. Write $B' = B^{(T_a)}$, so that, by Theorem 2.20, the process B' is a Brownian motion independent of $\mathscr{F}_{T_a}$ hence in particular of T_a. Since B' has the same distribution as $-B'$, the pair (T_a, B') also has the same distribution as $(T_a, -B')$ (this common distribution is just the product of the law of T_a with the Wiener measure). Let

$$H = \{(s, w) \in \mathbb{R}_+ \times C(\mathbb{R}_+, \mathbb{R}) : s \leq t, \, w(t-s) \leq b-a\}.$$

The preceding probability is equal to

$$\begin{aligned}
P((T_a, B') \in H) &= P((T_a, -B') \in H) \\
&= P(T_a \leq t, \, -B_{t-T_a}^{(T_a)} \leq b-a) \\
&= P(T_a \leq t, \, B_t \geq 2a-b) \\
&= P(B_t \geq 2a-b)
\end{aligned}$$

because the event $\{B_t \geq 2a-b\}$ is a.s. contained in $\{T_a \leq t\}$. This gives the first assertion (Fig. 2.2).

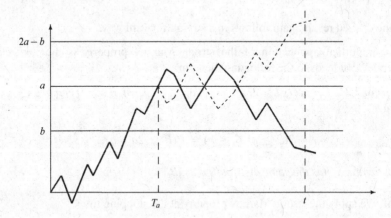

Fig. 2.2 Illustration of the reflection principle: the conditional probability, knowing that $\{T_a \leq t\}$, that the graph is below b at time t is the same as the conditional probability that the graph reflected at level a after time T_a (in *dashed lines*) is above $2a - b$ at time t

For the last assertion of the theorem, we observe that

$$P(S_t \geq a) = P(S_t \geq a, B_t \geq a) + P(S_t \geq a, B_t \leq a) = 2P(B_t \geq a) = P(|B_t| \geq a),$$

and the desired result follows. $\qquad\square$

It follows from the previous theorem that the law of the pair (S_t, B_t) has density

$$g(a, b) = \frac{2(2a - b)}{\sqrt{2\pi t^3}} \exp\left(-\frac{(2a - b)^2}{2t}\right) \mathbf{1}_{\{a>0, b<a\}}. \qquad (2.2)$$

Corollary 2.22 *For every $a > 0$, T_a has the same distribution as $\dfrac{a^2}{B_1^2}$ and has density*

$$f(t) = \frac{a}{\sqrt{2\pi t^3}} \exp\left(-\frac{a^2}{2t}\right) \mathbf{1}_{\{t>0\}}.$$

Proof Using Theorem 2.21 in the second equality, we have, for every $t \geq 0$,

$$P(T_a \leq t) = P(S_t \geq a) = P(|B_t| \geq a) = P(B_t^2 \geq a^2) = P(tB_1^2 \geq a^2) = P(\frac{a^2}{B_1^2} \leq t).$$

Furthermore, since B_1 is distributed according to $\mathcal{N}(0, 1)$, a straightforward calculation gives the density of a^2/B_1^2. $\qquad\square$

Remark From the form of the density of T_a, we immediately get that $E[T_a] = \infty$.

We finally extend the definition of Brownian motion to the case of an arbitrary (possibly random) initial value and to any dimension.

Definition 2.23 If Z is a real random variable, a process $(X_t)_{t\geq 0}$ is a *real Brownian motion* started from Z if we can write $X_t = Z + B_t$ where B is a real Brownian motion started from 0 and is *independent* of Z.

Definition 2.24 A random process $B_t = (B_t^1, \ldots, B_t^d)$ with values in $\mathbb{R}^d$ is a *d-dimensional Brownian motion* started from 0 if its components $B^1, \ldots, B^d$ are *independent* real Brownian motions started from 0. If Z a random variable with values in $\mathbb{R}^d$ and $X_t = (X_t^1, \ldots, X_t^d)$ is a process with values in $\mathbb{R}^d$, we say that X is a d-dimensional Brownian motion started from Z if we can write $X_t = Z + B_t$ where B is a d-dimensional Brownian motion started from 0 and is *independent* of Z.

Note that, if X is a d-dimensional Brownian motion and the initial value of X is random, the components of X may not be independent because the initial value may introduce some dependence (this does not occur if the initial value is deterministic).

In a way similar to the end of Sect. 2.2, the Wiener measure in dimension d is defined as the probability measure on $C(\mathbb{R}_+, \mathbb{R}^d)$ which is the law of a

d-dimensional Brownian motion started from 0. The canonical construction of Sect. 2.2 also applies to d-dimensional Brownian motion.

Many of the preceding results can be extended to d-dimensional Brownian motion with an arbitrary starting point. In particular, the invariance properties of Proposition 2.5 still hold with the obvious adaptations. Furthermore, property (i) of this proposition can be extended as follows. If X is a d-dimensional Brownian motion and Φ is an isometry of $\mathbb{R}^d$, the process $(\Phi(X_t))_{t\geq 0}$ is still a d-dimensional Brownian motion. The construction of the Wiener measure and Blumenthal's zero-one law are easily extended, and the strong Markov property also holds: One can adapt the proof of Theorem 2.20 to show that, if T is a stopping time – in the sense of the obvious extension of Definition 2.18 – which is finite with positive probability, then under the probability measure $P(\cdot \mid T < \infty)$, the process $X_t^{(T)} = X_{T+t} - X_T$, $t \geq 0$, is a d-dimensional Brownian motion started from 0 and is independent of $\mathscr{F}_T$.

Exercises

In all exercises below, $(B_t)_{t\geq 0}$ is a real Brownian motion started from 0, and $S_t = \sup_{0\leq s\leq t} B_s$.

Exercise 2.25 (*Time inversion*)

1. Show that the process $(W_t)_{t\geq 0}$ defined by $W_0 = 0$ and $W_t = tB_{1/t}$ for $t > 0$ is (indistinguishable of) a real Brownian motion started from 0. (*Hint:* First verify that W is a pre-Brownian motion.)

2. Infer that $\displaystyle\lim_{t\to\infty} \frac{B_t}{t} = 0$ a.s.

Exercise 2.26 For every real $a \geq 0$, we set $T_a = \inf\{t \geq 0 : B_t = a\}$. Show that the process $(T_a)_{a\geq 0}$ has stationary independent increments, in the sense that, for every $0 \leq a \leq b$, the variable $T_b - T_a$ is independent of the σ-field $\sigma(T_c, 0 \leq c \leq a)$ and has the same distribution as T_{b-a}.

Exercise 2.27 (*Brownian bridge*)
We set $W_t = B_t - tB_1$ for every $t \in [0, 1]$.

1. Show that $(W_t)_{t\in[0,1]}$ is a centered Gaussian process and give its covariance function.

2. Let $0 < t_1 < t_2 < \cdots < t_p < 1$. Show that the law of $(W_{t_1}, W_{t_2}, \ldots, W_{t_p})$ has density

$$g(x_1,\ldots,x_p) = \sqrt{2\pi}\, p_{t_1}(x_1)p_{t_2-t_1}(x_2 - x_1)\cdots p_{t_p-t_{p-1}}(x_p - x_{p-1})p_{1-t_p}(-x_p),$$

where $p_t(x) = \frac{1}{\sqrt{2\pi t}}\exp(-x^2/2t)$. Explain why the law of $(W_{t_1}, W_{t_2}, \ldots, W_{t_p})$ can be interpreted as the conditional law of $(B_{t_1}, B_{t_2}, \ldots, B_{t_p})$ knowing that $B_1 = 0$.

3. Verify that the two processes $(W_t)_{t\in[0,1]}$ and $(W_{1-t})_{t\in[0,1]}$ have the same distribution (similarly as in the definition of Wiener measure, this law is a probability measure on the space of all continuous functions from $[0, 1]$ into $\mathbb{R}$).

Exercise 2.28 (*Local maxima of Brownian paths*)   Show that, a.s., the local maxima of Brownian motion are distinct: a.s., for any choice of the rational numbers $p, q, r, s \geq 0$ such that $p < q < r < s$ we have

$$\sup_{p\leq t\leq q} B_t \neq \sup_{r\leq t\leq s} B_t.$$

Exercise 2.29 (*Non-differentiability*)   Using the zero-one law, show that, a.s.,

$$\limsup_{t\downarrow 0} \frac{B_t}{\sqrt{t}} = +\infty \quad , \quad \liminf_{t\downarrow 0} \frac{B_t}{\sqrt{t}} = -\infty .$$

Infer that, for every $s \geq 0$, the function $t \mapsto B_t$ has a.s. no right derivative at s.

Exercise 2.30 (*Zero set of Brownian motion*)   Let $H := \{t \in [0, 1] : B_t = 0\}$. Using Proposition 2.14 and the strong Markov property, show that H is a.s. a compact subset of $[0, 1]$ with no isolated point and zero Lebesgue measure.

Exercise 2.31 (*Time reversal*)   We set $B'_t = B_1 - B_{1-t}$ for every $t \in [0, 1]$. Show that the two processes $(B_t)_{t\in[0,1]}$ and $(B'_t)_{t\in[0,1]}$ have the same law (as in the definition of Wiener measure, this law is a probability measure on the space of all continuous functions from $[0, 1]$ into $\mathbb{R}$).

Exercise 2.32 (*Arcsine law*)
Set $T := \inf\{t \geq 0 : B_t = S_1\}$.

1. Show that $T < 1$ a.s. (one may use the result of the previous exercise) and then that T is not a stopping time.
2. Verify that the three variables S_t, $S_t - B_t$ and $|B_t|$ have the same law.
3. Show that T is distributed according to the so-called arcsine law, whose density is

$$g(t) = \frac{1}{\pi\sqrt{t(1-t)}} \mathbf{1}_{(0,1)}(t).$$

4. Show that the results of questions **1.** and **3.** remain valid if T is replaced by $L := \sup\{t \leq 1 : B_t = 0\}$.

Exercise 2.33 (*Law of the iterated logarithm*)
The goal of the exercise is to prove that

$$\limsup_{t\to\infty} \frac{B_t}{\sqrt{2t \log\log t}} = 1 \quad \text{a.s.}$$

We set $h(t) = \sqrt{2t \log \log t}$.

1. Show that, for every $t > 0$, $P(S_t > u\sqrt{t}) \sim \dfrac{2}{u\sqrt{2\pi}} e^{-u^2/2}$, when $u \to +\infty$.

2. Let r and c be two real numbers such that $1 < r < c^2$. From the behavior of the probabilities $P(S_{r^n} > c\, h(r^{n-1}))$ when $n \to \infty$, infer that, a.s.,

$$\limsup_{t \to \infty} \frac{B_t}{\sqrt{2t \log \log t}} \leq 1.$$

3. Show that a.s. there are infinitely many values of n such that

$$B_{r^n} - B_{r^{n-1}} \geq \sqrt{\frac{r-1}{r}} h(r^n).$$

Conclude that the statement given at the beginning of the exercise holds.

4. What is the value of $\displaystyle \liminf_{t \to \infty} \frac{B_t}{\sqrt{2t \log \log t}}$?

Notes and Comments

The first rigorous mathematical construction of Brownian motion is due to Wiener [81] in 1923. We use the nonstandard terminology of "pre-Brownian motion" to emphasize the necessity of choosing an appropriate modification in order to get a random process with continuous sample paths. There are several constructions of Brownian motion from a sequence of independent Gaussian random variables that directly yield the continuity property, and a very elegant one is Lévy's construction (see Exercise 1.18), which can be found in the books [49] or [62]. Lévy's construction avoids the use of Kolmogorov's lemma, but the latter will have other applications in this book. We refer to Talagrand's book [78] for far-reaching refinements of the "chaining method" used above in the proof of Kolmogorov's lemma. Much of what we know about linear Brownian motion comes from Lévy, see in particular [54, Chapitre VI]. Perhaps surprisingly, the strong Markov property of Brownian motion was proved only in the 1950s by Hunt [32] (see also Dynkin [19] for a more general version obtained independently of Hunt's work), but it had been used before by other authors, in particular by Lévy [54], without a precise justification. The reflection principle and its consequences already appeared, long before Brownian motion was rigorously constructed, in Bachelier's thesis [2], which was a pioneering work in financial mathematics. The book [62] by Mörters and Peres is an excellent source for various sample path properties of Brownian motion.

Chapter 3
Filtrations and Martingales

In this chapter, we provide a short introduction to the theory of continuous time random processes on a filtered probability space. On the way, we generalize several notions introduced in the previous chapter in the framework of Brownian motion, and we provide a thorough discussion of stopping times. In a second step, we develop the theory of continuous time martingales, and, in particular, we derive regularity results for sample paths of martingales. We finally discuss the optional stopping theorem for martingales and supermartingales, and we give applications to explicit calculations of distributions related to Brownian motion.

3.1 Filtrations and Processes

Throughout this chapter, we consider a probability space $(\Omega, \mathcal{F}, P)$. In this section, we introduce some general notions that will be of constant use later.

Definition 3.1 A *filtration* on $(\Omega, \mathcal{F}, P)$ is a collection $(\mathcal{F}_t)_{0 \leq t \leq \infty}$ indexed by $[0, \infty]$ of sub-σ-fields of $\mathcal{F}$, such that $\mathcal{F}_s \subset \mathcal{F}_t$ for every $s \leq t \leq \infty$.

We have thus, for every $0 \leq s < t$,

$$\mathcal{F}_0 \subset \mathcal{F}_s \subset \mathcal{F}_t \subset \mathcal{F}_\infty \subset \mathcal{F}.$$

We also say that $(\Omega, \mathcal{F}, (\mathcal{F}_t), P)$ is a filtered probability space

Example If B is a Brownian motion, we considered in Chap. 2 the filtration

$$\mathcal{F}_t = \sigma(B_s, 0 \leq s \leq t), \quad \mathcal{F}_\infty = \sigma(B_s, s \geq 0).$$

© Springer International Publishing Switzerland 2016
J.-F. Le Gall, *Brownian Motion, Martingales, and Stochastic Calculus*,
Graduate Texts in Mathematics 274, DOI 10.1007/978-3-319-31089-3_3

More generally, if $X = (X_t, t \geq 0)$ is any random process indexed by $\mathbb{R}_+$, the *canonical filtration* of X is defined by $\mathscr{F}_t^X = \sigma(X_s, 0 \leq s \leq t)$ and $\mathscr{F}_\infty^X = \sigma(X_s, s \geq 0)$.

Let $(\mathscr{F}_t)_{0 \leq t \leq \infty}$ be a filtration on $(\Omega, \mathscr{F}, P)$. We set, for every $t \geq 0$

$$\mathscr{F}_{t+} = \bigcap_{s > t} \mathscr{F}_s,$$

and $\mathscr{F}_{\infty+} = \mathscr{F}_\infty$. Note that $\mathscr{F}_t \subset \mathscr{F}_{t+}$ for every $t \in [0, \infty]$. The collection $(\mathscr{F}_{t+})_{0 \leq t \leq \infty}$ is also a filtration. We say that the filtration $(\mathscr{F}_t)$ is right-continuous if

$$\mathscr{F}_{t+} = \mathscr{F}_t, \qquad \forall t \geq 0.$$

By construction, the filtration $(\mathscr{F}_{t+})$ is right-continuous.

Let $(\mathscr{F}_t)$ be a filtration and let $\mathscr{N}$ be the class of all $(\mathscr{F}_\infty, P)$-negligible sets (i.e. $A \in \mathscr{N}$ if there exists an $A' \in \mathscr{F}_\infty$ such $A \subset A'$ and $P(A') = 0$). The filtration is said to be *complete* if $\mathscr{N} \subset \mathscr{F}_0$ (and thus $\mathscr{N} \subset \mathscr{F}_t$ for every t).

If $(\mathscr{F}_t)$ is not complete, it can be completed by setting $\mathscr{F}_t' = \mathscr{F}_t \vee \sigma(\mathscr{N})$, for every $t \in [0, \infty]$, using the notation $\mathscr{F}_t \vee \sigma(\mathscr{N})$ for the smallest σ-field that contains both $\mathscr{F}_t$ and $\sigma(\mathscr{N})$ (recall that $\sigma(\mathscr{N})$ is the σ-field generated by $\mathscr{N}$). We will often apply this completion procedure to the canonical filtration of a random process $(X_t)_{t \geq 0}$, and call the resulting filtration the *completed canonical filtration* of X. The reader will easily check that all results stated in Chap. 2, where we were considering the canonical filtration of a Brownian motion B, remain valid if instead we deal with the completed canonical filtration. The point is that augmenting a σ-field with negligible sets does not alter independence properties.

Let us turn to random processes, which in this chapter will be indexed by $\mathbb{R}_+$.

Definition 3.2 A process $X = (X_t)_{t \geq 0}$ with values in a measurable space $(E, \mathscr{E})$ is said to be *measurable* if the mapping

$$(\omega, t) \mapsto X_t(\omega)$$

defined on $\Omega \times \mathbb{R}_+$ equipped with the product σ-field $\mathscr{F} \otimes \mathscr{B}(\mathbb{R}_+)$ is measurable. (We recall that $\mathscr{B}(\mathbb{R}_+)$ stands for the Borel σ-field of $\mathbb{R}$.)

This is stronger than saying that, for every $t \geq 0$, X_t is $\mathscr{F}$-measurable. On the other hand, considering for instance the case where $E = \mathbb{R}$, it is easy to see that if the sample paths of X are continuous, or only right-continuous, the fact that X_t is $\mathscr{F}$-measurable for every t implies that the process is measurable in the previous sense – see the argument in the proof of Proposition 3.4 below.

In the remaining part of this chapter, we fix a filtration $(\mathscr{F}_t)$ on $(\Omega, \mathscr{F}, P)$, and the notions that will be introduced depend on the choice of this filtration.

Definition 3.3 A random process $(X_t)_{t \geq 0}$ with values in a measurable space $(E, \mathscr{E})$ is called *adapted* if, for every $t \geq 0$, X_t is $\mathscr{F}_t$-measurable. This process is said to be *progressive* if, for every $t \geq 0$, the mapping

$$(\omega, s) \mapsto X_s(\omega)$$

defined on $\Omega \times [0, t]$ is measurable for the σ-field $\mathcal{F}_t \otimes \mathcal{B}([0, t])$.

Note that a progressive process is both adapted and measurable (saying that a process is measurable is equivalent to saying that, for every $t \geq 0$, the mapping $(\omega, s) \mapsto X_s(\omega)$ defined on $\Omega \times [0, t]$ is measurable for $\mathcal{F} \otimes \mathcal{B}([0, t])$).

Proposition 3.4 *Let* $(X_t)_{t \geq 0}$ *be a random process with values in a metric space* (E, d) *(equipped with its Borel σ-field). Suppose that X is adapted and that the sample paths of X are right-continuous (i.e. for every $\omega \in \Omega$, $t \mapsto X_t(\omega)$ is right-continuous). Then X is progressive. The same conclusion holds if one replaces right-continuous by left-continuous.*

Proof We treat only the case of right-continuous sample paths, as the other case is similar. Fix $t > 0$. For every $n \geq 1$ and $s \in [0, t]$, define a random variable X_s^n by setting

$$X_s^n = X_{kt/n} \quad \text{if } s \in [(k-1)t/n, kt/n), \ k \in \{1, \ldots, n\},$$

and $X_t^n = X_t$. The right-continuity of sample paths ensures that, for every $s \in [0, t]$ and $\omega \in \Omega$,

$$X_s(\omega) = \lim_{n \to \infty} X_s^n(\omega).$$

On the other hand, for every Borel subset A of E,

$$\{(\omega, s) \in \Omega \times [0, t] : X_s^n(\omega) \in A\} = (\{X_t \in A\} \times \{t\})$$

$$\bigcup \left(\bigcup_{k=1}^{n} \left(\{X_{kt/n} \in A\} \times [\frac{(k-1)t}{n}, \frac{kt}{n}) \right) \right)$$

which belongs to the σ-field $\mathcal{F}_t \otimes \mathcal{B}([0, t])$. Hence, for every $n \geq 1$, the mapping $(\omega, s) \mapsto X_s^n(\omega)$, defined on $\Omega \times [0, t]$, is measurable for $\mathcal{F}_t \otimes \mathcal{B}([0, t])$. Since a pointwise limit of measurable functions is also measurable, the same measurability property holds for the mapping $(\omega, s) \mapsto X_s(\omega)$ defined on $\Omega \times [0, t]$. It follows that the process X is progressive. $\qquad \square$

The progressive σ-field The collection $\mathcal{P}$ of all sets $A \in \mathcal{F} \otimes \mathcal{B}(\mathbb{R}_+)$ such that the process $X_t(\omega) = \mathbf{1}_A(\omega, t)$ is progressive forms a σ-field on $\Omega \times \mathbb{R}_+$, which is called the progressive σ-field. A subset A of $\Omega \times \mathbb{R}_+$ belongs to $\mathcal{P}$ if and only if, for every $t \geq 0$, $A \cap (\Omega \times [0, t])$ belongs to $\mathcal{F}_t \otimes \mathcal{B}([0, t])$.

One then verifies that a process X is progressive if and only if the mapping $(\omega, t) \mapsto X_t(\omega)$ is measurable on $\Omega \times \mathbb{R}_+$ equipped with the σ-field $\mathcal{P}$.

3.2   Stopping Times and Associated σ-Fields

In this section, we extend certain notions that were already introduced in the previous chapter in the framework of Brownian motion. The following definition is just a repetition of the corresponding one in the previous chapter.

Definition 3.5 A random variable $T : \Omega \longrightarrow [0, \infty]$ is a *stopping time* of the filtration $(\mathscr{F}_t)$ if $\{T \leq t\} \in \mathscr{F}_t$, for every $t \geq 0$. The *σ-field of the past before T is* then defined by

$$\mathscr{F}_T = \{A \in \mathscr{F}_\infty : \forall t \geq 0, A \cap \{T \leq t\} \in \mathscr{F}_t\}.$$

The reader will verify that $\mathscr{F}_T$ is indeed a σ-field.

In what follows, "stopping time" will mean stopping time of the filtration $(\mathscr{F}_t)$ unless otherwise specified. If T is a stopping time, we also have $\{T < t\} \in \mathscr{F}_t$ for every $t > 0$, by the same argument as in the Brownian case, and moreover

$$\{T = \infty\} = \left(\bigcup_{n=1}^{\infty} \{T \leq n\} \right)^c \in \mathscr{F}_\infty.$$

Recall the definition of the filtration $(\mathscr{F}_{t+})$. A stopping time (of the filtration $(\mathscr{F}_t)$) is obviously also a stopping time of the filtration $(\mathscr{F}_{t+})$, but the converse need not be true in general.

Proposition 3.6 *Write* $\mathscr{G}_t = \mathscr{F}_{t+}$ *for every* $t \in [0, \infty]$.

(i) *A random variable* $T : \Omega \longrightarrow [0, \infty]$ *is a stopping time of the filtration* $(\mathscr{G}_t)$ *if and only if* $\{T < t\} \in \mathscr{F}_t$ *for every* $t > 0$. *This is also equivalent to saying that* $T \wedge t$ *is* $\mathscr{F}_t$-*measurable for every* $t > 0$.

(ii) *Let* T *be a stopping time of the filtration* $(\mathscr{G}_t)$. *Then*

$$\mathscr{G}_T = \{A \in \mathscr{F}_\infty : \forall t > 0, A \cap \{T < t\} \in \mathscr{F}_t\}.$$

We will write

$$\mathscr{F}_{T+} := \mathscr{G}_T.$$

Proof

(i) Suppose that T is a stopping time of the filtration $(\mathscr{G}_t)$. Then, for every $t > 0$,

$$\{T < t\} = \bigcup_{q \in \mathbb{Q}_+, q < t} \{T \leq q\} \in \mathscr{F}_t$$

because $\{T \leq q\} \in \mathcal{G}_q \subset \mathcal{F}_t$ if $q < t$. Conversely, assume that $\{T < t\} \in \mathcal{F}_t$ for every $t > 0$. Then, for every $t \geq 0$ and $s > t$,

$$\{T \leq t\} = \bigcap_{q \in \mathbb{Q}_+, t < q < s} \{T < q\} \in \mathcal{F}_s$$

and it follows that $\{T \leq t\} \in \mathcal{F}_{t+} = \mathcal{G}_t$.

Then, saying that $T \wedge t$ is $\mathcal{F}_t$-measurable for every $t > 0$ is equivalent to saying that, for every $s < t$, $\{T \leq s\} \in \mathcal{F}_t$. Taking a sequence of values of s that increases to t, we see that the latter property implies that $\{T < t\} \in \mathcal{F}_t$, and so T is a stopping time of the filtration $(\mathcal{G}_t)$. Conversely, if T is a stopping time of the filtration $(\mathcal{G}_t)$, we have $\{T \leq s\} \in \mathcal{G}_s \subset \mathcal{F}_t$ whenever $s < t$, and thus $T \wedge t$ is $\mathcal{F}_t$-measurable.

(ii) First, if $A \in \mathcal{G}_T$, we have $A \cap \{T \leq t\} \in \mathcal{G}_t$ for every $t \geq 0$. Hence, for $t > 0$,

$$A \cap \{T < t\} = \bigcup_{q \in \mathbb{Q}_+, q < t} \left(A \cap \{T \leq q\} \right) \in \mathcal{F}_t$$

since $A \cap \{T \leq q\} \in \mathcal{G}_q \subset \mathcal{F}_t$, for every $q < t$.

Conversely, assume that $A \cap \{T < t\} \in \mathcal{F}_t$ for every $t > 0$. Then, for every $t \geq 0$, and $s > t$,

$$A \cap \{T \leq t\} = \bigcap_{q \in \mathbb{Q}_+, t < q < s} \left(A \cap \{T < q\} \right) \in \mathcal{F}_s.$$

In this way, we get that $A \cap \{T \leq t\} \in \mathcal{F}_{t+} = \mathcal{G}_t$ and thus $A \in \mathcal{G}_T$.

□

Properties of stopping times and of the associated σ-fields

(a) For every stopping time T, we have $\mathcal{F}_T \subset \mathcal{F}_{T+}$. If the filtration $(\mathcal{F}_t)$ is right-continuous, we have $\mathcal{F}_{T+} = \mathcal{F}_T$.

(b) If $T = t$ is a constant stopping time, $\mathcal{F}_T = \mathcal{F}_t$ and $\mathcal{F}_{T+} = \mathcal{F}_{t+}$.

(c) Let T be a stopping time. Then T is $\mathcal{F}_T$-measurable.

(d) Let T be a stopping time and $A \in \mathcal{F}_\infty$. Set

$$T^A(\omega) = \begin{cases} T(\omega) & \text{if } \omega \in A, \\ +\infty & \text{if } \omega \notin A. \end{cases}$$

Then $A \in \mathcal{F}_T$ if and only if T^A is a stopping time.

(e) Let S, T be two stopping times such that $S \leq T$. Then $\mathcal{F}_S \subset \mathcal{F}_T$ and $\mathcal{F}_{S+} \subset \mathcal{F}_{T+}$.

(f) Let S, T be two stopping times. Then, $S \vee T$ and $S \wedge T$ are also stopping times and $\mathcal{F}_{S \wedge T} = \mathcal{F}_S \cap \mathcal{F}_T$. Furthermore, $\{S \leq T\} \in \mathcal{F}_{S \wedge T}$ and $\{S = T\} \in \mathcal{F}_{S \wedge T}$.

(g) If (S_n) is a monotone increasing sequence of stopping times, then $S = \lim \uparrow S_n$ is also a stopping time.

(h) If (S_n) is a monotone decreasing sequence of stopping times, then $S = \lim \downarrow S_n$ is a stopping time of the filtration $(\mathscr{F}_{t+})$, and

$$\mathscr{F}_{S+} = \bigcap_n \mathscr{F}_{S_n+}.$$

(i) If (S_n) is a monotone decreasing sequence of stopping times, which is also stationary (in the sense that, for every ω, there exists an integer $N(\omega)$ such that $S_n(\omega) = S(\omega)$ for every $n \geq N(\omega)$) then $S = \lim \downarrow S_n$ is also a stopping time, and

$$\mathscr{F}_S = \bigcap_n \mathscr{F}_{S_n}.$$

(j) Let T be a stopping time. A function $\omega \mapsto Y(\omega)$ defined on the set $\{T < \infty\}$ and taking values in the measurable set $(E, \mathscr{E})$ is $\mathscr{F}_T$-measurable if and only if, for every $t \geq 0$, the restriction of Y to the set $\{T \leq t\}$ is $\mathscr{F}_t$-measurable.

Remark In property (j) we use the (obvious) notion of $\mathscr{G}$-measurability for a random variable $\omega \mapsto Y(\omega)$ that is defined only on a $\mathscr{G}$-measurable subset of Ω (here $\mathscr{G}$ is a σ-field on Ω). This notion will be used again in Theorem 3.7 below.

Proof (a), (b) and (c) are almost immediate from our definitions. Let us prove the other statements.

(d) For every $t \geq 0$,

$$\{T^A \leq t\} = A \cap \{T \leq t\}$$

and the result follows from the definition of $\mathscr{F}_T$.

(e) It is enough to prove that $\mathscr{F}_S \subset \mathscr{F}_T$. If $A \in \mathscr{F}_S$, we have

$$A \cap \{T \leq t\} = (A \cap \{S \leq t\}) \cap \{T \leq t\} \in \mathscr{F}_t,$$

hence $A \in \mathscr{F}_T$.

(f) We have

$$\{S \wedge T \leq t\} = \{S \leq t\} \cup \{T \leq t\} \in \mathscr{F}_t,$$

$$\{S \vee T \leq t\} = \{S \leq t\} \cap \{T \leq t\} \in \mathscr{F}_t,$$

so that $S \wedge T$ and $S \vee T$ are stopping times.

It follows from (e) that $\mathscr{F}_{S \wedge T} \subset (\mathscr{F}_S \cap \mathscr{F}_T)$. Moreover, if $A \in \mathscr{F}_S \cap \mathscr{F}_T$,

$$A \cap \{S \wedge T \leq t\} = (A \cap \{S \leq t\}) \cup (A \cap \{T \leq t\}) \in \mathscr{F}_t,$$

hence $A \in \mathscr{F}_{S \wedge T}$.

Then, for every $t \geq 0$,

$$\{S \leq T\} \cap \{T \leq t\} = \{S \leq t\} \cap \{T \leq t\} \cap \{S \wedge t \leq T \wedge t\} \in \mathscr{F}_t,$$

$$\{S \leq T\} \cap \{S \leq t\} = \{S \wedge t \leq T \wedge t\} \cap \{S \leq t\} \in \mathscr{F}_t,$$

because $S \wedge t$ and $T \wedge t$ are both $\mathscr{F}_t$-measurable by Proposition 3.6 (i). It follows that $\{S \leq T\} \in \mathscr{F}_S \cap \mathscr{F}_T = \mathscr{F}_{S \wedge T}$. Then $\{S = T\} = \{S \leq T\} \cap \{T \leq S\}$.

(g) For every $t \geq 0$,

$$\{S \leq t\} = \bigcap_n \{S_n \leq t\} \in \mathscr{F}_t.$$

(h) Similarly

$$\{S < t\} = \bigcup_n \{S_n < t\} \in \mathscr{F}_t,$$

and we use Proposition 3.6 (i). Then, by (e), we have $\mathscr{F}_{S+} \subset \mathscr{F}_{S_n+}$ for every n, and conversely, if $A \in \bigcap_n \mathscr{F}_{S_n+}$,

$$A \cap \{S < t\} = \bigcup_n (A \cap \{S_n < t\}) \in \mathscr{F}_t,$$

hence $A \in \mathscr{F}_{S+}$.

(i) In that case, we also have

$$\{S \leq t\} = \bigcup_n \{S_n \leq t\} \in \mathscr{F}_t,$$

and if $A \in \bigcap_n \mathscr{F}_{S_n}$,

$$A \cap \{S \leq t\} = \bigcup_n (A \cap \{S_n \leq t\}) \in \mathscr{F}_t,$$

so that $A \in \mathscr{F}_S$.

(j) First assume that, for every $t \geq 0$, the restriction of Y to $\{T \leq t\}$ is $\mathscr{F}_t$-measurable. Then, for every measurable subset A of E,

$$\{Y \in A\} \cap \{T \leq t\} \in \mathscr{F}_t.$$

Letting $t \to \infty$, we first obtain that $\{Y \in A\} \in \mathscr{F}_\infty$, and then we deduce from the previous display that $\{Y \in A\} \in \mathscr{F}_T$.

Conversely, if Y is $\mathscr{F}_T$-measurable, $\{Y \in A\} \in \mathscr{F}_T$ and thus $\{Y \in A\} \cap \{T \leq t\} \in \mathscr{F}_t$, giving the desired result. $\qquad\square$

Theorem 3.7 *Let $(X_t)_{t\geq 0}$ be a progressive process with values in a measurable space $(E, \mathscr{E})$, and let T be a stopping time. Then the function $\omega \mapsto X_T(\omega) := X_{T(\omega)}(\omega)$, which is defined on the event $\{T < \infty\}$, is $\mathscr{F}_T$-measurable.*

Proof We use property (j) above. Let $t \geq 0$. The restriction to $\{T \leq t\}$ of the function $\omega \mapsto X_T(\omega)$ is the composition of the two mappings

$$\{T \leq t\} \ni \omega \mapsto (\omega, T(\omega) \wedge t)$$
$$\mathscr{F}_t \qquad \mathscr{F}_t \otimes \mathscr{B}([0, t])$$

and

$$\Omega \times [0, t] \ni (\omega, s) \mapsto X_s(\omega)$$
$$\mathscr{F}_t \otimes \mathscr{B}([0, t]) \qquad \mathscr{E}$$

which are both measurable (the first one since $T \wedge t$ is $\mathscr{F}_t$-measurable, by Proposition 3.6 (i), and the second one by the definition of a progressive process). It follows that the restriction to $\{T \leq t\}$ of the function $\omega \mapsto X_T(\omega)$ is $\mathscr{F}_t$-measurable, which gives the desired result by property (j). $\square$

Proposition 3.8 *Let T be a stopping time and let S be an $\mathscr{F}_T$-measurable random variable with values in $[0, \infty]$, such that $S \geq T$. Then S is also a stopping time.*
    In particular, if T is a stopping time,

$$T_n = \sum_{k=0}^{\infty} \frac{k+1}{2^n} \mathbf{1}_{\{k2^{-n} < T \leq (k+1)2^{-n}\}} + \infty \cdot \mathbf{1}_{\{T=\infty\}}, \qquad n = 0, 1, 2, \ldots$$

defines a sequence of stopping times that decreases to T.

Proof For the first assertion, we write, for every $t \geq 0$,

$$\{S \leq t\} = \{S \leq t\} \cap \{T \leq t\} \in \mathscr{F}_t$$

since $\{S \leq t\}$ is $\mathscr{F}_T$-measurable. The second assertion follows since $T_n \geq T$, and T_n is a function of T, hence $\mathscr{F}_T$-measurable, and $T_n \downarrow T$ as $n \uparrow \infty$ by construction. $\square$

The following proposition will be our main tool to construct stopping times associated with random processes.

Proposition 3.9 *Let $(X_t)_{t\geq 0}$ be an adapted process with values in a metric space (E, d).*

(i) *Assume that the sample paths of X are right-continuous, and let O be an open subset of E. Then*

$$T_O = \inf\{t \geq 0 : X_t \in O\}$$

is a stopping time of the filtration $(\mathscr{F}_{t+})$.

(ii) *Assume that the sample paths of X are continuous, and let F be a closed subset of E. Then*

$$T_F = \inf\{t \geq 0 : X_t \in F\}$$

is a stopping time (of the filtration $(\mathscr{F}_t)$).

Proof

(i) For every $t > 0$,

$$\{T_O < t\} = \bigcup_{s \in [0,t) \cap \mathbb{Q}} \{X_s \in O\} \in \mathscr{F}_t,$$

and we use Proposition 3.6 (i).

(ii) For every $t \geq 0$,

$$\{T_F \leq t\} = \left\{ \inf_{0 \leq s \leq t} d(X_s, F) = 0 \right\} = \left\{ \inf_{s \in [0,t] \cap \mathbb{Q}} d(X_s, F) = 0 \right\} \in \mathscr{F}_t.$$

$\square$

3.3 Continuous Time Martingales and Supermartingales

Recall that we have fixed a filtered probability space $(\Omega, \mathscr{F}, (\mathscr{F}_t), P)$. In the remaining part of this chapter, all processes take values in $\mathbb{R}$. The following is an obvious analog of the corresponding definition in discrete time (see Appendix A2 below).

Definition 3.10 An adapted real-valued process $(X_t)_{t \geq 0}$ such that $X_t \in L^1$ for every $t \geq 0$ is called

- a *martingale* if, for every $0 \leq s < t$, $E[X_t \mid \mathscr{F}_s] = X_s$;
- a *supermartingale* if, for every $0 \leq s < t$, $E[X_t \mid \mathscr{F}_s] \leq X_s$;
- a *submartingale* if, for every $0 \leq s < t$, $E[X_t \mid \mathscr{F}_s] \geq X_s$.

If $(X_t)_{t \geq 0}$ is a submartingale, $(-X_t)_{t \geq 0}$ is a supermartingale. For this reason, some of the results below are stated for supermartingales only, but the analogous results for submartingales immediately follow.

If $(X_t)_{t \geq 0}$ is a martingale (resp. a supermartingale, resp. a submartingale), we have $E[X_s] = E[X_t]$ (resp. $E[X_s] \geq E[X_t]$, resp. $E[X_s] \leq E[X_t]$) whenever $0 \leq s \leq t$.

A simple way to construct a martingale is to take a random variable $Z \in L^1$ and to set $X_t = E[Z \mid \mathscr{F}_t]$ for every $t \geq 0$. Not all martingales are of this type, however. Let us turn to an important class of examples.

Important example We say that a process $(Z_t)_{t \geq 0}$ with values in $\mathbb{R}$ or in $\mathbb{R}^d$ has independent increments with respect to the filtration $(\mathscr{F}_t)$ if Z is adapted and if, for every $0 \leq s < t$, $Z_t - Z_s$ is independent of $\mathscr{F}_s$ (for instance, a Brownian motion has independent increments with respect to its canonical filtration). If Z is a real-valued process having independent increments with respect to $(\mathscr{F}_t)$, then

(i) if $Z_t \in L^1$ for every $t \geq 0$, then $\widetilde{Z}_t = Z_t - E[Z_t]$ is a martingale;
(ii) if $Z_t \in L^2$ for every $t \geq 0$, then $Y_t = \widetilde{Z}_t^2 - E[\widetilde{Z}_t^2]$ is a martingale;
(iii) if, for some $\theta \in \mathbb{R}$, we have $E[e^{\theta Z_t}] < \infty$ for every $t \geq 0$, then

$$X_t = \frac{e^{\theta Z_t}}{E[e^{\theta Z_t}]}$$

is a martingale.

Proofs of these facts are very easy. In the second case, we have for every $0 \leq s < t$,

$$\begin{aligned}
E[(\widetilde{Z}_t)^2 \mid \mathscr{F}_s] &= E[(\widetilde{Z}_s + \widetilde{Z}_t - \widetilde{Z}_s)^2 \mid \mathscr{F}_s] \\
&= \widetilde{Z}_s^2 + 2\widetilde{Z}_s E[\widetilde{Z}_t - \widetilde{Z}_s \mid \mathscr{F}_s] + E[(\widetilde{Z}_t - \widetilde{Z}_s)^2 \mid \mathscr{F}_s] \\
&= \widetilde{Z}_s^2 + E[(\widetilde{Z}_t - \widetilde{Z}_s)^2] \\
&= \widetilde{Z}_s^2 + E[\widetilde{Z}_t^2] - 2E[\widetilde{Z}_s \widetilde{Z}_t] + E[\widetilde{Z}_s^2] \\
&= \widetilde{Z}_s^2 + E[\widetilde{Z}_t^2] - E[\widetilde{Z}_s^2],
\end{aligned}$$

because $E[\widetilde{Z}_s \widetilde{Z}_t] = E[\widetilde{Z}_s E[\widetilde{Z}_t \mid \mathscr{F}_s]] = E[\widetilde{Z}_s^2]$. The desired result follows. In the third case,

$$E[X_t \mid \mathscr{F}_s] = \frac{e^{\theta Z_s} E[e^{\theta(Z_t - Z_s)} \mid \mathscr{F}_s]}{E[e^{\theta Z_s}] E[e^{\theta(Z_t - Z_s)}]} = \frac{e^{\theta Z_s}}{E[e^{\theta Z_s}]} = X_s,$$

using the fact that $E[e^{\theta(Z_t - Z_s)} \mid \mathscr{F}_s] = E[e^{\theta(Z_t - Z_s)}]$ by independence.

Consider the special case of Brownian motion.

Definition 3.11 A real-valued process $B = (B_t)_{t \geq 0}$ is an $(\mathscr{F}_t)$-Brownian motion if B is a Brownian motion and if B is adapted and has independent increments with respect to $(\mathscr{F}_t)$. Similarly, a process $B = (B_t)_{t \geq 0}$ with values in $\mathbb{R}^d$ is a d-dimensional $(\mathscr{F}_t)$-Brownian motion if B is a d-dimensional Brownian motion and if B is adapted and has independent increments with respect to $(\mathscr{F}_t)$.

Note that if B is a (d-dimensional) Brownian motion and $(\mathscr{F}_t^B)$ is the (possibly completed) canonical filtration of B, then B is a (d-dimensional) $(\mathscr{F}_t^B)$-Brownian motion.

Let B be an $(\mathscr{F}_t)$-Brownian motion started from 0 (or from any $a \in \mathbb{R}$). Then it follows from the above observations that the processes

$$B_t \, , \ B_t^2 - t \, , \ e^{\theta B_t - \frac{\theta^2}{2} t}$$

are martingales with continuous sample paths. The processes $e^{\theta B_t - \frac{\theta^2}{2}t}$ are called exponential martingales of Brownian motion

We can also take, for $f \in L^2(\mathbb{R}_+, \mathscr{B}(\mathbb{R}_+), dt)$,

$$Z_t = \int_0^t f(s)\,dB_s\,.$$

Properties of Gaussian white noise imply that Z has independent increments with respect to the canonical filtration of B, and thus

$$\int_0^t f(s)dB_s\,, \left(\int_0^t f(s)dB_s\right)^2 - \int_0^t f(s)^2 ds\,, \exp\left(\theta \int_0^t f(s)dB_s - \frac{\theta^2}{2}\int_0^t f(s)^2 ds\right)$$

are martingales (with respect to this filtration). One can prove that these martingales have a modification with continuous sample paths – it is enough to do it for the first one, and this will follow from the more general results in Chap. 5 below.

Finally, if $Z = N$ is a Poisson process with parameter λ (and $(\mathscr{F}_t)$ is the canonical filtration of N), it is well known that Z has independent increments, and we get that

$$N_t - \lambda t\,, \ (N_t - \lambda t)^2 - \lambda t,\ \exp(\theta N_t - \lambda t(e^\theta - 1))$$

are martingales. In contrast with the previous examples, these martingales do not have a modification with continuous sample paths.

Proposition 3.12 *Let $(X_t)_{t \geq 0}$ be an adapted process and let $f : \mathbb{R} \longrightarrow \mathbb{R}_+$ be a convex function such that $E[f(X_t)] < \infty$ for every $t \geq 0$.*

(i) *If $(X_t)_{t \geq 0}$ is a martingale, then $(f(X_t))_{t \geq 0}$ is a submartingale.*
(ii) *If $(X_t)_{t \geq 0}$ is a submartingale, and if in addition f is nondecreasing, then $(f(X_t))_{t \geq 0}$ is a submartingale.*

Proof By Jensen's inequality, we have, for $s < t$,

$$E[f(X_t) \mid \mathscr{F}_s] \geq f(E[X_t \mid \mathscr{F}_s]) \geq f(X_s).$$

In the last inequality, we need the fact that f is nondecreasing when (X_t) is only a submartingale. $\qquad\square$

Consequences If $(X_t)_{t \geq 0}$ is a martingale, $|X_t|$ is a submartingale and more generally, for every $p \geq 1$, $|X_t|^p$ is a submartingale, provided that we have $E[|X_t|^p] < \infty$ for every $t \geq 0$. If $(X_t)_{t \geq 0}$ is a submartingale, $(X_t)^+ = X_t \vee 0$ is also a submartingale.

Remark If $(X_t)_{t \geq 0}$ is any martingale, Jensen's inequality shows that $E[|X_t|^p]$ is a nondecreasing function of t with values in $[0, \infty]$, for every $p \geq 1$.

Proposition 3.13 *Let $(X_t)_{t\geq 0}$ be a submartingale or a supermartingale. Then, for every $t > 0$,*

$$\sup_{0\leq s\leq t} E[|X_s|] < \infty.$$

Proof It is enough to treat the case where $(X_t)_{t\geq 0}$ is a submartingale. Since $(X_t)^+$ is also a submartingale, we have for every $s \in [0, t]$,

$$E[(X_s)^+] \leq E[(X_t)^+].$$

On the other hand, since X is a submartingale, we also have for $s \in [0, t]$,

$$E[X_s] \geq E[X_0].$$

By combining these two bounds, and noting that $|x| = 2x^+ - x$, we get

$$\sup_{s\in[0,t]} E[|X_s|] \leq 2\,E[(X_t)^+] - E[X_0] < \infty,$$

giving the desired result. $\square$

The next proposition will be very useful in the study of square integrable martingales.

Proposition 3.14 *Let $(M_t)_{t\geq 0}$ be a square integrable martingale (that is, $M_t \in L^2$ for every $t \geq 0$). Let $0 \leq s < t$ and let $s = t_0 < t_1 < \cdots < t_p = t$ be a subdivision of the interval $[s, t]$. Then,*

$$E\Big[\sum_{i=1}^{p}(M_{t_i} - M_{t_{i-1}})^2 \,\Big|\, \mathscr{F}_s\Big] = E[M_t^2 - M_s^2 \mid \mathscr{F}_s] = E[(M_t - M_s)^2 \mid \mathscr{F}_s].$$

In particular,

$$E\Big[\sum_{i=1}^{p}(M_{t_i} - M_{t_{i-1}})^2\Big] = E[M_t^2 - M_s^2] = E[(M_t - M_s)^2].$$

Proof For every $i = 1, \ldots, p$,

$$E[(M_{t_i} - M_{t_{i-1}})^2 \mid \mathscr{F}_s] = E[E[(M_{t_i} - M_{t_{i-1}})^2 \mid \mathscr{F}_{t_{i-1}}] \mid \mathscr{F}_s]$$

$$= E\Big[E[M_{t_i}^2 \mid \mathscr{F}_{t_{i-1}}] - 2M_{t_{i-1}}\,E[M_{t_i} \mid \mathscr{F}_{t_{i-1}}] + M_{t_{i-1}}^2 \,\Big|\, \mathscr{F}_s\Big]$$

$$= E\Big[E[M_{t_i}^2 \mid \mathscr{F}_{t_{i-1}}] - M_{t_{i-1}}^2 \,\Big|\, \mathscr{F}_s\Big]$$

$$= E[M_{t_i}^2 - M_{t_{i-1}}^2 \mid \mathscr{F}_s]$$

and the desired result follows by summing over i. $\square$

Our next goal is to study the regularity properties of sample paths of martingales and supermartingales. We first establish continuous time- analogs of classical inequalities in the discrete time setting.

Proposition 3.15

(i) (Maximal inequality) *Let* $(X_t)_{t\geq 0}$ *be a supermartingale with right-continuous sample paths. Then, for every $t > 0$ and every $\lambda > 0$,*

$$\lambda P\Big(\sup_{0\leq s\leq t} |X_s| > \lambda \Big) \leq E[|X_0|] + 2E[|X_t|].$$

(ii) (Doob's inequality in L^p) *Let* $(X_t)_{t\geq 0}$ *be a martingale with right-continuous sample paths. Then, for every $t > 0$ and every $p > 1$,*

$$E\Big[\sup_{0\leq s\leq t} |X_s|^p \Big] \leq \Big(\frac{p}{p-1}\Big)^p E[|X_t|^p].$$

Note that part (ii) of the proposition is useful only if $E[|X_t|^p] < \infty$.

Proof

(i) Fix $t > 0$ and consider a countable dense subset D of $\mathbb{R}_+$ such that $0 \in D$ and $t \in D$. Then $D \cap [0, t]$ is the increasing union of a sequence $(D_m)_{m\geq 1}$ of finite subsets $[0, t]$ of the form $D_m = \{t_0^m, t_1^m, \ldots, t_m^m\}$ where $0 = t_0^m < t_1^m < \cdots < t_m^m = t$. For every fixed m, we can apply the discrete time maximal inequality (see Appendix A2) to the sequence $Y_n = X_{t_n \wedge m}$, which is a discrete supermartingale with respect to the filtration $\mathscr{G}_n = \mathscr{F}_{t_n \wedge m}$. We get

$$\lambda P\Big(\sup_{s\in D_m} |X_s| > \lambda \Big) \leq E[|X_0|] + 2E[|X_t|].$$

Then, we observe that

$$P\Big(\sup_{s\in D_m} |X_s| > \lambda \Big) \uparrow P\Big(\sup_{s\in D\cap[0,t]} |X_s| > \lambda \Big)$$

when $m \uparrow \infty$. We have thus

$$\lambda P\Big(\sup_{s\in D\cap[0,t]} |X_s| > \lambda \Big) \leq E[|X_0|] + 2E[|X_t|].$$

Finally, the right-continuity of sample paths (and the fact that $t \in D$) ensures that

$$\sup_{s\in D\cap[0,t]} |X_s| = \sup_{s\in[0,t]} |X_s|. \tag{3.1}$$

Assertion (i) now follows.

(ii) Following the same strategy as in the proof of (i), and using now Doob's inequality in L^p for discrete martingales (see Appendix A2), we get, for every $m \geq 1$,

$$E\left[\sup_{s \in D_m} |X_s|^p\right] \leq \left(\frac{p}{p-1}\right)^p E[|X_t|^p].$$

Now we just have to let m tend to infinity, using the monotone convergence theorem and then the identity (3.1).

$\square$

Remark If we no longer assume that the sample paths of the supermartingale X are right-continuous, the preceding proof shows that, for every countable dense subset D of $\mathbb{R}_+$, and every $t > 0$,

$$P\left(\sup_{s \in D \cap [0,t]} |X_s| > \lambda\right) \leq \frac{1}{\lambda}(E[|X_0|] + 2E[|X_t|]).$$

Letting $\lambda \to \infty$, we have in particular

$$\sup_{s \in D \cap [0,t]} |X_s| < \infty, \quad \text{a.s.}$$

Upcrossing numbers Let $f : I \longrightarrow \mathbb{R}$ be a function defined on a subset I of $\mathbb{R}_+$. If $a < b$, the upcrossing number of f along $[a, b]$, denoted by $M_{ab}^f(I)$, is the maximal integer $k \geq 1$ such that there exists a finite increasing sequence $s_1 < t_1 < \cdots < s_k < t_k$ of elements of I such that $f(s_i) \leq a$ and $f(t_i) \geq b$ for every $i \in \{1, \ldots, k\}$ (if, even for $k = 1$, there is no such subsequence, we take $M_{ab}^f(I) = 0$, and if such a subsequence exists for every $k \geq 1$, we take $M_{ab}^f(I) = \infty$). Upcrossing numbers are a convenient tool to study the regularity of functions.

In the next lemma, the notation

$$\lim_{s \downarrow \downarrow t} f(s) \qquad (\text{resp. } \lim_{s \uparrow \uparrow t} f(s))$$

means

$$\lim_{s \downarrow t, s > t} f(s) \qquad (\text{resp. } \lim_{s \uparrow t, s < t} f(s)).$$

We say that $g : \mathbb{R}_+ \to \mathbb{R}$ is càdlàg (for the French "continue à droite avec des limites à gauche") if g is right-continuous and has left-limits at every $t > 0$.

Lemma 3.16 Let D be a countable dense subset of $\mathbb{R}_+$ and let f be a real function defined on D. We assume that, for every $T \in D$,

(i) the function f is bounded on $D \cap [0, T]$;

(ii) for all rationals a and b such that $a < b$,

$$M_{ab}^f(D \cap [0, T]) < \infty.$$

Then, the right-limit

$$f(t+) := \lim_{s \downarrow \downarrow t, s \in D} f(s)$$

exists for every real $t \geq 0$, and similarly the left-limit

$$f(t-) := \lim_{s \uparrow \uparrow t, s \in D} f(s)$$

exists for every real $t > 0$. Furthermore, the function $g : \mathbb{R}_+ \longrightarrow \mathbb{R}$ defined by $g(t) = f(t+)$ is càdlàg.

We omit the proof of this analytic lemma. It is important to note that the right and left-limits $f(t+)$ and $f(t-)$ are defined for every $t \geq 0$ ($t > 0$ in the case of $f(t-)$) and not only for $t \in D$.

Theorem 3.17 *Let $(X_t)_{t \geq 0}$ be a supermartingale, and let D be a countable dense subset of $\mathbb{R}_+$.*

(i) *For almost every $\omega \in \Omega$, the restriction of the function $s \mapsto X_s(\omega)$ to the set D has a right-limit*

$$X_{t+}(\omega) := \lim_{s \downarrow \downarrow t, s \in D} X_s(\omega) \qquad (3.2)$$

at every $t \in [0, \infty)$, and a left-limit

$$X_{t-}(\omega) := \lim_{s \uparrow \uparrow t, s \in D} X_s(\omega)$$

at every $t \in (0, \infty)$.

(ii) *For every $t \in \mathbb{R}_+$, $X_{t+} \in L^1$ and*

$$X_t \geq E[X_{t+} \mid \mathscr{F}_t],$$

with equality if the function $t \longrightarrow E[X_t]$ is right-continuous (in particular if X is a martingale). The process $(X_{t+})_{t \geq 0}$ is a supermartingale with respect to the filtration $(\mathscr{F}_{t+})$. It is a martingale if X is a martingale.

Remark For the last assertions of (ii), we need $X_{t+}(\omega)$ to be defined for *every* $\omega \in \Omega$ and not only outside a negligible set. As we will see in the proof, we can just take $X_{t+}(\omega) = 0$ when the limit in (3.2) does not exist.

Proof

(i) Fix $T \in D$. By the remark following Proposition 3.15, we have

$$\sup_{s \in D \cap [0,T]} |X_s| < \infty, \quad \text{a.s.}$$

As in the proof of Proposition 3.15, we can choose a sequence $(D_m)_{m \geq 1}$ of finite subsets of D that increase to $D \cap [0, T]$ and are such that $0, T \in D_m$. Doob's upcrossing inequality for discrete supermartingales (see Appendix A2) gives, for every $a < b$ and every $m \geq 1$,

$$E[M_{ab}^X(D_m)] \leq \frac{1}{b-a} E[(X_T - a)^-].$$

We let $m \to \infty$ and get by monotone convergence

$$E[M_{ab}^X(D \cap [0, T])] \leq \frac{1}{b-a} E[(X_T - a)^-] < \infty.$$

We thus have

$$M_{ab}^X([0, T] \cap D) < \infty , \quad \text{a.s.}$$

Set

$$N = \bigcup_{T \in D} \left(\left\{ \sup_{t \in D \cap [0,T]} |X_t| = \infty \right\} \bigcup \left(\bigcup_{a,b \in \mathbb{Q}, a<b} \{ M_{ab}^X(D \cap [0, T]) = \infty \} \right) \right).$$

(3.3)

Then $P(N) = 0$ by the preceding considerations. On the other hand, if $\omega \notin N$, the function $D \ni t \mapsto X_t(\omega)$ satisfies all assumptions of Lemma 3.16. Assertion (i) now follows from this lemma.

(ii) To define $X_{t+}(\omega)$ for every $\omega \in \Omega$ and not only on $\Omega \backslash N$, we set

$$X_{t+}(\omega) = \begin{cases} \lim_{s \downarrow \downarrow t, s \in D} X_s(\omega) & \text{if the limit exists} \\ \qquad 0 & \text{otherwise.} \end{cases}$$

With this definition, X_{t+} is $\mathscr{F}_{t+}$-measurable.

Fix $t \geq 0$ and choose a sequence $(t_n)_{n \geq 0}$ in D such that t_n decreases strictly to t as $n \to \infty$. Then, by construction, we have a.s.

$$X_{t+} = \lim_{n \to \infty} X_{t_n}.$$

Set $Y_k = X_{t-k}$ for every integer $k \leq 0$. Then Y is a backward supermartingale with respect to the (backward) discrete filtration $\mathscr{H}_k = \mathscr{F}_{t-k}$ (see Appendix A2). From Proposition 3.13, we have $\sup_{k \leq 0} E[|Y_k|] < \infty$. The convergence theorem for backward supermartingales (see Appendix A2) then implies that the sequence X_{t_n} converges to X_{t+} in L^1. In particular, $X_{t+} \in L^1$.

Thanks to the L^1-convergence, we can pass to the limit $n \to \infty$ in the inequality $X_t \geq E[X_{t_n} \mid \mathscr{F}_t]$, and we get

$$X_t \geq E[X_{t+} \mid \mathscr{F}_t]$$

(we use the fact that the conditional expectation is continuous for the L^1-norm, and it is important to realize that an a.s. convergence would not be sufficient to warrant this passage to the limit). Furthermore, thanks again to the L^1-convergence, we have $E[X_{t+}] = \lim E[X_{t_n}]$. Thus, if the function $s \longrightarrow E[X_s]$ is right-continuous, we must have $E[X_t] = E[X_{t+}] = E[E[X_{t+} \mid \mathscr{F}_t]]$, and the inequality $X_t \geq E[X_{t+} \mid \mathscr{F}_t]$ then forces $X_t = E[X_{t+} \mid \mathscr{F}_t]$.

We already noticed that X_{t+} is $\mathscr{F}_{t+}$-measurable. Let $s < t$ and let $(s_n)_{n \geq 0}$ be a sequence in D that decreases strictly to s. We may assume that $s_n \leq t_n$ for every n. Then as previously X_{s_n} converges to X_{s+} in L^1, and thus, if $A \in \mathscr{F}_{s+}$, which implies $A \in \mathscr{F}_{s_n}$ for every n, we have

$$E[X_{s+}\mathbf{1}_A] = \lim_{n\to\infty} E[X_{s_n}\mathbf{1}_A] \geq \lim_{n\to\infty} E[X_{t_n}\mathbf{1}_A] = E[X_{t+}\mathbf{1}_A] = E[E[X_{t+} \mid \mathscr{F}_{s+}]\mathbf{1}_A].$$

Since this inequality holds for every $A \in \mathscr{F}_{s+}$, and since X_{s+} and $E[X_{t+} \mid \mathscr{F}_{s+}]$ are both $\mathscr{F}_{s+}$-measurable, it follows that $X_{s+} \geq E[X_{t+} \mid \mathscr{F}_{s+}]$. Finally, if X is a martingale, inequalities can be replaced by equalities in the previous considerations.

$\square$

Theorem 3.18 *Assume that the filtration $(\mathscr{F}_t)$ is right-continuous and complete. Let $X = (X_t)_{t \geq 0}$ be a supermartingale, such that the function $t \longrightarrow E[X_t]$ is right-continuous. Then X has a modification with càdlàg sample paths, which is also an $(\mathscr{F}_t)$-supermartingale.*

Proof Let D be a countable dense subset of $\mathbb{R}_+$ as in Theorem 3.17. Let N be the negligible set defined in (3.3). We set, for every $t \geq 0$,

$$Y_t(\omega) = \begin{cases} X_{t+}(\omega) & \text{if } \omega \notin N \\ 0 & \text{if } \omega \in N. \end{cases}$$

Lemma 3.16 then shows that the sample paths of Y are càdlàg.

The random variable X_{t+} is $\mathscr{F}_{t+}$-measurable, and thus $\mathscr{F}_t$-measurable since the filtration is right-continuous. As the negligible set N belongs to $\mathscr{F}_\infty$, the completeness of the filtration ensures that Y_t is $\mathscr{F}_t$-measurable. By Theorem 3.17 (ii), we have for every $t \geq 0$,

$$X_t = E[X_{t+} \mid \mathscr{F}_t] = X_{t+} = Y_t, \quad \text{a.s.}$$

because X_{t+} is $\mathscr{F}_t$-measurable. Consequently, Y is a modification of X. The process Y is adapted to the filtration $(\mathscr{F}_t)$. Since Y is a modification of X the inequality $E[X_t \mid \mathscr{F}_s] \leq X_s$, for $0 \leq s < t$, implies that the same inequality holds for Y. $\square$

Remarks

(i) Let us comment on the assumptions of the theorem. A simple example shows that our assumption that the filtration is right-continuous is necessary. Take

$\Omega = \{-1, 1\}$, with the probability measure P defined by $P(\{1\}) = P(\{-1\}) = 1/2$. Let ε be the random variable $\varepsilon(\omega) = \omega$, and let the process $(X_t)_{t \geq 0}$ be defined by $X_t = 0$ if $0 \leq t \leq 1$, and $X_t = \varepsilon$ if $t > 1$. Then it is easy to verify that X is a martingale with respect to its canonical filtration $(\mathscr{F}_t^X)$ (which is complete since there are no nonempty negligible sets!). On the other hand, no modification of X can be right-continuous at $t = 1$. This does not contradict the theorem since the filtration is not right-continuous ($\mathscr{F}_{1+}^X \neq \mathscr{F}_1^X$).

(ii) Similarly, to show that the right-continuity of the mapping $t \longrightarrow E[X_t]$ is needed, we can just take $X_t = f(t)$, where f is any nonincreasing deterministic function. If f is not right-continuous, no modification of X can have right-continuous sample paths.

3.4 Optional Stopping Theorems

We start with a convergence theorem for supermartingales.

Theorem 3.19 *Let X be a supermartingale with right-continuous sample paths. Assume that the collection $(X_t)_{t \geq 0}$ is bounded in L^1. Then there exists a random variable $X_\infty \in L^1$ such that*

$$\lim_{t \to \infty} X_t = X_\infty, \quad a.s.$$

Proof Let D be a countable dense subset of $\mathbb{R}_+$. From the proof of Theorem 3.17, we have, for every $T \in D$ and $a < b$,

$$E[M_{ab}^X(D \cap [0, T])] \leq \frac{1}{b - a} E[(X_T - a)^-].$$

By monotone convergence, we get, for every $a < b$,

$$E[M_{ab}^X(D)] \leq \frac{1}{b - a} \sup_{t \geq 0} E[(X_t - a)^-] < \infty,$$

since the collection $(X_t)_{t \geq 0}$ is bounded in L^1. Hence, a.s. for all rationals $a < b$, we have $M_{ab}^X(D) < \infty$. This implies that the limit

$$X_\infty := \lim_{D \ni t \to \infty} X_t \tag{3.4}$$

exists a.s. in $[-\infty, \infty]$. We can in fact exclude the values $+\infty$ and $-\infty$, since Fatou's lemma gives

$$E[|X_\infty|] \leq \liminf_{D \ni t \to \infty} E[|X_t|] < \infty,$$

and we get that $X_\infty \in L^1$. The right-continuity of sample paths (which we have not yet used) allows us to remove the restriction $t \in D$ in the limit (3.4). □

Under the assumptions of Theorem 3.19, the convergence of X_t towards X_∞ may not hold in L^1. The next result gives, in the case of a martingale, necessary and sufficient conditions for the convergence to also hold in L^1.

Definition 3.20 A martingale $(X_t)_{t\geq0}$ is said to be *closed* if there exists a random variable $Z \in L^1$ such that, for every $t \geq 0$,

$$X_t = E[Z \mid \mathcal{F}_t].$$

Theorem 3.21 *Let X be a martingale with right-continuous sample paths. Then the following properties are equivalent:*

 (i) *X is closed;*
 (ii) *the collection $(X_t)_{t\geq0}$ is uniformly integrable;*
(iii) *X_t converges a.s. and in L^1 as $t \to \infty$.*

Moreover, if these properties hold, we have $X_t = E[X_\infty \mid \mathcal{F}_t]$ for every $t \geq 0$, where $X_\infty \in L^1$ is the a.s. limit of X_t as $t \to \infty$.

Proof The fact that (i)⇒(ii) is easy: If $Z \in L^1$, the collection of all random variables $E[Z \mid \mathcal{G}]$, when $\mathcal{G}$ varies over sub-σ-fields of $\mathcal{F}$, is uniformly integrable. If (ii) holds, in particular the collection $(X_t)_{t\geq0}$ is bounded in L^1 and Proposition 3.19 implies that X_t converges a.s. to X_∞. By uniform integrability, the latter convergence also holds in L^1. Finally, if (iii) holds, for every $s \geq 0$, we can pass to the limit $t \to \infty$ in the equality $X_s = E[X_t \mid \mathcal{F}_s]$ (using the fact that the conditional expectation is continuous for the L^1-norm), and we get $X_s = E[X_\infty \mid \mathcal{F}_s]$. □

We will now use the optional stopping theorems for discrete martingales and supermartingales in order to establish similar results in the continuous time setting. Let $(X_t)_{t\geq0}$ be a martingale or a supermartingale with right-continuous sample paths, and such that X_t converges a.s. as $t \to \infty$ to a random variable denoted by X_∞. Then, for every stopping time T, we write X_T for the random variable

$$X_T(\omega) = \mathbf{1}_{\{T(\omega)<\infty\}}X_{T(\omega)}(\omega) + \mathbf{1}_{\{T(\omega)=\infty\}}X_\infty(\omega).$$

Compare with Theorem 3.7, where the random variable X_T was only defined on the subset $\{T < \infty\}$ of Ω. With this definition, the random variable X_T is still $\mathcal{F}_T$-measurable: Use Theorem 3.7 and the easily verified fact that $\mathbf{1}_{\{T=\infty\}}X_\infty$ is $\mathcal{F}_T$-measurable.

Theorem 3.22 (Optional stopping theorem for martingales) *Let $(X_t)_{t\geq0}$ be a uniformly integrable martingale with right-continuous sample paths. Let S and T be two stopping times with $S \leq T$. Then X_S and X_T are in L^1 and*

$$X_S = E[X_T \mid \mathcal{F}_S].$$

In particular, for every stopping time S, we have

$$X_S = E[X_\infty \mid \mathscr{F}_S],$$

and

$$E[X_S] = E[X_\infty] = E[X_0].$$

Proof Set, for every integer $n \geq 0$,

$$T_n = \sum_{k=0}^{\infty} \frac{k+1}{2^n} \mathbf{1}_{\{k2^{-n} < T \leq (k+1)2^{-n}\}} + \infty \cdot \mathbf{1}_{\{T=\infty\}},$$

and similarly

$$S_n = \sum_{k=0}^{\infty} \frac{k+1}{2^n} \mathbf{1}_{\{k2^{-n} < S \leq (k+1)2^{-n}\}} + \infty \cdot \mathbf{1}_{\{S=\infty\}}.$$

By Proposition 3.8, (T_n) and (S_n) are two sequences of stopping times that decrease respectively to T and to S. Moreover, we have $S_n \leq T_n$ for every $n \geq 0$.

Now observe that, for every fixed n, $2^n S_n$ and $2^n T_n$ are stopping times of the discrete filtration $\mathscr{H}_k^{(n)} := \mathscr{F}_{k/2^n}$, and $Y_k^{(n)} := X_{k/2^n}$ is a discrete martingale with respect to this filtration. From the optional stopping theorem for uniformly integrable discrete martingales (see Appendix A2) we get that $Y_{2^n S_n}^{(n)}$ and $Y_{2^n T_n}^{(n)}$ are in L^1, and

$$X_{S_n} = Y_{2^n S_n}^{(n)} = E[Y_{2^n T_n}^{(n)} \mid \mathscr{H}_{2^n S_n}^{(n)}] = E[X_{T_n} \mid \mathscr{F}_{S_n}]$$

(here we need to verify that $\mathscr{H}_{2^n S_n}^{(n)} = \mathscr{F}_{S_n}$, but this is straightforward).

Let $A \in \mathscr{F}_S$. Since $\mathscr{F}_S \subset \mathscr{F}_{S_n}$, we have $A \in \mathscr{F}_{S_n}$ and thus

$$E[\mathbf{1}_A X_{S_n}] = E[\mathbf{1}_A X_{T_n}].$$

By the right-continuity of sample paths, we get a.s.

$$X_S = \lim_{n \to \infty} X_{S_n} , \ X_T = \lim_{n \to \infty} X_{T_n}.$$

These limits also hold in L^1. Indeed, thanks again to the optional stopping theorem for uniformly integrable discrete martingales, we have $X_{S_n} = E[X_\infty \mid \mathscr{F}_{S_n}]$ for every n, and thus the sequence (X_{S_n}) is uniformly integrable (and the same holds for the sequence (X_{T_n})).

The L^1-convergence implies that X_S and X_T belong to L^1, and also allows us to pass to the limit $n \to \infty$ in the equality $E[1_A X_{S_n}] = E[1_A X_{T_n}]$ in order to get

$$E[1_A X_S] = E[1_A X_T].$$

Since this holds for every $A \in \mathscr{F}_S$, and since the variable X_S is $\mathscr{F}_S$-measurable (by the remarks before the theorem), we conclude that

$$X_S = E[X_T \mid \mathscr{F}_S],$$

which completes the proof. $\qquad\qquad\square$

We now give two corollaries of Theorem 3.22.

Corollary 3.23 *Let $(X_t)_{t \geq 0}$ be a martingale with right-continuous sample paths, and let $S \leq T$ be two bounded stopping times. Then X_S and X_T are in L^1 and*

$$X_S = E[X_T \mid \mathscr{F}_S].$$

Proof Let $a \geq 0$ such that $S \leq T \leq a$. We apply Theorem 3.22 to the martingale $(X_{t \wedge a})_{t \geq 0}$ which is closed by X_a. $\qquad\qquad\square$

The second corollary shows that a martingale (resp. a uniformly integrable martingale) stopped at an arbitrary stopping time remains a martingale (resp. a uniformly integrable martingale). This result will play an important role in the next chapters.

Corollary 3.24 *Let $(X_t)_{t \geq 0}$ be a martingale with right-continuous sample paths, and let T be a stopping time.*

 (i) *The process $(X_{t \wedge T})_{t \geq 0}$ is still a martingale.*
 (ii) *Suppose in addition that the martingale $(X_t)_{t \geq 0}$ is uniformly integrable. Then the process $(X_{t \wedge T})_{t \geq 0}$ is also a uniformly integrable martingale, and more precisely we have for every $t \geq 0$,*

$$X_{t \wedge T} = E[X_T \mid \mathscr{F}_t]. \tag{3.5}$$

Proof We start with the proof of (ii). Note that $t \wedge T$ is a stopping time by property (f) of stopping times. By Theorem 3.22, $X_{t \wedge T}$ and X_T are in L^1, and we also know that $X_{t \wedge T}$ is $\mathscr{F}_{t \wedge T}$-measurable, hence $\mathscr{F}_t$-measurable since $\mathscr{F}_{t \wedge T} \subset \mathscr{F}_t$. So in order to get (3.5), it is enough to prove that, for every $A \in \mathscr{F}_t$,

$$E[1_A X_T] = E[1_A X_{t \wedge T}].$$

Let us fix $A \in \mathscr{F}_t$. First, we have trivially

$$E[1_{A \cap \{T \leq t\}} X_T] = E[1_{A \cap \{T \leq t\}} X_{t \wedge T}]. \tag{3.6}$$

On the other hand, by Theorem 3.22, we have

$$X_{t \wedge T} = E[X_T \mid \mathscr{F}_{t \wedge T}],$$

and we notice that we have both $A \cap \{T > t\} \in \mathscr{F}_t$ and $A \cap \{T > t\} \in \mathscr{F}_T$ (the latter as a straightforward consequence of the definition of $\mathscr{F}_T$), so that $A \cap \{T > t\} \in \mathscr{F}_t \cap \mathscr{F}_T = \mathscr{F}_{t \wedge T}$. Using the preceding display, we obtain

$$E[\mathbf{1}_{A \cap \{T > t\}} X_T] = E[\mathbf{1}_{A \cap \{T > t\}} X_{t \wedge T}].$$

By adding this equality to (3.6), we get the desired result.

To prove (i), we just need to apply (ii) to the (uniformly integrable) martingale $(X_{t \wedge a})_{a \geq 0}$, for any choice of $a \geq 0$.   □

Applications Above all, the optional stopping theorem is a powerful tool for explicit calculations of probability distributions. Let us give a few important and typical examples of such applications (several other examples can be found in the exercises of this and the following chapters). Let B be a real Brownian motion started from 0. We know that B is a martingale with continuous sample paths with respect to its canonical filtration. For every real a, set $T_a = \inf\{t \geq 0 : B_t = a\}$. Recall that $T_a < \infty$ a.s.

(a) *Law of the exit point from an interval.* For every $a < 0 < b$, we have

$$P(T_a < T_b) = \frac{b}{b-a} \quad , \quad P(T_b < T_a) = \frac{-a}{b-a}.$$

To get this result, consider the stopping time $T = T_a \wedge T_b$ and the stopped martingale $M_t = B_{t \wedge T}$ (this is a martingale by Corollary 3.24). Then, $|M|$ is bounded above by $b \vee |a|$, and the martingale M is thus uniformly integrable. We can apply Theorem 3.22 and we get

$$0 = E[M_0] = E[M_T] = b\, P(T_b < T_a) + a\, P(T_a < T_b).$$

Since we also have $P(T_b < T_a) + P(T_a < T_b) = 1$, the desired result follows. In fact the proof shows that the result remains valid if we replace Brownian motion by a martingale with continuous sample paths and initial value 0, provided we know that this process exits (a, b) a.s.

(b) *First moment of exit times.* For every $a > 0$, consider the stopping time $U_a = \inf\{t \geq 0 : |B_t| = a\}$. Then

$$E[U_a] = a^2.$$

To verify this, consider the martingale $M_t = B_t^2 - t$. By Corollary 3.24, $M_{t \wedge U_a}$ is still a martingale, and therefore $E[M_{t \wedge U_a}] = E[M_0] = 0$, giving $E[(B_{t \wedge U_a})^2] = E[t \wedge U_a]$. Then, on one hand, $E[t \wedge U_a]$ converges to $E[U_a]$ as

$t \to \infty$ by monotone convergence, on the other hand, $E[(B_{t \wedge U_a})^2]$ converges to $E[(B_{U_a})^2] = a^2$ as $t \to \infty$, by dominated convergence (note that $(B_{t \wedge U_a})^2 \leq a^2$). The stated result follows. We may observe that we have $E[U_a] < \infty$ in contrast with the property $E[T_a] = \infty$, which was noticed in Chap. 2.

(c) *Laplace transform of hitting times.* We now fix $a > 0$ and our goal is to compute the Laplace transform of T_a. For every $\lambda \in \mathbb{R}$, we can consider the exponential martingale

$$N_t^\lambda = \exp(\lambda B_t - \frac{\lambda^2}{2} t).$$

Suppose first that $\lambda > 0$. By Corollary 3.24, the stopped process $N_{t \wedge T_a}^\lambda$ is still a martingale, and we immediately see that this martingale is bounded above by $e^{\lambda a}$, hence uniformly integrable. By applying the last assertion of Theorem 3.22 to this martingale and to the stopping time $S = T_a$ (or to $S = \infty$) we get

$$e^{\lambda a} E[e^{-\frac{\lambda^2}{2} T_a}] = E[N_{T_a}^\lambda] = E[N_0^\lambda] = 1 .$$

Replacing λ by $\sqrt{2\lambda}$, we conclude that, for every $\lambda > 0$,

$$E[e^{-\lambda T_a}] = e^{-a\sqrt{2\lambda}}. \tag{3.7}$$

(This formula could also be deduced from the knowledge of the density of T_a, see Corollary 2.22.) As an instructive example, one may try to reproduce the preceding line of reasoning, using now the martingale N_t^λ for $\lambda < 0$: one gets an absurd result, which can be explained by the fact that the stopped martingale $N_{t \wedge T_a}^\lambda$ is **not uniformly integrable** when $\lambda < 0$. When applying Theorem 3.22, it is crucial to always verify the uniform integrability of the martingale. In most cases, this is done by verifying that the (stopped) martingale is bounded.

(d) *Laplace transform of exit times from an interval.* With the notation of (b), we have for every $a > 0$ and every $\lambda > 0$,

$$E[\exp(-\lambda U_a)] = \frac{1}{\cosh(a\sqrt{2\lambda})}.$$

To see this, first note that U_a and B_{U_a} are independent since, using the symmetry property of Brownian motion,

$$E[\mathbf{1}_{\{B_{U_a}=a\}} \exp(-\lambda U_a)] = E[\mathbf{1}_{\{B_{U_a}=-a\}} \exp(-\lambda U_a)] = \frac{1}{2} E[\exp(-\lambda U_a)].$$

Then the claimed formula is proved by the same method as in (c), writing $E[N_{U_a}^\lambda] = E[N_0^\lambda] = 1$ and noting that the application of the optional stopping theorem is justified by the fact that $N^\lambda_{t\wedge U_a}$ is bounded above by $e^{\lambda a}$. See also Exercise 3.27 for a more general formula.

We end this chapter with the optional stopping theorem for nonnegative supermartingales. This result will be useful in later applications to Markov processes. We first note that, if $(Z_t)_{t\geq 0}$ is a nonnegative supermartingale with right-continuous sample paths, $(Z_t)_{t\geq 0}$ is automatically bounded in L^1 since $E[Z_t] \leq E[Z_0]$, and by Theorem 3.19, Z_t converges a.s. to a random variable $Z_\infty \in L^1$ as $t \to \infty$. As explained before Theorem 3.22, we can thus make sense of Z_T for any (finite or not) stopping time T.

Theorem 3.25 *Let $(Z_t)_{t\geq 0}$ be a nonnegative supermartingale with right-continuous sample paths. Let U and V be two stopping times such that $U \leq V$. Then, Z_U and Z_V are in L^1, and*

$$Z_U \geq E[Z_V \mid \mathscr{F}_U].$$

Remark This implies that $E[Z_U] \geq E[Z_V]$, and since $Z_U = Z_V = Z_\infty$ on the event $\{U = \infty\}$, it also follows that

$$E[1_{\{U<\infty\}} Z_U] \geq E[1_{\{U<\infty\}} Z_V] \geq E[1_{\{V<\infty\}} Z_V].$$

Proof In the first step of the proof, we make the extra assumption that U and V are bounded and we verify that we then have $E[Z_U] \geq E[Z_V]$. Let $p \geq 1$ be an integer such that $U \leq p$ and $V \leq p$. For every integer $n \geq 0$, set

$$U_n = \sum_{k=0}^{p2^n-1} \frac{k+1}{2^n} 1_{\{k2^{-n}<U\leq(k+1)2^{-n}\}}, \quad V_n = \sum_{k=0}^{p2^n-1} \frac{k+1}{2^n} 1_{\{k2^{-n}<V\leq(k+1)2^{-n}\}}$$

in such a way (by Proposition 3.8) that (U_n) and (V_n) are two sequences of bounded stopping times that decrease respectively to U and V, and additionally we have $U_n \leq V_n$ for every $n \geq 0$. The right-continuity of sample paths ensures that $Z_{U_n} \longrightarrow Z_U$ and $Z_{V_n} \longrightarrow Z_V$ a.s. as $n \to \infty$. Then, by the optional stopping theorem for discrete supermartingales in the case of bounded stopping times (see Appendix A2), with respect to the filtration $(\mathscr{F}_{k/2^{n+1}})_{k\geq 0}$, we have for every $n \geq 0$,

$$Z_{U_{n+1}} \geq E[Z_{U_n} \mid \mathscr{F}_{U_{n+1}}].$$

Setting $Y_n = Z_{U_{-n}}$ and $\mathscr{H}_n = \mathscr{F}_{U_{-n}}$, for every integer $n \leq 0$, we get that the sequence $(Y_n)_{n\leq 0}$ is a backward supermartingale with respect to the filtration $(\mathscr{H}_n)_{n\leq 0}$. Since, for every $n \geq 0$, $E[Z_{U_n}] \leq E[Z_0]$ (by another application of the discrete optional stopping theorem), the sequence $(Y_n)_{n\leq 0}$ is bounded in L^1, and by the convergence theorem for backward supermartingales (see Appendix A2), it converges in L^1. Hence the convergence of Z_{U_n} to Z_U also holds in L^1 and similarly the convergence of Z_{V_n} to Z_V holds in L^1. Since $U_n \leq V_n$, by yet another application of the discrete optional stopping theorem, we have $E[Z_{U_n}] \geq E[Z_{V_n}]$. Using the L^1-convergence of Z_{U_n} and Z_{V_n} we can pass to the limit $n \to \infty$ and obtain that $E[Z_U] \geq E[Z_V]$ as claimed.

Let us prove the statement of the theorem (no longer assuming that U and V are bounded). By the first step of the proof applied to the stopping times 0 and $U \wedge p$, we have $E[Z_{U \wedge p}] \leq E[Z_0]$ for every $p \geq 1$, and Fatou's lemma gives $E[Z_U] \leq E[Z_0] < \infty$ and similarly $E[Z_V] < \infty$. Fix $A \in \mathscr{F}_U \subset \mathscr{F}_V$ and recall our notation U^A for the stopping time defined by $U^A(\omega) = U(\omega)$ if $\omega \in A$ and $U^A(\omega) = \infty$ otherwise (cf. property (d) of stopping times). By the first part of the proof, we have, for every $p \geq 1$,

$$E[Z_{U^A \wedge p}] \geq E[Z_{V^A \wedge p}].$$

By writing each of these two expectations as a sum of expectations over the sets A^c, $A \cap \{U \leq p\}$ and $A \cap \{U > p\}$, and noting that $U > p$ implies $V > p$, we get

$$E[Z_U \, \mathbf{1}_{A \cap \{U \leq p\}}] \geq E[Z_V \, \mathbf{1}_{A \cap \{U \leq p\}}].$$

Letting $p \to \infty$ then gives

$$E[Z_U \, \mathbf{1}_{A \cap \{U < \infty\}}] \geq E[Z_V \, \mathbf{1}_{A \cap \{U < \infty\}}].$$

On the other hand, the equality $E[Z_U \, \mathbf{1}_{A \cap \{U = \infty\}}] = E[Z_V \, \mathbf{1}_{A \cap \{U = \infty\}}]$ is trivial and by adding it to the preceding display, we obtain

$$E[Z_U \, \mathbf{1}_A] \geq E[Z_V \, \mathbf{1}_A] = E[E[Z_V \mid \mathscr{F}_U] \, \mathbf{1}_A].$$

Since this holds for every $A \in \mathscr{F}_U$ and Z_U is $\mathscr{F}_U$-measurable, the desired result follows. $\square$

Exercises

In the following exercises, processes are defined on a probability space $(\Omega, \mathscr{F}, P)$ equipped with a complete filtration $(\mathscr{F}_t)_{t \in [0, \infty]}$.

Exercise 3.26

1. Let M be a martingale with continuous sample paths such that $M_0 = x \in \mathbb{R}_+$. We assume that $M_t \geq 0$ for every $t \geq 0$, and that $M_t \to 0$ when $t \to \infty$, a.s. Show that, for every $y > x$,

$$P\left(\sup_{t \geq 0} M_t \geq y \right) = \frac{x}{y}.$$

2. Give the law of

$$\sup_{t \leq T_0} B_t$$

when B is a Brownian motion started from $x > 0$ and $T_0 = \inf\{t \geq 0 : B_t = 0\}$.

3. Assume now that B is a Brownian motion started from 0, and let $\mu > 0$. Using an appropriate exponential martingale, show that

$$\sup_{t \geq 0}(B_t - \mu t)$$

is exponentially distributed with parameter 2μ.

Exercise 3.27 Let B be an $(\mathscr{F}_t)$-Brownian motion started from 0. Recall the notation $T_x = \inf\{t \geq 0 : B_t = x\}$, for every $x \in \mathbb{R}$. We fix two real numbers a and b with $a < 0 < b$, and we set

$$T = T_a \wedge T_b \,.$$

1. Show that, for every $\lambda > 0$,

$$E[\exp(-\lambda T)] = \frac{\cosh(\frac{b+a}{2}\sqrt{2\lambda})}{\cosh(\frac{b-a}{2}\sqrt{2\lambda})}\,.$$

(*Hint*: One may consider a martingale of the form

$$M_t = \exp\left(\sqrt{2\lambda}(B_t - \alpha) - \lambda t\right) + \exp\left(-\sqrt{2\lambda}(B_t - \alpha) - \lambda t\right)$$

with a suitable choice of α.)

2. Show similarly that, for every $\lambda > 0$,

$$E[\exp(-\lambda T)\,\mathbf{1}_{\{T=T_a\}}] = \frac{\sinh(b\sqrt{2\lambda})}{\sinh((b-a)\sqrt{2\lambda})}\,.$$

3. Recover the formula for $P(T_a < T_b)$ from question (**2**).

Exercise 3.28 Let $(B_t)_{t \geq 0}$ be an $(\mathscr{F}_t)$-Brownian motion started from 0. Let $a > 0$ and

$$\sigma_a = \inf\{t \geq 0 : B_t \leq t - a\}.$$

1. Show that σ_a is a stopping time and that $\sigma_a < \infty$ a.s.

2. Using an appropriate exponential martingale, show that, for every $\lambda \geq 0$,

$$E[\exp(-\lambda \sigma_a)] = \exp(-a(\sqrt{1 + 2\lambda} - 1)).$$

The fact that this formula remains valid for $\lambda \in [-\frac{1}{2}, 0[$ can be obtained via an argument of analytic continuation.

3. Let $\mu \in \mathbb{R}$ and $M_t = \exp(\mu B_t - \frac{\mu^2}{2} t)$. Show that the stopped martingale $M_{\sigma_a \wedge t}$ is closed if and only if $\mu \leq 1$. (*Hint*: This martingale is closed if and only if $E[M_{\sigma_a}] = 1$.)

Exercise 3.29 Let $(Y_t)_{t \geq 0}$ be a uniformly integrable martingale with continuous sample paths, such that $Y_0 = 0$. We set $Y_\infty = \lim_{t \to \infty} Y_t$. Let $p \geq 1$ be a fixed real number. We say that Property (P) holds for the martingale Y if there exists a constant C such that, for every stopping time T, we have

$$E[|Y_\infty - Y_T|^p \mid \mathscr{F}_T] \leq C.$$

1. Show that Property (P) holds for Y if Y_∞ is bounded.
2. Let B be an $(\mathscr{F}_t)$-Brownian motion started from 0. Show that Property (P) holds for the martingale $Y_t = B_{t \wedge 1}$. (*Hint*: One may observe that the random variable $\sup_{t \leq 1} |B_t|$ is in L^p.)
3. Show that Property (P) holds for Y, with the constant C, if and only if, for any stopping time T,

$$E[|Y_T - Y_\infty|^p] \leq C P[T < \infty].$$

   (*Hint*: It may be useful to consider the stopping times T^A defined for $A \in \mathscr{F}_T$ in property (d) of stopping times.)
4. We assume that Property (P) holds for Y with the constant C. Let S be a stopping time and let Y^S be the stopped martingale defined by $Y_t^S = Y_{t \wedge S}$ (see Corollary 3.24). Show that Property (P) holds for Y^S with the same constant C. One may start by observing that, if S and T are stopping times, one has $Y_T^S = Y_{S \wedge T} = Y_S^T = E[Y_T \mid \mathscr{F}_S]$.
5. We assume in this question and the next one that Property (P) holds for Y with the constant $C = 1$. Let $a > 0$, and let $(R_n)_{n \geq 0}$ be the sequence of stopping times defined by induction by

$$R_0 = 0, \qquad R_{n+1} = \inf\{t \geq R_n : |Y_t - Y_{R_n}| \geq a\} \qquad (\inf \varnothing = \infty).$$

   Show that, for every integer $n \geq 0$,

$$a^p P(R_{n+1} < \infty) \leq P(R_n < \infty).$$

6. Infer that, for every $x > 0$,

$$P\left(\sup_{t \geq 0} Y_t > x\right) \leq 2^p \, 2^{-px/2}.$$

Notes and Comments

This chapter gives a brief presentation of the so-called general theory of processes. We limited ourselves to the notions that are needed in the remaining part of this book, but the interested reader can consult the treatise of Dellacherie and Meyer [13, 14] for more about this subject. Most of the martingale theory presented in Sections 3 and 4 goes back to Doob [15]. A comprehensive study of the theory of continuous time martingales can be found in [14]. The applications of the optional stopping theorem to Brownian motion are very classical. For other applications in the same vein, see in particular the book [70] by Revuz and Yor. Exercise 3.29 is taken from the theory of BMO martingales, see e.g. [18, Chapter 7].

Chapter 4
Continuous Semimartingales

Continuous semimartingales provide the general class of processes with continuous sample paths for which we will develop the theory of stochastic integration in the next chapter. By definition, a continuous semimartingale is the sum of a continuous local martingale and a (continuous) finite variation process. In the present chapter, we study separately these two classes of processes. We start with some preliminaries about deterministic functions with finite variation, before considering the corresponding random processes. We then define (continuous) local martingales and we construct the quadratic variation of a local martingale, which will play a fundamental role in the construction of stochastic integrals. We explain how properties of a local martingale are related to those of its quadratic variation. Finally, we introduce continuous semimartingales and their quadratic variation processes.

4.1 Finite Variation Processes

In this chapter, all processes are indexed by $\mathbb{R}_+$ and take real values. The first section provides a brief presentation of finite variation processes. We start by discussing functions with finite variation in a deterministic setting.

4.1.1 Functions with Finite Variation

In our discussion of functions with finite variation, we restrict our attention to *continuous* functions, as this is the case of interest in the subsequent developments. Recall that a signed measure on a compact interval $[0, T]$ is the difference of two finite positive measures on $[0, T]$.

© Springer International Publishing Switzerland 2016
J.-F. Le Gall, *Brownian Motion, Martingales, and Stochastic Calculus*,
Graduate Texts in Mathematics 274, DOI 10.1007/978-3-319-31089-3_4

Definition 4.1 Let $T \geq 0$. A continuous function $a : [0, T] \longrightarrow \mathbb{R}$ such that $a(0) = 0$ is said to have *finite variation* if there exists a signed measure μ on $[0, T]$ such that $a(t) = \mu([0, t])$ for every $t \in [0, T]$.

The measure μ is then determined uniquely by a. Since a is continuous and $a(0) = 0$, it follows that μ has no atoms.

Remark The general definition of a function with finite variation does not require continuity nor the condition $a(0) = 0$. We impose these two conditions for convenience.

The decomposition of μ as a difference of two finite positive measures on $[0, T]$ is not unique, but there exists a unique decomposition $\mu = \mu_+ - \mu_-$ such that μ_+ and μ_- are supported on disjoint Borel sets. To get the existence of such a decomposition, start from an arbitrary decomposition $\mu = \mu_1 - \mu_2$, set $\nu = \mu_1 + \mu_2$ and then use the Radon–Nikodym theorem to find two nonnegative Borel functions h_1 and h_2 on $[0, T]$ such that

$$\mu_1(dt) = h_1(t)\nu(dt), \quad \mu_2(dt) = h_2(t)\nu(dt).$$

Then, if $h(t) = h_1(t) - h_2(t)$, we have

$$\mu(dt) = h(t)\nu(dt) = h(t)^+\nu(dt) - h(t)^-\nu(dt),$$

which gives the decomposition $\mu = \mu_+ - \mu_-$ with $\mu_+(dt) = h(t)^+\nu(dt)$, $\mu_-(dt) = h(t)^-\nu(dt)$, and the measures μ_+ and μ_- are supported respectively on the disjoint Borel sets $D_+ = \{t : h(t) > 0\}$ and $D_- = \{t : h(t) < 0\}$. The uniqueness of this decomposition $\mu = \mu_+ - \mu_-$ follows from the fact that we have necessarily, for every $A \in \mathscr{B}([0, T])$,

$$\mu_+(A) = \sup\{\mu(C) : C \in \mathscr{B}([0, T]), \ C \subset A\}.$$

We write $|\mu|$ for the (finite) positive measure $|\mu| = \mu_+ + \mu_-$. The measure $|\mu|$ is called the *total variation* of a. We have $|\mu(A)| \leq |\mu|(A)$ for every $A \in \mathscr{B}([0, T])$. Moreover, the Radon–Nikodym derivative of μ with respect to $|\mu|$ is

$$\frac{d\mu}{d|\mu|} = 1_{D_+} - 1_{D_-}.$$

The fact that $a(t) = \mu_+([0, t]) - \mu_-([0, t])$ shows that a is the difference of two monotone nondecreasing continuous functions that vanish at 0 (since μ has no atoms, the same holds for μ_+ of μ_-). Conversely, the difference of two monotone nondecreasing continuous functions that vanish at 0 has finite variation in the sense of the previous definition. Indeed, this follows from the well-known fact that the formula $g(t) = \theta([0, t])$, $t \in [0, T]$ induces a bijection between monotone nondecreasing right-continuous functions $g : [0, T] \longrightarrow \mathbb{R}_+$ and finite positive measures θ on $[0, T]$.

Let $f : [0, T] \longrightarrow \mathbb{R}$ be a measurable function such that $\int_{[0,T]} |f(s)| \, |\mu|(ds) < \infty$. We set

$$\int_0^T f(s) \, da(s) = \int_{[0,T]} f(s) \, \mu(ds),$$

$$\int_0^T f(s) \, |da(s)| = \int_{[0,T]} f(s) \, |\mu|(ds).$$

Then the bound

$$\left| \int_0^T f(s) \, da(s) \right| \le \int_0^T |f(s)| \, |da(s)|$$

holds. By restricting a to $[0, t]$ (which amounts to restricting μ, μ_+, μ_-), we can define $\int_0^t f(s) \, da(s)$ for every $t \in [0, T]$, and we observe that the function $t \mapsto \int_0^t f(s) \, da(s)$ also has finite variation on $[0, T]$ (the associated measure is just $\mu'(ds) = f(s)\mu(ds)$).

Proposition 4.2 *For every* $t \in (0, T]$,

$$\int_0^t |da(s)| = \sup \left\{ \sum_{i=1}^p |a(t_i) - a(t_{i-1})| \right\},$$

where the supremum is over all subdivisions $0 = t_0 < t_1 < \cdots < t_p = t$ *of* $[0, t]$. *More precisely, for any increasing sequence* $0 = t_0^n < t_1^n < \cdots < t_{p_n}^n = t$ *of subdivisions of* $[0, t]$ *whose mesh tends to* 0, *we have*

$$\lim_{n \to \infty} \sum_{i=1}^{p_n} |a(t_i^n) - a(t_{i-1}^n)| = \int_0^t |da(s)|.$$

Remark In the usual presentation of functions with finite variation, one starts from the property that the supremum in the first display of the proposition is finite.

Proof Clearly, it is enough to treat the case $t = T$. The inequality $\ge$ in the first assertion is very easy since, for any subdivision $0 = t_0 < t_1 < \cdots < t_p = T$ of $[0, T]$,

$$|a(t_i) - a(t_{i-1})| = |\mu((t_{i-1}, t_i])| \le |\mu|((t_{i-1}, t_i]), \qquad \forall i \in \{1, \ldots, p\},$$

and

$$\sum_{i=1}^p |\mu|((t_{i-1}, t_i]) = |\mu|([0, t]) = \int_0^T |da(s)|.$$

In order to get the reverse inequality, it suffices to prove the second assertion. So we consider an increasing sequence $0 = t_0^n < t_1^n < \cdots < t_{p_n}^n = T$ of subdivisions of $[0, T]$, whose mesh $\max\{t_i^n - t_{i-1}^n : 1 \leq i \leq p_n\}$ tends to 0. Although we are proving a "deterministic" result, we will use a martingale argument. Leaving aside the trivial case where $|\mu| = 0$, we introduce the probability space $\Omega = [0, T]$, which is equipped with the Borel σ-field $\mathscr{B} = \mathscr{B}([0, T])$ and the probability measure $P(\mathrm{d}s) = (|\mu|([0, T]))^{-1} |\mu|(\mathrm{d}s)$. On this probability space, we consider the discrete filtration $(\mathscr{B}_n)_{n \geq 0}$ such that, for every integer $n \geq 0$, $\mathscr{B}_n$ is the σ-field generated by the intervals $(t_{i-1}^n, t_i^n]$, $1 \leq i \leq p_n$. We then set

$$X(s) = \mathbf{1}_{D_+}(s) - \mathbf{1}_{D_-}(s) = \frac{\mathrm{d}\mu}{\mathrm{d}|\mu|}(s),$$

and, for every $n \geq 0$,

$$X_n = E[X \mid \mathscr{B}_n].$$

Properties of conditional expectation show that X_n is constant on every interval $(t_{i-1}^n, t_i^n]$ and takes the value

$$\frac{\mu((t_{i-1}^n, t_i^n])}{|\mu|((t_{i-1}^n, t_i^n])} = \frac{a(t_i^n) - a(t_{i-1}^n)}{|\mu|((t_{i-1}^n, t_i^n])}$$

on this interval. On the other hand, the sequence (X_n) is a closed martingale, with respect to the discrete filtration $(\mathscr{B}_n)$. Since X is measurable with respect to $\mathscr{B} = \bigvee_n \mathscr{B}_n$, this martingale converges to X in L^1, by the convergence theorem for closed discrete martingales (see Appendix A2). In particular,

$$\lim_{n \to \infty} E[|X_n|] = E[|X|] = 1,$$

where the last equality is clear since $|X(s)| = 1$, $|\mu|(\mathrm{d}s)$ a.e. The desired result follows by noting that

$$E[|X_n|] = (|\mu|([0, T]))^{-1} \sum_{i=1}^{p_n} |a(t_i^n) - a(t_{i-1}^n)|,$$

and recalling that $|\mu|([0, T]) = \int_0^T |\mathrm{d}a(s)|$.   □

We now give a useful approximation lemma for the integral of a continuous function with respect to a function with finite variation.

Lemma 4.3 *If $f : [0, T] \longrightarrow \mathbb{R}$ is a continuous function, and if $0 = t_0^n < t_1^n < \cdots < t_{p_n}^n = T$ is a sequence of subdivisions of $[0, T]$ whose mesh tends to 0, we have*

$$\int_0^T f(s) \, da(s) = \lim_{n \to \infty} \sum_{i=1}^{p_n} f(t_{i-1}^n) \, (a(t_i^n) - a(t_{i-1}^n)).$$

Proof Let f_n be defined on $[0, T]$ by $f_n(s) = f(t_{i-1}^n)$ if $s \in (t_{i-1}^n, t_i^n]$, $1 \le i \le p_n$, and $f_n(0) = f(0)$. Then,

$$\sum_{i=1}^{p_n} f(t_{i-1}^n) \, (a(t_i^n) - a(t_{i-1}^n)) = \int_{[0,T]} f_n(s) \, \mu(ds),$$

and the desired result follows by dominated convergence since $f_n(s) \longrightarrow f(s)$ as $n \to \infty$, for every $s \in [0, T]$. □

We say that a function $a : \mathbb{R}_+ \longrightarrow \mathbb{R}$ is a finite variation function on $\mathbb{R}_+$ if the restriction of a to $[0, T]$ has finite variation on $[0, T]$, for every $T > 0$. Then there is a unique σ-finite (positive) measure on $\mathbb{R}_+$ whose restriction to every interval $[0, T]$ is the total variation measure of the restriction of a to $[0, T]$, and we write

$$\int_0^\infty f(s) \, |da(s)|$$

for the integral of a nonnegative Borel function f on $\mathbb{R}_+$ with respect to this σ-finite measure. Furthermore, we can define

$$\int_0^\infty f(s) da(s) = \lim_{T \to \infty} \int_0^T f(s) da(s) \in (-\infty, \infty)$$

for any real Borel function f on $\mathbb{R}_+$ such that $\int_0^\infty |f(s)||da(s)| < \infty$.

4.1.2 Finite Variation Processes

We now consider random variables and processes defined on a filtered probability space $(\Omega, \mathscr{F}, (\mathscr{F}_t), P)$.

Definition 4.4 An adapted process $A = (A_t)_{t \ge 0}$ is called a *finite variation process* if all its sample paths are finite variation functions on $\mathbb{R}_+$. If in addition the sample paths are nondecreasing functions, the process A is called an *increasing process*.

Remark In particular, $A_0 = 0$ and the sample paths of A are continuous – one can define finite variation processes with càdlàg sample paths, but in this book we consider **only the case of continuous sample paths**. Our special convention that the initial value of a finite variation process is 0 will be convenient for certain uniqueness statements.

If A is a finite variation process, the process

$$V_t = \int_0^t |dA_s|$$

is an increasing process. Indeed, it is clear that the sample paths of V are nondecreasing functions (as well as continuous functions that vanish at $t = 0$). The fact that V_t is an $\mathscr{F}_t$-measurable random variable can be deduced from the second part of Proposition 4.2. Writing $A_t = \frac{1}{2}(V_t + A_t) - \frac{1}{2}(V_t - A_t)$ shows that any finite variation process can be written as the difference of two increasing processes (the converse is obvious).

Proposition 4.5 *Let A be a finite variation process, and let H be a progressive process such that*

$$\forall t \geq 0, \ \forall \omega \in \Omega, \ \int_0^t |H_s(\omega)| \, |dA_s(\omega)| < \infty.$$

Then the process $H \cdot A = ((H \cdot A)_t)_{t \geq 0}$ defined by

$$(H \cdot A)_t = \int_0^t H_s \, dA_s$$

is also a finite variation process.

Proof By the observations preceding the statement of Proposition 4.2, we know that the sample paths of $H \cdot A$ are finite variation functions. It remains to verify that the process $H \cdot A$ is adapted. To this end, it is enough to check that, if $t > 0$ is fixed, if $h : \Omega \times [0, t] \longrightarrow \mathbb{R}$ is measurable for the σ-field $\mathscr{F}_t \otimes \mathscr{B}([0, t])$, and if $\int_0^t |h(\omega, s)| |dA_s(\omega)| < \infty$ for every ω, then the variable $\int_0^t h(\omega, s) dA_s(\omega)$ is $\mathscr{F}_t$-measurable.

If $h(\omega, s) = \mathbf{1}_{(u,v]}(s) \mathbf{1}_\Gamma(\omega)$ with $(u, v] \subset [0, t]$ and $\Gamma \in \mathscr{F}_t$, the result is immediate since $\int_0^t h(\omega, s) dA_s(\omega) = \mathbf{1}_\Gamma(\omega) (A_v(\omega) - A_u(\omega))$ in that case. A monotone class argument (see Appendix A1) then gives the case $h = \mathbf{1}_G$, $G \in \mathscr{F}_t \otimes \mathscr{B}([0, t])$. Finally, in the general case, we observe that we can write h as a pointwise limit of a sequence of simple functions (i.e. finite linear combinations of indicator functions of measurable sets) h_n such that $|h_n| \leq |h|$ for every n, and that we then have $\int_0^t h_n(\omega, s) dA_s(\omega) \longrightarrow \int_0^t h(\omega, s) dA_s(\omega)$ by dominated convergence, for every $\omega \in \Omega$. $\square$

Remarks

(i) It happens frequently that instead of the assumption of the proposition we have the weaker assumption

$$\text{a.s.} \quad \forall t \geq 0, \quad \int_0^t |H_s(\omega)| \, |dA_s(\omega)| < \infty.$$

If the filtration is complete, we can still define $H \cdot A$ as a finite variation process under this weaker assumption. We replace H by the process H' defined by

$$H'_t(\omega) = \begin{cases} H_t(\omega) & \text{if } \int_0^n |H_s(\omega)| \, |dA_s(\omega)| < \infty, \ \forall n, \\ 0 & \text{otherwise.} \end{cases}$$

Thanks to the fact that the filtration is complete, the process H' is still progressive, which allows us to define $H \cdot A = H' \cdot A$. We will use this extension of Proposition 4.5 implicitly in what follows.

(ii) Under appropriate assumptions (if H and K are progressive and $\int_0^t |H_s| \, |dA_s| < \infty$, $\int_0^t |H_s K_s| \, |dA_s| < \infty$ for every $t \geq 0$), we have the "associativity" property

$$K \cdot (H \cdot A) = (KH) \cdot A. \tag{4.1}$$

This indeed follows from the analogous deterministic result saying informally that $k(s) \, (h(s) \, \mu(ds)) = (k(s)h(s)) \, \mu(ds)$ if $h(s)$ and $k(s)h(s)$ are integrable with respect to the signed measure μ on $[0, t]$.

An important special case of the proposition is the case where $A_t = t$. If H is a progressive process such that

$$\forall t \geq 0, \ \forall \omega \in \Omega, \quad \int_0^t |H_s(\omega)| \, ds < \infty,$$

the process $\int_0^t H_s \, ds$ is a finite variation process.

4.2 Continuous Local Martingales

We consider again a filtered probability space $(\Omega, \mathscr{F}, (\mathscr{F}_t), P)$. If T is a stopping time, and if $X = (X_t)_{t \geq 0}$ is an adapted process with continuous sample paths, we will write X^T for process X stopped at T, defined by $X_t^T = X_{t \wedge T}$ for every $t \geq 0$. It is useful to observe that, if S is another stopping time,

$$(X^T)^S = (X^S)^T = X^{S \wedge T}.$$

Definition 4.6 An adapted process $M = (M_t)_{t \geq 0}$ with continuous sample paths and such that $M_0 = 0$ a.s. is called a *continuous local martingale* if there exists a nondecreasing sequence $(T_n)_{n \geq 0}$ of stopping times such that $T_n \uparrow \infty$ (i.e. $T_n(\omega) \uparrow \infty$ for every ω) and, for every n, the stopped process M^{T_n} is a uniformly integrable martingale.

More generally, when we do not assume that $M_0 = 0$ a.s., we say that M is a continuous local martingale if the process $N_t = M_t - M_0$ is a continuous local martingale.

In all cases, we say that the sequence of stopping times (T_n) *reduces M* if $T_n \uparrow \infty$ and, for every n, the stopped process M^{T_n} is a uniformly integrable martingale.

Remarks

(i) We do not require in the definition of a continuous local martingale that the variables M_t are in L^1 (compare with the definition of martingales). In particular, the variable M_0 may be any $\mathscr{F}_0$-measurable random variable.

(ii) Any martingale with continuous sample paths is a continuous local martingale (see property (a) below) but the converse is false, and for this reason we will sometimes speak of "true martingales" to emphasize the difference with local martingales. Let us give a few examples of continuous local martingales which are not (true) martingales. If B is an $(\mathscr{F}_t)$-Brownian motion started from 0, and Z is an $\mathscr{F}_0$-measurable random variable, the process $M_t = Z + B_t$ is always a continuous local martingale, but is not a martingale if $E[|Z|] = \infty$. If we require the property $M_0 = 0$, we can also consider $M_t = ZB_t$, which is always a continuous local martingale (see Exercise 4.22) but is not a martingale if $E[|Z|] = \infty$. For a less artificial example, we refer to question (**8**) of Exercise 5.33.

(iii) One can define a notion of local martingale with càdlàg sample paths. In this course, however, we consider only continuous local martingales.

The following properties are easily established.

Properties of continuous local martingales.

(a) A martingale with continuous sample paths is a continuous local martingale, and the sequence $T_n = n$ reduces M.

(b) In the definition of a continuous local martingale starting from 0, one can replace "uniformly integrable martingale" by "martingale" (indeed, one can then observe that $M^{T_n \wedge n}$ is uniformly integrable, and we still have $T_n \wedge n \uparrow \infty$).

(c) If M is a continuous local martingale, then, for every stopping time T, M^T is a continuous local martingale (this follows from Corollary 3.24).

(d) If (T_n) reduces M and if (S_n) is a sequence of stopping times such that $S_n \uparrow \infty$, then the sequence $(T_n \wedge S_n)$ also reduces M (use Corollary 3.24 again).

(e) The space of all continuous local martingales is a vector space (to check stability under addition, note that if M and M' are two continuous local martingales such that $M_0 = 0$ and $M'_0 = 0$, if the sequence (T_n) reduces M and if the sequence (T'_n) reduces M', property (d) shows that the sequence $T_n \wedge T'_n$ reduces $M + M'$).

The next proposition gives three other useful properties of local martingales.

Proposition 4.7

(i) *A nonnegative continuous local martingale M such that $M_0 \in L^1$ is a supermartingale.*

(ii) *A continuous local martingale M such that there exists a random variable $Z \in L^1$ with $|M_t| \leq Z$ for every $t \geq 0$ (in particular a bounded continuous local martingale) is a uniformly integrable martingale.*

(iii) *If M is a continuous local martingale and $M_0 = 0$ (or more generally $M_0 \in L^1$), the sequence of stopping times*

$$T_n = \inf\{t \geq 0 : |M_t| \geq n\}$$

reduces M.

Proof

(i) Write $M_t = M_0 + N_t$. By definition, there exists a sequence (T_n) of stopping times that reduces N. Then, if $s \leq t$, we have for every n,

$$N_{s \wedge T_n} = E[N_{t \wedge T_n} \mid \mathscr{F}_s].$$

We can add on both sides the random variable M_0 (which is $\mathscr{F}_0$-measurable and in L^1 by assumption), and we get

$$M_{s \wedge T_n} = E[M_{t \wedge T_n} \mid \mathscr{F}_s].$$

Since M takes nonnegative values, we can now let n tend to ∞ and apply the version of Fatou's lemma for conditional expectations, which gives

$$M_s \geq E[M_t \mid \mathscr{F}_s].$$

Taking $s = 0$, we get $E[M_t] \leq E[M_0] < \infty$, hence $M_t \in L^1$ for every $t \geq 0$. The previous inequality now shows that M is a supermartingale.

(ii) By the same argument as in (i), we get for $0 \leq s \leq t$,

$$M_{s \wedge T_n} = E[M_{t \wedge T_n} \mid \mathscr{F}_s]. \tag{4.2}$$

Since $|M_{t \wedge T_n}| \leq Z$, we can use dominated convergence to obtain that the sequence $M_{t \wedge T_n}$ converges to M_t in L^1. We can thus pass to the limit $n \to \infty$ in (4.2), and get that $M_s = E[M_t \mid \mathscr{F}_s]$.

(iii) Suppose that $M_0 = 0$. The random times T_n are stopping times by Proposition 3.9. The desired result is an immediate consequence of (ii) since M^{T_n} is a continuous local martingale and $|M^{T_n}| \leq n$. If we only assume that $M_0 \in L^1$, we observe that M^{T_n} is dominated by $n + |M_0|$. $\square$

Remark Considering property (ii) of the proposition, one might expect that a continuous local martingale M such that the collection $(M_t)_{t\geq 0}$ is uniformly integrable (or even a continuous local martingale satisfying the stronger property of being bounded in L^p for some $p > 1$) is automatically a martingale. This is incorrect!! For instance, if B is a three-dimensional Brownian motion started from $x \neq 0$, the process $M_t = 1/|B_t|$ is a continuous local martingale bounded in L^2, but is not a martingale: see Exercise 5.33.

Theorem 4.8 *Let M be a continuous local martingale. Assume that M is also a finite variation process (in particular $M_0 = 0$). Then $M_t = 0$ for every $t \geq 0$, a.s.*

Proof Set

$$\tau_n = \inf\{t \geq 0 : \int_0^t |dM_s| \geq n\}$$

for every integer $n \geq 0$. By Proposition 3.9, τ_n is a stopping time (recall that $\int_0^t |dM_s|$ is an increasing process if M is a finite variation process).

Fix $n \geq 0$ and set $N = M^{\tau_n}$. Note that, for every $t \geq 0$,

$$|N_t| = |M_{t\wedge\tau_n}| \leq \int_0^{t\wedge\tau_n} |dM_s| \leq n.$$

By Proposition 4.7, N is a (bounded) martingale. Let $t > 0$ and let $0 = t_0 < t_1 < \cdots < t_p = t$ be any subdivision of $[0, t]$. Then, from Proposition 3.14, we have

$$E[N_t^2] = \sum_{i=1}^{p} E[(N_{t_i} - N_{t_{i-1}})^2]$$

$$\leq E\left[\left(\sup_{1\leq i\leq p} |N_{t_i} - N_{t_{i-1}}|\right) \sum_{i=1}^{p} |N_{t_i} - N_{t_{i-1}}|\right]$$

$$\leq n E\left[\sup_{1\leq i\leq p} |N_{t_i} - N_{t_{i-1}}|\right]$$

noting that $\int_0^t |dN_s| \leq n$ by the definition of τ_n, and using Proposition 4.2.

We now apply the preceding bound to a sequence $0 = t_0^k < t_1^k < \cdots < t_{p_k}^k = t$ of subdivisions of $[0, t]$ whose mesh tends to 0. Using the continuity of sample paths, and the fact that N is bounded (to justify dominated convergence), we get

$$\lim_{k \to \infty} E\left[\sup_{1 \le i \le p_k} |N_{t_i^k} - N_{t_{i-1}^k}| \right] = 0.$$

We then conclude that $E[N_t^2] = 0$, hence $M_{t \wedge \tau_n} = 0$ a.s. Letting n tend to ∞, we get that $M_t = 0$ a.s. $\qquad \square$

4.3 The Quadratic Variation of a Continuous Local Martingale

From now on until the end of this chapter (and in the next chapter), we assume that the filtration $(\mathscr{F}_t)$ is complete. The next theorem will play a very important role in forthcoming developments.

Theorem 4.9 *Let $M = (M_t)_{t \ge 0}$ be a continuous local martingale. There exists an increasing process denoted by $(\langle M, M \rangle_t)_{t \ge 0}$, which is unique up to indistinguishability, such that $M_t^2 - \langle M, M \rangle_t$ is a continuous local martingale. Furthermore, for every fixed $t > 0$, if $0 = t_0^n < t_1^n < \cdots < t_{p_n}^n = t$ is an increasing sequence of subdivisions of $[0, t]$ with mesh tending to 0, we have*

$$\langle M, M \rangle_t = \lim_{n \to \infty} \sum_{i=1}^{p_n} (M_{t_i^n} - M_{t_{i-1}^n})^2 \qquad (4.3)$$

in probability. The process $\langle M, M \rangle$ is called the quadratic variation of M.

Let us immediately mention an important special case. If $M = B$ is an $(\mathscr{F}_t)$-Brownian motion (see Definition 3.11) then B is a martingale with continuous sample paths, hence a continuous local martingale. Then by comparing (4.3) with Proposition 2.16, we get that, for every $t \ge 0$,

$$\langle B, B \rangle_t = t.$$

So the quadratic variation of a Brownian motion is the simplest increasing process one can imagine.

Remarks

(i) We observe that the process $\langle M, M \rangle$ does not depend on the initial value M_0, but only on the increments of M: if $M_t = M_0 + N_t$, we have $\langle M, M \rangle = \langle N, N \rangle$. This is obvious from the second assertion of the theorem, and this will also be clear from the proof that follows.

(ii) In the second assertion of the theorem, it is in fact not necessary to assume that the sequence of subdivisions is increasing.

Proof We start by proving the first assertion. Uniqueness is an easy consequence of Theorem 4.8. Indeed, let A and A' be two increasing processes satisfying the condition given in the statement. Then the process $A_t - A'_t = (M_t^2 - A'_t) - (M_t^2 - A_t)$ is both a continuous local martingale and a finite variation process. It follows that $A - A' = 0$.

In order to prove existence, consider first the case where $M_0 = 0$ and M is bounded (hence M is a true martingale, by Proposition 4.7 (ii)). Fix $K > 0$ and an increasing sequence $0 = t_0^n < t_1^n < \cdots < t_{p_n}^n = K$ of subdivisions of $[0, K]$ with mesh tending to 0.

We observe that, for every $0 \leq r < s$ and for every bounded $\mathscr{F}_r$-measurable variable Z, the process

$$N_t = Z(M_{s \wedge t} - M_{r \wedge t})$$

is a martingale (the reader is invited to write down the easy proof!). It follows that, for every n, the process

$$X_t^n = \sum_{i=1}^{p_n} M_{t_{i-1}^n} (M_{t_i^n \wedge t} - M_{t_{i-1}^n \wedge t})$$

is a (bounded) martingale. The reason for considering these martingales comes from the following identity, which results from a simple calculation: for every n, for every $j \in \{0, 1, \ldots, p_n\}$,

$$M_{t_j^n}^2 - 2X_{t_j^n}^n = \sum_{i=1}^{j} (M_{t_i^n} - M_{t_{i-1}^n})^2. \tag{4.4}$$

Lemma 4.10 *We have*

$$\lim_{n,m \to \infty} E[(X_K^n - X_K^m)^2] = 0.$$

Proof of the lemma Let us fix $n \leq m$ and evaluate the product $E[X_K^n X_K^m]$. This product is equal to

$$\sum_{i=1}^{p_n} \sum_{j=1}^{p_m} E[M_{t_{i-1}^n} (M_{t_i^n} - M_{t_{i-1}^n}) M_{t_{j-1}^m} (M_{t_j^m} - M_{t_{j-1}^m})].$$

In this double sum, the only terms that may be nonzero are those corresponding to indices i and j such that the interval $(t_{j-1}^m, t_j^m]$ is contained in $(t_{i-1}^n, t_i^n]$. Indeed, suppose that $t_i^n \leq t_{j-1}^m$ (the symmetric case $t_j^m \leq t_{i-1}^n$ is treated in an analogous way).

Then, conditioning on the σ-field $\mathscr{F}_{t^m_{j-1}}$, we have

$$E[M_{t^n_{i-1}}(M_{t^n_i} - M_{t^n_{i-1}}) M_{t^m_{j-1}}(M_{t^m_j} - M_{t^m_{j-1}})]$$
$$= E[M_{t^n_{i-1}}(M_{t^n_i} - M_{t^n_{i-1}}) M_{t^m_{j-1}} E[M_{t^m_j} - M_{t^m_{j-1}} \mid \mathscr{F}_{t^m_{j-1}}]] = 0.$$

For every $j = 1, \ldots, p_m$, write $i_{n,m}(j)$ for the unique index i such that $(t^m_{j-1}, t^m_j] \subset (t^n_{i-1}, t^n_i]$. It follows from the previous considerations that

$$E[X^n_K X^m_K] = \sum_{1 \le j \le p_m,\, i=i_{n,m}(j)} E[M_{t^n_{i-1}}(M_{t^n_i} - M_{t^n_{i-1}}) M_{t^m_{j-1}}(M_{t^m_j} - M_{t^m_{j-1}})].$$

In each term $E[M_{t^n_{i-1}}(M_{t^n_i} - M_{t^n_{i-1}}) M_{t^m_{j-1}}(M_{t^m_j} - M_{t^m_{j-1}})]$, we can now decompose

$$M_{t^n_i} - M_{t^n_{i-1}} = \sum_{k:i_{n,m}(k)=i} (M_{t^m_k} - M_{t^m_{k-1}})$$

and we observe that, if k is such that $i_{n,m}(k) = i$ but $k \ne j$,

$$E[M_{t^n_{i-1}}(M_{t^m_k} - M_{t^m_{k-1}}) M_{t^m_{j-1}}(M_{t^m_j} - M_{t^m_{j-1}})] = 0$$

(condition on $\mathscr{F}_{t^m_{k-1}}$ if $k > j$ and on $\mathscr{F}_{t^m_{j-1}}$ if $k < j$). The only case that remains is $k = j$, and we have thus obtained

$$E[X^n_K X^m_K] = \sum_{1 \le j \le p_m,\, i=i_{n,m}(j)} E[M_{t^n_{i-1}} M_{t^m_{j-1}} (M_{t^m_j} - M_{t^m_{j-1}})^2].$$

As a special case of this relation, we have

$$E[(X^m_K)^2] = \sum_{1 \le j \le p_m} E[M^2_{t^m_{j-1}} (M_{t^m_j} - M_{t^m_{j-1}})^2].$$

Furthermore,

$$E[(X^n_K)^2] = \sum_{1 \le i \le p_n} E[M^2_{t^n_{i-1}} (M_{t^n_i} - M_{t^n_{i-1}})^2]$$
$$= \sum_{1 \le i \le p_n} E[M^2_{t^n_{i-1}} E[(M_{t^n_i} - M_{t^n_{i-1}})^2 \mid \mathscr{F}_{t^n_{i-1}}]]$$
$$= \sum_{1 \le i \le p_n} E\Big[M^2_{t^n_{i-1}} \sum_{j:i_{n,m}(j)=i} E[(M_{t^m_j} - M_{t^m_{j-1}})^2 \mid \mathscr{F}_{t^n_{i-1}}]\Big]$$
$$= \sum_{1 \le j \le p_m,\, i=i_{n,m}(j)} E[M^2_{t^n_{i-1}} (M_{t^m_j} - M_{t^m_{j-1}})^2],$$

where we have used Proposition 3.14 in the third equality.

If we combine the last three displays, we get

$$E[(X_K^n - X_K^m)^2] = E\Big[\sum_{1 \le j \le p_m,\, i = i_{n,m}(j)} (M_{t_{i-1}^n} - M_{t_{j-1}^m})^2 (M_{t_j^m} - M_{t_{j-1}^m})^2 \Big].$$

Using the Cauchy–Schwarz inequality, we then have

$$E[(X_K^n - X_K^m)^2] \le E\Big[\sup_{1 \le j \le p_m,\, i = i_{n,m}(j)} (M_{t_{i-1}^n} - M_{t_{j-1}^m})^4 \Big]^{1/2}$$

$$\times E\Big[\Big(\sum_{1 \le j \le p_m} (M_{t_j^m} - M_{t_{j-1}^m})^2 \Big)^2 \Big]^{1/2}.$$

By the continuity of sample paths (together with the fact that the mesh of our subdivisions tends to 0) and dominated convergence, we have

$$\lim_{n,m \to \infty,\, n \le m} E\Big[\sup_{1 \le j \le p_m,\, i = i_{n,m}(j)} (M_{t_{i-1}^n} - M_{t_{j-1}^m})^4 \Big] = 0.$$

To complete the proof of the lemma, it is then enough to prove the existence of a finite constant C such that, for every m,

$$E\Big[\Big(\sum_{1 \le j \le p_m} (M_{t_j^m} - M_{t_{j-1}^m})^2 \Big)^2 \Big] \le C. \tag{4.5}$$

Let A be a constant such that $|M_t| \le A$ for every $t \ge 0$. Expanding the square and using Proposition 3.14 twice, we have

$$E\Big[\Big(\sum_{1 \le j \le p_m} (M_{t_j^m} - M_{t_{j-1}^m})^2 \Big)^2 \Big]$$

$$= E\Big[\sum_{1 \le j \le p_m} (M_{t_j^m} - M_{t_{j-1}^m})^4 \Big] + 2E\Big[\sum_{1 \le j < k \le p_m} (M_{t_j^m} - M_{t_{j-1}^m})^2 (M_{t_k^m} - M_{t_{k-1}^m})^2 \Big]$$

$$\le 4A^2 E\Big[\sum_{1 \le j \le p_m} (M_{t_j^m} - M_{t_{j-1}^m})^2 \Big]$$

$$+ 2 \sum_{j=1}^{p_m - 1} E\Big[(M_{t_j^m} - M_{t_{j-1}^m})^2 E\Big[\sum_{k=j+1}^{p_m} (M_{t_k^m} - M_{t_{k-1}^m})^2 \,\Big|\, \mathscr{F}_{t_j^m} \Big] \Big]$$

$$= 4A^2 E\Big[\sum_{1 \le j \le p_m} (M_{t_j^m} - M_{t_{j-1}^m})^2 \Big]$$

$$+ 2 \sum_{j=1}^{p_m - 1} E\Big[(M_{t_j^m} - M_{t_{j-1}^m})^2 E[(M_K - M_{t_j^m})^2 \,|\, \mathscr{F}_{t_j^m}] \Big]$$

$$\leq 12A^2 E\Big[\sum_{1\leq j\leq p_m} (M_{t_j^m} - M_{t_{j-1}^m})^2 \Big]$$

$$= 12A^2 E[(M_K - M_0)^2]$$

$$\leq 48A^4$$

which gives the bound (4.5) with $C = 48A^4$. This completes the proof. □

We now return to the proof of the theorem. Thanks to Doob's inequality in L^2 (Proposition 3.15 (ii)), and to Lemma 4.10, we have

$$\lim_{n,m\to\infty} E\Big[\sup_{t\leq K}(X_t^n - X_t^m)^2 \Big] = 0. \tag{4.6}$$

In particular, for every $t \in [0, K]$, $(X_t^n)_{n\geq 1}$ is a Cauchy sequence in L^2 and thus converges in L^2. We want to argue that the limit yields a process Y indexed by $[0, K]$ with continuous sample paths. To see this, we note that (4.6) allows us find a strictly increasing sequence $(n_k)_{k\geq 1}$ of positive integers such that, for every $k \geq 1$,

$$E\Big[\sup_{t\leq K}(X_t^{n_{k+1}} - X_t^{n_k})^2 \Big] \leq 2^{-k}.$$

This implies that

$$E\Big[\sum_{k=1}^{\infty} \sup_{t\leq K} |X_t^{n_{k+1}} - X_t^{n_k}| \Big] < \infty$$

and thus

$$\sum_{k=1}^{\infty} \sup_{t\leq K} |X_t^{n_{k+1}} - X_t^{n_k}| < \infty, \quad \text{a.s.}$$

Consequently, except on the negligible set $\mathcal{N}$ where the series in the last display diverges, the sequence of random functions $(X_t^{n_k}, 0 \leq t \leq K)$ converges uniformly on $[0, K]$ as $k \to \infty$, and the limiting random function is continuous by uniform convergence. We can thus set $Y_t(\omega) = \lim X_t^{n_k}(\omega)$, for every $t \in [0, K]$, if $\omega \in \Omega\backslash\mathcal{N}$, and $Y_t(\omega) = 0$, for every $t \in [0, K]$, if $\omega \in \mathcal{N}$. The process $(Y_t)_{0\leq t\leq K}$ has continuous sample paths and Y_t is $\mathcal{F}_t$-measurable for every $t \in [0, K]$ (here we use the fact that the filtration is complete, which ensures that $\mathcal{N} \in \mathcal{F}_t$ for every $t \geq 0$). Furthermore, since the L^2-limit of $(X_t^n)_{n\geq 1}$ must coincide with the a.s. limit of a subsequence, Y_t is also the limit of X_t^n in L^2, for every $t \in [0, K]$, and we can pass to the limit in the martingale property for X^n, to obtain that $E[Y_t \mid \mathcal{F}_s] = Y_s$ for every $0 \leq s \leq t \leq K$. It follows that $(Y_{t\wedge K})_{t\geq 0}$ is a martingale with continuous sample paths.

On the other hand, the identity (4.4) shows that the sample paths of the process $M_t^2 - 2X_t^n$ are nondecreasing along the finite sequence $(t_i^n, 0 \leq i \leq p_n)$. By passing to the limit $n \to \infty$ along the sequence $(n_k)_{k \geq 1}$, we get that the sample paths of $M_t^2 - 2Y_t$ are nondecreasing on $[0, K]$, except maybe on the negligible set $\mathcal{N}$. For every $t \in [0, K]$, we set $A_t^{(K)} = M_t^2 - 2Y_t$ on $\Omega \setminus \mathcal{N}$, and $A_t^{(K)} = 0$ on $\mathcal{N}$. Then $A_0^{(K)} = 0$, $A_t^{(K)}$ is $\mathscr{F}_t$-measurable for every $t \in [0, K]$, $A^{(K)}$ has nondecreasing continuous sample paths, and $(M_{t \wedge K}^2 - A_{t \wedge K}^{(K)})_{t \geq 0}$ is a martingale.

We apply the preceding considerations with $K = \ell$, for every integer $\ell \geq 1$, and we get a process $(A_t^{(\ell)})_{0 \leq t \leq \ell}$. We then observe that, for every $\ell \geq 1$, $A_{t \wedge \ell}^{(\ell+1)} = A_{t \wedge \ell}^{(\ell)}$ for every $t \geq 0$, a.s., by the uniqueness argument explained at the beginning of the proof. It follows that we can define an increasing process $\langle M, M \rangle$ such that $\langle M, M \rangle_t = A_t^{(\ell)}$ for every $t \in [0, \ell]$ and every $\ell \geq 1$, a.s., and clearly $M_t^2 - \langle M, M \rangle_t$ is a martingale.

In order to get (4.3), we observe that, if $K > 0$ and the sequence of subdivisions $0 = t_0^n < t_1^n < \cdots < t_{p_n}^n = K$ are fixed as in the beginning of the proof, the process $A_{t \wedge K}^{(K)}$ must be indistinguishable from $\langle M, M \rangle_{t \wedge K}$, again by the uniqueness argument (we know that both $M_{t \wedge K}^2 - A_{t \wedge K}^{(K)}$ and $M_{t \wedge K}^2 - \langle M, M \rangle_{t \wedge K}$ are martingales). In particular, we have $\langle M, M \rangle_K = A_K^{(K)}$ a.s. Then, from (4.4) with $j = p_n$, and the fact that X_K^n converges in L^2 to $Y_K = \frac{1}{2}(M_K^2 - A_K^{(K)})$, we get that

$$\lim_{n \to \infty} \sum_{j=1}^{p_n} (M_{t_j^n} - M_{t_{j-1}^n})^2 = \langle M, M \rangle_K$$

in L^2. This completes the proof of the theorem in the case when $M_0 = 0$ and M is bounded.

Let us consider the general case. Writing $M_t = M_0 + N_t$, so that $M_t^2 = M_0^2 + 2M_0 N_t + N_t^2$, and noting that $M_0 N_t$ is a continuous local martingale (see Exercise 4.22), we see that we may assume that $M_0 = 0$. We then set

$$T_n = \inf\{t \geq 0 : |M_t| \geq n\}$$

and we can apply the bounded case to the stopped martingales M^{T_n}. Set $A^{[n]} = \langle M^{T_n}, M^{T_n} \rangle$. The uniqueness part of the theorem shows that the processes $A_{t \wedge T_n}^{[n+1]}$ and $A_t^{[n]}$ are indistinguishable. It follows that there exists an increasing process A such that, for every n, the processes $A_{t \wedge T_n}$ and $A_t^{[n]}$ are indistinguishable. By construction, $M_{t \wedge T_n}^2 - A_{t \wedge T_n}$ is a martingale for every n, which precisely implies that $M_t^2 - A_t$ is a continuous local martingale. We take $\langle M, M \rangle_t = A_t$, which completes the proof of the existence part of the theorem.

Finally, to get (4.3), it suffices to consider the case $M_0 = 0$. The bounded case then shows that (4.3) holds if M and $\langle M, M \rangle_t$ are replaced respectively by M^{T_n} and $\langle M, M \rangle_{t \wedge T_n}$ (even with convergence in L^2). Then it is enough to observe that, for every $t > 0$, $P(t \leq T_n)$ converges to 1 when $n \to \infty$. □

Proposition 4.11 *Let M be a continuous local martingale and let T be a stopping time. Then we have a.s. for every t ≥ 0,*

$$\langle M^T, M^T \rangle_t = \langle M, M \rangle_{t \wedge T}.$$

This follows from the fact that $M^2_{t \wedge T} - \langle M, M \rangle_{t \wedge T}$ is a continuous local martingale (cf. property (c) of continuous local martingales).

Proposition 4.12 *Let M be a continuous local martingale such that $M_0 = 0$. Then we have $\langle M, M \rangle = 0$ if and only if $M = 0$.*

Proof Suppose that $\langle M, M \rangle = 0$. Then M_t^2 is a nonnegative continuous local martingale and, by Proposition 4.7 (i), M_t^2 is a supermartingale, hence $E[M_t^2] \leq E[M_0^2] = 0$, so that $M_t = 0$ for every t. The converse is obvious. $\qquad\square$

The next theorem shows that properties of a continuous local martingale are closely related to those of its quadratic variation. If A is an increasing process, A_∞ denotes the increasing limit of A_t as $t \to \infty$ (this limit always exists in $[0, \infty]$).

Theorem 4.13 *Let M be a continuous local martingale with $M_0 \in L^2$.*

(i) *The following are equivalent:*

    (a) *M is a (true) martingale bounded in L^2.*
    (b) $E[\langle M, M \rangle_\infty] < \infty$.

    Furthermore, if these properties hold, the process $M_t^2 - \langle M, M \rangle_t$ is a uniformly integrable martingale, and in particular $E[M_\infty^2] = E[M_0^2] + E[\langle M, M \rangle_\infty]$.

(ii) *The following are equivalent:*

    (a) *M is a (true) martingale and $M_t \in L^2$ for every $t \geq 0$.*
    (b) $E[\langle M, M \rangle_t] < \infty$ *for every $t \geq 0$.*

Furthermore, if these properties hold, the process $M_t^2 - \langle M, M \rangle_t$ is a martingale.

Remark In property (a) of (i) (or of (ii)), it is essential to suppose that M is a martingale, and not only a continuous local martingale. Doob's inequality used in the following proof is not valid in general for a continuous local martingale!

Proof

(i) Replacing M by $M - M_0$, we may assume that $M_0 = 0$ in the proof. Let us first assume that M is a martingale bounded in L^2. Doob's inequality in L^2 (Proposition 3.15 (ii)) shows that, for every $T > 0$,

$$E\left[\sup_{0 \leq t \leq T} M_t^2 \right] \leq 4 E[M_T^2].$$

By letting T go to ∞, we have

$$E\left[\sup_{t \geq 0} M_t^2 \right] \leq 4 \sup_{t \geq 0} E[M_t^2] =: C < \infty.$$

Set $S_n = \inf\{t \geq 0 : \langle M, M \rangle_t \geq n\}$. Then the continuous local martingale $M^2_{t \wedge S_n} - \langle M, M \rangle_{t \wedge S_n}$ is dominated by the variable

$$\sup_{s \geq 0} M^2_s + n,$$

which is integrable. From Proposition 4.7 (ii), we get that this continuous local martingale is a uniformly integrable martingale, hence

$$E[\langle M, M \rangle_{t \wedge S_n}] = E[M^2_{t \wedge S_n}] \leq E\Big[\sup_{s \geq 0} M^2_s \Big] \leq C.$$

By letting n, and then t tend to infinity, and using monotone convergence, we get $E[\langle M, M \rangle_\infty] \leq C < \infty$.

Conversely, assume that $E[\langle M, M \rangle_\infty] < \infty$. Set $T_n = \inf\{t \geq 0 : |M_t| \geq n\}$. Then the continuous local martingale $M^2_{t \wedge T_n} - \langle M, M \rangle_{t \wedge T_n}$ is dominated by the variable

$$n^2 + \langle M, M \rangle_\infty,$$

which is integrable. From Proposition 4.7 (ii) again, this continuous local martingale is a uniformly integrable martingale, hence, for every $t \geq 0$,

$$E[M^2_{t \wedge T_n}] = E[\langle M, M \rangle_{t \wedge T_n}] \leq E[\langle M, M \rangle_\infty] =: C' < \infty.$$

By letting $n \to \infty$ and using Fatou's lemma, we get $E[M^2_t] \leq C'$, so that the collection $(M_t)_{t \geq 0}$ is bounded in L^2. We have not yet verified that $(M_t)_{t \geq 0}$ is a martingale. However, the previous bound on $E[M^2_{t \wedge T_n}]$ shows that the sequence $(M_{t \wedge T_n})_{n \geq 1}$ is uniformly integrable, and therefore converges both a.s. and in L^1 to M_t, for every $t \geq 0$. Recalling that M^{T_n} is a martingale (Proposition 4.7 (iii)), the L^1-convergence allows us to pass to the limit $n \to \infty$ in the martingale property $E[M_{t \wedge T_n} \mid \mathscr{F}_s] = M_{s \wedge T_n}$, for $0 \leq s < t$, and to get that M is a martingale.

Finally, if properties (a) and (b) hold, the continuous local martingale $M^2 - \langle M, M \rangle$ is dominated by the integrable variable

$$\sup_{t \geq 0} M^2_t + \langle M, M \rangle_\infty$$

and is therefore (by Proposition 4.7 (ii)) a uniformly integrable martingale.

(ii) It suffices to apply (i) to $(M_{t \wedge a})_{t \geq 0}$ for every choice of $a \geq 0$.   □

4.4 The Bracket of Two Continuous Local Martingales

Definition 4.14 If M and N are two continuous local martingales, the *bracket* $\langle M, N \rangle$ is the finite variation process defined by setting, for every $t \geq 0$,

$$\langle M, N \rangle_t = \frac{1}{2} (\langle M + N, M + N \rangle_t - \langle M, M \rangle_t - \langle N, N \rangle_t).$$

Let us state a few easy properties of the bracket.

Proposition 4.15

(i) $\langle M, N \rangle$ is the unique (up to indistinguishability) finite variation process such that $M_t N_t - \langle M, N \rangle_t$ is a continuous local martingale.
(ii) The mapping $(M, N) \mapsto \langle M, N \rangle$ is bilinear and symmetric.
(iii) If $0 = t_0^n < t_1^n < \cdots < t_{p_n}^n = t$ is an increasing sequence of subdivisions of $[0, t]$ with mesh tending to 0, we have

$$\lim_{n \to \infty} \sum_{i=1}^{p_n} (M_{t_i^n} - M_{t_{i-1}^n})(N_{t_i^n} - N_{t_{i-1}^n}) = \langle M, N \rangle_t$$

in probability.
(iv) For every stopping time T, $\langle M^T, N^T \rangle_t = \langle M^T, N \rangle_t = \langle M, N \rangle_{t \wedge T}$.
(v) If M and N are two martingales (with continuous sample paths) bounded in L^2, $M_t N_t - \langle M, N \rangle_t$ is a uniformly integrable martingale. Consequently, $\langle M, N \rangle_\infty$ is well defined as the almost sure limit of $\langle M, N \rangle_t$ as $t \to \infty$, is integrable, and satisfies

$$E[M_\infty N_\infty] = E[M_0 N_0] + E[\langle M, N \rangle_\infty].$$

Proof (i) follows from the analogous characterization in Theorem 4.9 (uniqueness follows from Theorem 4.8). Similarly (iii) is a consequence of the analogous assertion in Theorem 4.9. (ii) follows from (iii), or can be proved directly via the uniqueness argument. We can then get (iv) as a consequence of property (iii), noting that this property implies, for every $0 \leq s \leq t$, a.s.,

$$\langle M^T, N^T \rangle_t = \langle M^T, N \rangle_t = \langle M, N \rangle_t \qquad \text{on } \{T \geq t\},$$
$$\langle M^T, N^T \rangle_t - \langle M^T, N^T \rangle_s = \langle M^T, N \rangle_t - \langle M^T, N \rangle_s = 0 \text{ on } \{T \leq s < t\}.$$

Finally, (v) follows as a consequence of Theorem 4.13 (i). $\qquad \square$

Remark A consequence of (iv) is the fact that $M^T(N - N^T)$ is a continuous local martingale, which is not so easy to prove directly.

Proposition 4.16 *Let B and B' be two independent $(\mathscr{F}_t)$-Brownian motions. Then $\langle B, B' \rangle_t = 0$ for every $t \geq 0$.*

Proof By subtracting the initial values, we may assume that $B_0 = B'_0 = 0$. We then observe that the process $X_t = \frac{1}{\sqrt{2}}(B_t + B'_t)$ is a martingale, as a linear combination of martingales. By checking the finite-dimensional marginals of X, we verify that X is also a Brownian motion (notice that we do not claim that X is an $(\mathscr{F}_t)$-Brownian motion). Proposition 2.16 implies that $\langle X, X \rangle_t = t$, and, using the bilinearity of the bracket, it follows that $\langle B, B' \rangle_t = 0$. $\qquad\qquad\qquad\qquad\qquad\qquad\square$

Definition 4.17 Two continuous local martingales M and N are said to be *orthogonal* if $\langle M, N \rangle = 0$, which holds if and only if MN is a continuous local martingale.

In particular, two independent $(\mathscr{F}_t)$-Brownian motions are orthogonal martingales, by Proposition 4.16.

If M and N are two orthogonal martingales bounded in L^2, we have $E[M_t N_t] = E[M_0 N_0]$, and even $E[M_S N_S] = E[M_0 N_0]$ for any stopping time S. This follows from Theorem 3.22, using property (v) of Proposition 4.15.

Proposition 4.18 (Kunita–Watanabe) *Let M and N be two continuous local martingales and let H and K be two measurable processes. Then, a.s.,*

$$\int_0^\infty |H_s|\,|K_s|\,|\mathrm{d}\langle M, N \rangle_s| \leq \left(\int_0^\infty H_s^2 \, \mathrm{d}\langle M, M \rangle_s \right)^{1/2} \left(\int_0^\infty K_s^2 \, \mathrm{d}\langle N, N \rangle_s \right)^{1/2}.$$

Proof Only in this proof, we use the special notation $\langle M, N \rangle_s^t = \langle M, N \rangle_t - \langle M, N \rangle_s$ for $0 \leq s \leq t$. The first step of the proof is to observe that we have a.s. for every choice of the rationals $s < t$ (and also by continuity for every reals $s < t$),

$$|\langle M, N \rangle_s^t| \leq \sqrt{\langle M, M \rangle_s^t}\,\sqrt{\langle N, N \rangle_s^t}.$$

Indeed, this follows from the approximations of $\langle M, M \rangle$ and $\langle M, N \rangle$ given in Theorem 4.9 and in Proposition 4.15 respectively (note that these approximations are easily extended to the increments of $\langle M, M \rangle$ and $\langle M, N \rangle$), together with the Cauchy–Schwarz inequality. From now on, we fix ω such that the inequality of the last display holds for every $s < t$, and we argue with this value of ω (the remaining part of the argument is "deterministic").

We then observe that we also have, for every $0 \leq s \leq t$,

$$\int_s^t |\mathrm{d}\langle M, N \rangle_u| \leq \sqrt{\langle M, M \rangle_s^t}\,\sqrt{\langle N, N \rangle_s^t}. \tag{4.7}$$

Indeed, we use Proposition 4.2, noting that, for any subdivision $s = t_0 < t_1 < \cdots < t_p = t$, we can bound

$$\sum_{i=1}^{p} |\langle M, N \rangle_{t_{i-1}}^{t_i}| \leq \sum_{i=1}^{p} \sqrt{\langle M, M \rangle_{t_{i-1}}^{t_i}} \sqrt{\langle N, N \rangle_{t_{i-1}}^{t_i}}$$

$$\leq \left(\sum_{i=1}^{p} \langle M, M \rangle_{t_{i-1}}^{t_i} \right)^{1/2} \left(\sum_{i=1}^{p} \langle N, N \rangle_{t_{i-1}}^{t_i} \right)^{1/2}$$

$$= \sqrt{\langle M, M \rangle_s^t} \sqrt{\langle N, N \rangle_s^t}.$$

We then get that, for every bounded Borel subset A of $\mathbb{R}_+$,

$$\int_A |d\langle M, N \rangle_u| \leq \sqrt{\int_A d\langle M, M \rangle_u} \sqrt{\int_A d\langle N, N \rangle_u}.$$

When $A = [s, t]$, this is the bound (4.7). If A is a finite union of intervals, this follows from (4.7) and another application of the Cauchy–Schwarz inequality. A monotone class argument shows that the inequality of the last display remains valid for any bounded Borel set A (here we use a version of the monotone class lemma that is different from the one in Appendix A1: precisely, a class of sets which is stable under increasing and decreasing sequential limits and which contains an algebra of sets must contain the σ-field generated by this algebra – see the first chapter of [64]).

Next let $h = \sum_{i=1}^{p} \lambda_i \mathbf{1}_{A_i}$ and $k = \sum_{i=1}^{p} \mu_i \mathbf{1}_{A_i}$ be two nonnegative simple functions on $\mathbb{R}_+$ with bounded support contained in $[0, K]$, for some $K > 0$. Here $A_1, \ldots, A_p$ is a measurable partition of $[0, K]$, and $\lambda_1, \ldots, \lambda_p, \mu_1, \ldots, \mu_p$ are reals (we can always assume that h and k are expressed in terms of the same partition). Then,

$$\int h(s) k(s) |d\langle M, N \rangle_s| = \sum_{i=1}^{p} \lambda_i \mu_i \int_{A_i} |d\langle M, N \rangle_s|$$

$$\leq \left(\sum_{i=1}^{p} \lambda_i^2 \int_{A_i} d\langle M, M \rangle_s \right)^{1/2} \left(\sum_{i=1}^{p} \mu_i^2 \int_{A_i} d\langle N, N \rangle_s \right)^{1/2}$$

$$= \left(\int h(s)^2 d\langle M, M \rangle_s \right)^{1/2} \left(\int k(s)^2 d\langle N, N \rangle_s \right)^{1/2},$$

which gives the desired inequality for simple functions. Since every nonnegative Borel function is a monotone increasing limit of simple functions with bounded support, an application of the monotone convergence theorem completes the proof.

□

4.5   Continuous Semimartingales

We now introduce the class of processes for which we will develop the theory of stochastic integrals.

Definition 4.19 A process $X = (X_t)_{t \geq 0}$ is a *continuous semimartingale* if it can be written in the form

$$X_t = M_t + A_t,$$

where M is a continuous local martingale and A is a finite variation process.

The decomposition $X = M + A$ is then unique up to indistinguishability thanks to Theorem 4.8. We say that this is the *canonical decomposition* of X.

By construction, continuous semimartingales have continuous sample paths. It is possible to define a notion of semimartingale with càdlàg sample paths, but in this book, we will only deal with *continuous* semimartingales, and for this reason we sometimes omit the word continuous.

Definition 4.20 Let $X = M + A$ and $Y = M' + A'$ be the canonical decompositions of two continuous semimartingales X and Y. The *bracket* $\langle X, Y \rangle$ is the finite variation process defined by

$$\langle X, Y \rangle_t = \langle M, M' \rangle_t.$$

In particular, we have $\langle X, X \rangle_t = \langle M, M \rangle_t$.

Proposition 4.21 *Let $0 = t_0^n < t_1^n < \cdots < t_{p_n}^n = t$ be an increasing sequence of subdivisions of $[0, t]$ whose mesh tends to 0. Then,*

$$\lim_{n \to \infty} \sum_{i=1}^{p_n} (X_{t_i^n} - X_{t_{i-1}^n})(Y_{t_i^n} - Y_{t_{i-1}^n}) = \langle X, Y \rangle_t$$

in probability.

Proof We treat the case where $X = Y$ and leave the general case to the reader. We have

$$\sum_{i=1}^{p_n} (X_{t_i^n} - X_{t_{i-1}^n})^2 = \sum_{i=1}^{p_n} (M_{t_i^n} - M_{t_{i-1}^n})^2 + \sum_{i=1}^{p_n} (A_{t_i^n} - A_{t_{i-1}^n})^2$$

$$+ 2 \sum_{i=1}^{p_n} (M_{t_i^n} - M_{t_{i-1}^n})(A_{t_i^n} - A_{t_{i-1}^n}).$$

By Theorem 4.9,

$$\lim_{n\to\infty} \sum_{i=1}^{p_n} (M_{t_i^n} - M_{t_{i-1}^n})^2 = \langle M, M \rangle_t = \langle X, X \rangle_t,$$

in probability. On the other hand,

$$\sum_{i=1}^{p_n} (A_{t_i^n} - A_{t_{i-1}^n})^2 \leq \left(\sup_{1 \leq i \leq p_n} |A_{t_i^n} - A_{t_{i-1}^n}| \right) \sum_{i=1}^{p_n} |A_{t_i^n} - A_{t_{i-1}^n}|$$

$$\leq \left(\int_0^t |dA_s| \right) \sup_{1 \leq i \leq p_n} |A_{t_i^n} - A_{t_{i-1}^n}|,$$

which tends to 0 a.s. when $n \to \infty$ by the continuity of sample paths of A. The same argument shows that

$$\left| \sum_{i=1}^{p_n} (A_{t_i^n} - A_{t_{i-1}^n})(M_{t_i^n} - M_{t_{i-1}^n}) \right| \leq \left(\int_0^t |dA_s| \right) \sup_{1 \leq i \leq p_n} |M_{t_i^n} - M_{t_{i-1}^n}|$$

tends to 0 a.s.   □

Exercises

In the following exercises, processes are defined on a probability space $(\Omega, \mathscr{F}, P)$ equipped with a complete filtration $(\mathscr{F}_t)_{t\in[0,\infty]}$.

Exercise 4.22 Let U be an $\mathscr{F}_0$-measurable real random variable, and let M be a continuous local martingale. Show that the process $N_t = UM_t$ is a continuous local martingale. (*This result was used in the construction of the quadratic variation of a continuous local martingale.*)

Exercise 4.23

1. Let M be a (true) martingale with continuous sample paths, such that $M_0 = 0$. We assume that $(M_t)_{t\geq0}$ is also a Gaussian process. Show that, for every $t \geq 0$ and every $s > 0$, the random variable $M_{t+s} - M_t$ is independent of $\sigma(M_r, 0 \leq r \leq t)$.
2. Under the assumptions of question 1., show that there exists a continuous monotone nondecreasing function $f : \mathbb{R}_+ \to \mathbb{R}_+$ such that $\langle M, M \rangle_t = f(t)$ for every $t \geq 0$.

Exercise 4.24 Let M be a continuous local martingale with $M_0 = 0$.

1. For every integer $n \geq 1$, we set $T_n = \inf\{t \geq 0 : |M_t| = n\}$. Show that, a.s.

$$\left\{ \lim_{t\to\infty} M_t \text{ exists and is finite} \right\} = \bigcup_{n=1}^{\infty} \{T_n = \infty\} \subset \{\langle M, M \rangle_\infty < \infty\}.$$

2. We set $S_n = \inf\{t \geq 0 : \langle M, M \rangle_t = n\}$ for every $n \geq 1$. Show that, a.s.,

$$\{\langle M, M \rangle_\infty < \infty\} = \bigcup_{n=1}^{\infty} \{S_n = \infty\} \subset \left\{ \lim_{t\to\infty} M_t \text{ exists and is finite} \right\},$$

and conclude that

$$\left\{ \lim_{t\to\infty} M_t \text{ exists and is finite} \right\} = \{\langle M, M \rangle_\infty < \infty\} \quad , \text{ a.s.}$$

Exercise 4.25 For every integer $n \geq 1$, let $M^n = (M_t^n)_{t\geq 0}$ be a continuous local martingale with $M_0 = 0$. We assume that

$$\lim_{n\to\infty} \langle M^n, M^n \rangle_\infty = 0$$

in probability.

1. Let $\varepsilon > 0$, and, for every $n \geq 1$, let

$$T_\varepsilon^n = \inf\{t \geq 0 : \langle M^n, M^n \rangle_t \geq \varepsilon\}.$$

Justify the fact that T_ε^n is a stopping time, then prove that the stopped continuous local martingale

$$M_t^{n,\varepsilon} = M_{t\wedge T_\varepsilon^n}^n , \qquad \forall t \geq 0 ,$$

is a true martingale bounded in L^2.

2. Show that

$$E\left[\sup_{t\geq 0} |M_t^{n,\varepsilon}|^2 \right] \leq 4\varepsilon^2.$$

3. Writing, for every $a > 0$,

$$P\left(\sup_{t\geq 0} |M_t^n| \geq a \right) \leq P\left(\sup_{t\geq 0} |M_t^{n,\varepsilon}| \geq a \right) + P(T_\varepsilon^n < \infty),$$

show that

$$\lim_{n \to \infty} \left(\sup_{t \geq 0} |M_t^n| \right) = 0$$

in probability.

Exercise 4.26

1. Let A be an increasing process (adapted, with continuous sample paths and such that $A_0 = 0$) such that $A_\infty < \infty$ a.s., and let Z be an integrable random variable. We assume that, for every stopping time T,

$$E[A_\infty - A_T] \leq E[Z \mathbf{1}_{\{T < \infty\}}].$$

Show, by introducing an appropriate stopping time, that, for every $\lambda > 0$,

$$E[(A_\infty - \lambda) \mathbf{1}_{\{A_\infty > \lambda\}}] \leq E[Z \mathbf{1}_{\{A_\infty > \lambda\}}].$$

2. Let $f : \mathbb{R}_+ \longrightarrow \mathbb{R}$ be a continuously differentiable monotone increasing function such that $f(0) = 0$ and set $F(x) = \int_0^x f(t) dt$ for every $x \geq 0$. Show that, under the assumptions of question 1., one has

$$E[F(A_\infty)] \leq E[Zf(A_\infty)].$$

(*Hint*: It may be useful to observe that $F(x) = xf(x) - \int_0^x \lambda f'(\lambda) d\lambda$ for every $x \geq 0$.)

3. Let M be a (true) martingale with continuous sample paths and bounded in L^2 such that $M_0 = 0$, and let M_∞ be the almost sure limit of M_t as $t \to \infty$. Show that the assumptions of question 1. hold when $A_t = \langle M, M \rangle_t$ and $Z = M_\infty^2$. Infer that, for every real $q \geq 1$,

$$E[(\langle M, M \rangle_\infty)^{q+1}] \leq (q+1) E[(\langle M, M \rangle_\infty)^q M_\infty^2].$$

4. Let $p \geq 2$ be a real number such that $E[(\langle M, M \rangle_\infty)^p] < \infty$. Show that

$$E[(\langle M, M \rangle_\infty)^p] \leq p^p E[|M_\infty|^{2p}].$$

5. Let N be a continuous local martingale such that $N_0 = 0$, and let T be a stopping time such that the stopped martingale N^T is uniformly integrable. Show that, for every real $p \geq 2$,

$$E[(\langle N, N \rangle_T)^p] \leq p^p E[|N_T|^{2p}].$$

Give an example showing that this result may fail if N^T is not uniformly integrable.

Exercise 4.27 Let $(X_t)_{t\geq 0}$ be an adapted process with continuous sample paths and taking nonnegative values. Let $(A_t)_{t\geq 0}$ be an increasing process (adapted, with continuous sample paths and such that $A_0 = 0$). We consider the following condition:

(D)   For every bounded stopping time T, we have $E[X_T] \leq E[A_T]$.

1. Show that, if M is a square integrable martingale with continuous sample paths and $M_0 = 0$, the condition (D) holds for $X_t = M_t^2$ and $A_t = \langle M, M \rangle_t$.
2. Show that the conclusion of the previous question still holds if one only assumes that M is a continuous local martingale with $M_0 = 0$.
3. We set $X_t^* = \sup_{s\leq t} X_s$. Show that, under the condition (D), we have, for every bounded stopping time S and every $c > 0$,

$$P(X_S^* \geq c) \leq \frac{1}{c}E[A_S].$$

(*Hint*: One may apply (D) to $T = S \wedge R$, where $R = \inf\{t \geq 0 : X_t \geq c\}$.)
4. Infer that, still under the condition (D), one has, for every (finite or not) stopping time S,

$$P(X_S^* > c) \leq \frac{1}{c}E[A_S]$$

(when S takes the value ∞, we of course define $X_\infty^* = \sup_{s\geq 0} X_s$).
5. Let $c > 0$ and $d > 0$, and $S = \inf\{t \geq 0 : A_t \geq d\}$. Let T be a stopping time. Noting that

$$\{X_T^* > c\} \subset \left(\{X_{T\wedge S}^* > c\} \cup \{A_T \geq d\} \right)$$

show that, under the condition (D), one has

$$P(X_T^* > c) \leq \frac{1}{c}E[A_T \wedge d] + P(A_T \geq d).$$

6. Use questions (2) and (5) to verify that, if $M^{(n)}$ is a sequence of continuous local martingales and T is a stopping time such that $\langle M^{(n)}, M^{(n)} \rangle_T$ converges in probability to 0 as $n \to \infty$, then,

$$\lim_{n\to\infty} \left(\sup_{s\leq T} |M_s^{(n)}| \right) = 0 , \quad \text{in probability.}$$

Notes and Comments

The book [14] of Dellacherie and Meyer is again an excellent reference for the topics of this chapter, in the more general setting of local martingales and semimartingales with càdlàg sample paths. See also [72] and [49] (in particular, a discussion of the elementary theory of finite variation processes can be found in [72]). The notion of a local martingale appeared in Itô and Watanabe [43] in 1965. The notion of a semimartingale seems to be due to Fisk [25] in 1965, who used the name "quasimartingales". See also Meyer [60]. The classical approach to the quadratic variation of a continuous (local) martingale is based on the Doob–Meyer decomposition theorem [58], see e.g. [49]. Our more elementary presentation is inspired by [70].

Chapter 5
Stochastic Integration

This chapter is at the core of the present book. We start by defining the stochastic integral with respect to a square-integrable continuous martingale, considering first the integral of elementary processes (which play a role analogous to step functions in the theory of the Riemann integral) and then using an isometry between Hilbert spaces to deal with the general case. It is easy to extend the definition of stochastic integrals to continuous local martingales and semimartingales. We then derive the celebrated Itô's formula, which shows that the image of one or several continuous semimartingales under a smooth function is still a continuous semimartingale, whose canonical decomposition is given in terms of stochastic integrals. Itô's formula is the main technical tool of stochastic calculus, and we discuss several important applications of this formula, including Lévy's theorem characterizing Brownian motion as a continuous local martingale with quadratic variation process equal to t, the Burkholder–Davis–Gundy inequalities and the representation of martingales as stochastic integrals in a Brownian filtration. The end of the chapter is devoted to Girsanov's theorem, which deals with the stability of the notions of a martingale and a semimartingale under an absolutely continuous change of probability measure. As an application of Girsanov's theorem, we establish the famous Cameron–Martin formula giving the image of the Wiener measure under a translation by a deterministic function.

5.1 The Construction of Stochastic Integrals

Throughout this chapter, we argue on a filtered probability space $(\Omega, \mathscr{F}, (\mathscr{F}_t), P)$, and we assume that the filtration $(\mathscr{F}_t)$ is complete. Unless otherwise specified, all processes in this chapter are indexed by $\mathbb{R}_+$ and take real values. We often say "continuous martingale" instead of "martingale with continuous sample paths".

© Springer International Publishing Switzerland 2016
J.-F. Le Gall, *Brownian Motion, Martingales, and Stochastic Calculus*,
Graduate Texts in Mathematics 274, DOI 10.1007/978-3-319-31089-3_5

5.1.1 Stochastic Integrals for Martingales Bounded in L^2

We write $\mathbb{H}^2$ for the space of all continuous martingales M which are bounded in L^2 and such that $M_0 = 0$, with the usual convention that two indistinguishable processes are identified. Equivalently, $M \in \mathbb{H}^2$ if and only if M is a continuous local martingale such that $M_0 = 0$ and $E[\langle M, M \rangle_\infty] < \infty$ (Proposition 4.13). By Proposition 3.21, if $M \in \mathbb{H}^2$, we have $M_t = E[M_\infty \mid \mathscr{F}_t]$ where $M_\infty \in L^2$ is the almost sure limit of M_t as $t \to \infty$.

Proposition 4.15 (v) shows that, if $M, N \in \mathbb{H}^2$, the random variable $\langle M, N \rangle_\infty$ is well defined, and we have $E[|\langle M, N \rangle_\infty|] < \infty$. This allows us to define a symmetric bilinear form on $\mathbb{H}^2$ via the formula

$$(M, N)_{\mathbb{H}^2} = E[\langle M, N \rangle_\infty] = E[M_\infty N_\infty],$$

where the second equality comes from Proposition 4.15 (v). Clearly $(M, M)_{\mathbb{H}^2} = 0$ if and only if $M = 0$. The scalar product $(M, N)_{\mathbb{H}^2}$ thus yields a norm on $\mathbb{H}^2$ given by

$$\|M\|_{\mathbb{H}^2} = (M, M)_{\mathbb{H}^2}^{1/2} = E[\langle M, M \rangle_\infty]^{1/2} = E[(M_\infty)^2]^{1/2}.$$

Proposition 5.1 *The space $\mathbb{H}^2$ equipped with the scalar product $(M, N)_{\mathbb{H}^2}$ is a Hilbert space.*

Proof We need to verify that the vector space $\mathbb{H}^2$ is complete for the norm $\| \ \|_{\mathbb{H}^2}$. Let $(M^n)_{n \geq 1}$ be a sequence in $\mathbb{H}^2$ which is Cauchy for that norm. We have then

$$\lim_{m,n \to \infty} E[(M_\infty^n - M_\infty^m)^2] = \lim_{m,n \to \infty} (M^n - M^m, M^n - M^m)_{\mathbb{H}^2} = 0.$$

Consequently, the sequence (M_∞^n) converges in L^2 to a limit, which we denote by Z. On the other hand, Doob's inequality in L^2 (Proposition 3.15 (ii)) and a straightforward passage to the limit show that, for every m, n,

$$E\left[\sup_{t \geq 0} (M_t^n - M_t^m)^2 \right] \leq 4 \, E[(M_\infty^n - M_\infty^m)^2].$$

We thus obtain that

$$\lim_{m,n \to \infty} E\left[\sup_{t \geq 0} (M_t^n - M_t^m)^2 \right] = 0. \tag{5.1}$$

Hence, for every $t \geq 0$, M_t^n converges in L^2, and we want to argue that the limit yields a process with continuous sample paths. To this end, we use (5.1) to find an increasing sequence $n_k \uparrow \infty$ such that

$$E\Big[\sum_{k=1}^{\infty} \sup_{t\geq 0} |M_t^{n_k} - M_t^{n_{k+1}}| \Big] \leq \sum_{k=1}^{\infty} E\Big[\sup_{t\geq 0} (M_t^{n_k} - M_t^{n_{k+1}})^2 \Big]^{1/2} < \infty.$$

The last display implies that, a.s.,

$$\sum_{k=1}^{\infty} \sup_{t\geq 0} |M_t^{n_k} - M_t^{n_{k+1}}| < \infty,$$

and thus the sequence $(M_t^{n_k})_{t\geq 0}$ converges uniformly on $\mathbb{R}_+$, a.s., to a limit denoted by $(M_t)_{t\geq 0}$. On the zero probability set where the uniform convergence does not hold, we take $M_t = 0$ for every $t \geq 0$. Clearly the limiting process M has continuous sample paths and is adapted (here we use the fact that the filtration is complete). Furthermore, from the L^2-convergence of (M_∞^n) to Z, we immediately get by passing to the limit in the identity $M_t^{n_k} = E[M_\infty^{n_k} \mid \mathscr{F}_t]$ that $M_t = E[Z \mid \mathscr{F}_t]$. Hence $(M_t)_{t\geq 0}$ is a continuous martingale and is bounded in L^2, so that $M \in \mathbb{H}^2$. The a.s. uniform convergence of $(M_t^{n_k})_{t\geq 0}$ to $(M_t)_{t\geq 0}$ then ensures that $M_\infty = \lim M_\infty^{n_k} = Z$ a.s. Finally, the L^2-convergence of (M_∞^n) to $Z = M_\infty$ shows that the sequence (M^n) converges to M in $\mathbb{H}^2$.   □

We denote the progressive σ-field on $\Omega \times \mathbb{R}_+$ by $\mathscr{P}$ (see the end of Sect. 3.1), and, if $M \in \mathbb{H}^2$, we let $L^2(M)$ be the set of all progressive processes H such that

$$E\Big[\int_0^{\infty} H_s^2 \, d\langle M, M \rangle_s \Big] < \infty,$$

with the convention that two progressive processes H and H' satisfying this integrability condition are identified if $H_s = H'_s = 0$, $d\langle M, M \rangle_s$ a.e., a.s. We can view $L^2(M)$ as an ordinary L^2 space, namely

$$L^2(M) = L^2(\Omega \times \mathbb{R}_+, \mathscr{P}, dP \, d\langle M, M \rangle_s)$$

where $dP \, d\langle M, M \rangle_s$ refers to the finite measure on $(\Omega \times \mathbb{R}_+, \mathscr{P})$ that assigns the mass

$$E\Big[\int_0^{\infty} 1_A(\omega, s) \, d\langle M, M \rangle_s \Big]$$

to a set $A \in \mathscr{P}$ (the total mass of this measure is $E[\langle M, M \rangle_\infty] = \|M\|_{\mathbb{H}^2}^2$).

Just like any L^2 space, the space $L^2(M)$ is a Hilbert space for the scalar product

$$(H, K)_{L^2(M)} = E\Big[\int_0^{\infty} H_s K_s \, d\langle M, M \rangle_s \Big],$$

and the associated norm is

$$\|H\|_{L^2(M)} = \left(E\left[\int_0^\infty H_s^2 \, d\langle M, M\rangle_s \right] \right)^{1/2}.$$

Definition 5.2 An *elementary process* is a progressive process of the form

$$H_s(\omega) = \sum_{i=0}^{p-1} H_{(i)}(\omega) \, \mathbf{1}_{(t_i, t_{i+1}]}(s),$$

where $0 = t_0 < t_1 < t_2 < \cdots < t_p$ and for every $i \in \{0, 1, \ldots, p-1\}$, $H_{(i)}$ is a bounded $\mathscr{F}_{t_i}$-measurable random variable.

The set $\mathscr{E}$ of all elementary processes forms a linear subspace of $L^2(M)$. To be precise, we should here say "equivalence classes of elementary processes" (recall that H and H' are identified in $L^2(M)$ if $\|H - H'\|_{L^2(M)} = 0$).

Proposition 5.3 *For every $M \in \mathbb{H}^2$, $\mathscr{E}$ is dense in $L^2(M)$.*

Proof By elementary Hilbert space theory, it is enough to verify that, if $K \in L^2(M)$ is orthogonal to $\mathscr{E}$, then $K = 0$. Assume that $K \in L^2(M)$ is orthogonal to $\mathscr{E}$, and set, for every $t \geq 0$,

$$X_t = \int_0^t K_u \, d\langle M, M\rangle_u.$$

To see that the integral in the right-hand side makes sense, and defines a finite variation process $(X_t)_{t \geq 0}$, we use the Cauchy–Schwarz inequality to observe that

$$E\left[\int_0^t |K_u| \, d\langle M, M\rangle_u \right] \leq \left(E\left[\int_0^t (K_u)^2 \, d\langle M, M\rangle_u \right] \right)^{1/2} \times (E[\langle M, M\rangle_\infty])^{1/2}.$$

The right-hand side is finite since $M \in \mathbb{H}^2$ and $K \in L^2(M)$, and thus we have in particular

$$\text{a.s.} \quad \forall t \geq 0, \quad \int_0^t |K_u| \, d\langle M, M\rangle_u < \infty.$$

By Proposition 4.5 (and Remark (i) following this proposition), $(X_t)_{t \geq 0}$ is well defined as a finite variation process. The preceding bound also shows that $X_t \in L^1$ for every $t \geq 0$.

Let $0 \leq s < t$, let F be a bounded $\mathscr{F}_s$-measurable random-variable, and let $H \in \mathscr{E}$ be the elementary process defined by $H_r(\omega) = F(\omega) \, \mathbf{1}_{(s,t]}(r)$. Writing $(H, K)_{L^2(M)} = 0$, we get

$$E\left[F \int_s^t K_u \, d\langle M, M\rangle_u \right] = 0.$$

It follows that $E[F(X_t - X_s)] = 0$ for every $s < t$ and every bounded $\mathscr{F}_s$-measurable variable F. Since the process X is adapted and we know that $X_r \in L^1$ for every $r \geq 0$, this implies that X is a (continuous) martingale. On the other hand, X is also a finite variation process and, by Theorem 4.8, this is only possible if $X = 0$. We have thus proved that

$$\int_0^t K_u \, d\langle M, M\rangle_u = 0 \qquad \forall t \geq 0, \quad \text{a.s.}$$

which implies that, a.s., the signed measure having density K_u with respect to $d\langle M, M\rangle_u$ is the zero measure, which is only possible if

$$K_u = 0, \qquad d\langle M, M\rangle_u \text{ a.e.,} \quad \text{a.s.}$$

or equivalently $K = 0$ in $L^2(M)$. $\square$

Recall our notation X^T for the process X stopped at the stopping time T: $X_t^T = X_{t \wedge T}$. If $M \in \mathbb{H}^2$, the fact that $\langle M^T, M^T\rangle_\infty = \langle M, M\rangle_T$ immediately implies that M^T also belongs to $\mathbb{H}^2$. Furthermore, if $H \in L^2(M)$, the process $\mathbf{1}_{[0,T]} H$ defined by $(\mathbf{1}_{[0,T]} H)_s(\omega) = \mathbf{1}_{\{0 \leq s \leq T(\omega)\}} H_s(\omega)$ also belongs to $L^2(M)$ (note that $\mathbf{1}_{[0,T]}$ is progressive since it is adapted with left-continuous sample paths).

Theorem 5.4 *Let $M \in \mathbb{H}^2$. For every $H \in \mathscr{E}$ of the form*

$$H_s(\omega) = \sum_{i=0}^{p-1} H_{(i)}(\omega) \, \mathbf{1}_{(t_i, t_{i+1}]}(s),$$

the formula

$$(H \cdot M)_t = \sum_{i=0}^{p-1} H_{(i)} \, (M_{t_{i+1} \wedge t} - M_{t_i \wedge t})$$

defines a process $H \cdot M \in \mathbb{H}^2$. The mapping $H \mapsto H \cdot M$ extends to an isometry from $L^2(M)$ into $\mathbb{H}^2$. Furthermore, $H \cdot M$ is the unique martingale of $\mathbb{H}^2$ that satisfies the property

$$\langle H \cdot M, N\rangle = H \cdot \langle M, N\rangle, \qquad \forall N \in \mathbb{H}^2. \tag{5.2}$$

If T is a stopping time, we have

$$(\mathbf{1}_{[0,T]} H) \cdot M = (H \cdot M)^T = H \cdot M^T. \tag{5.3}$$

We often use the notation

$$(H \cdot M)_t = \int_0^t H_s \, dM_s$$

and call $H \cdot M$ the stochastic integral of H with respect to M.

Remark The quantity $H \cdot \langle M, N \rangle$ in the right-hand side of (5.2) is an integral with respect to a finite variation process, as defined in Sect. 4.1. The fact that we use a similar notation $H \cdot A$ and $H \cdot M$ for the integrals with respect to a finite variation process A and with respect to a martingale M creates no ambiguity since these two classes of processes are essentially disjoint.

Proof As a preliminary observation, we note that the definition of $H \cdot M$ when $H \in \mathscr{E}$ does not depend on the decomposition chosen for H in the first display of the theorem. Using this remark, one then checks that the mapping $H \mapsto H \cdot M$ is linear. We next verify that this mapping is an isometry from $\mathscr{E}$ (viewed as a subspace of $L^2(M)$) into $\mathbb{H}^2$.

Fix $H \in \mathscr{E}$ of the form given in the theorem, and for every $i \in \{0, 1, \ldots, p-1\}$, set

$$M_t^i = H_{(i)} (M_{t_{i+1} \wedge t} - M_{t_i \wedge t}),$$

for every $t \geq 0$. Then a simple verification shows that M^i is a continuous martingale (this was already used in the beginning of the proof of Theorem 4.9), and that this martingale belongs to $\mathbb{H}^2$. It follows that $H \cdot M = \sum_{i=0}^{p-1} M^i$ is also a martingale in $\mathbb{H}^2$. Then, we note that the continuous martingales M^i are orthogonal, and their respective quadratic variations are given by

$$\langle M^i, M^i \rangle_t = H_{(i)}^2 \big(\langle M, M \rangle_{t_{i+1} \wedge t} - \langle M, M \rangle_{t_i \wedge t} \big)$$

(the orthogonality of the martingales M^i as well as the formula of the last display are easily checked, for instance by using the approximations of $\langle M, N \rangle$). We conclude that

$$\langle H \cdot M, H \cdot M \rangle_t = \sum_{i=0}^{p-1} H_{(i)}^2 \big(\langle M, M \rangle_{t_{i+1} \wedge t} - \langle M, M \rangle_{t_i \wedge t} \big) = \int_0^t H_s^2 \, d\langle M, M \rangle_s.$$

Consequently,

$$\|H \cdot M\|_{\mathbb{H}^2}^2 = E[\langle H \cdot M, H \cdot M \rangle_\infty] = E\left[\int_0^\infty H_s^2 \, d\langle M, M \rangle_s \right] = \|H\|_{L^2(M)}^2.$$

By linearity, this implies that $H \cdot M = H' \cdot M$ if H' is another elementary process that is identified with H in $L^2(M)$. Therefore the mapping $H \mapsto H \cdot M$ makes sense from

$\mathscr{E}$ viewed as a subspace of $L^2(M)$ into $\mathbb{H}^2$. The latter mapping is linear, and, since it preserves the norm, it is an isometry from $\mathscr{E}$ (equipped with the norm of $L^2(M)$) into $\mathbb{H}^2$. Since $\mathscr{E}$ is dense in $L^2(M)$ (Proposition 5.3) and $\mathbb{H}^2$ is a Hilbert space (Proposition 5.1), this mapping can be extended in a unique way to an isometry from $L^2(M)$ into $\mathbb{H}^2$.

Let us verify property (5.2). We fix $N \in \mathbb{H}^2$. We first note that, if $H \in L^2(M)$, the Kunita–Watanabe inequality (Proposition 4.18) shows that

$$E\left[\int_0^\infty |H_s|\,|\mathrm{d}\langle M, N\rangle_s|\right] \le \|H\|_{L^2(M)}\,\|N\|_{\mathbb{H}^2} < \infty$$

and thus the variable $\int_0^\infty H_s \mathrm{d}\langle M, N\rangle_s = (H \cdot \langle M, N\rangle)_\infty$ is well defined and in L^1. Consider first the case where H is an elementary process of the form given in the theorem, and define the continuous martingales M^i, $0 \le i \le p-1$, as previously. Then, for every $i \in \{0, 1, \ldots, p-1\}$,

$$\langle H \cdot M, N\rangle = \sum_{i=0}^{p-1} \langle M^i, N\rangle$$

and we have

$$\langle M^i, N\rangle_t = H_{(i)}\big(\langle M, N\rangle_{t_{i+1}\wedge t} - \langle M, N\rangle_{t_i\wedge t}\big).$$

It follows that

$$\langle H \cdot M, N\rangle_t = \sum_{i=0}^{p-1} H_{(i)}\big(\langle M, N\rangle_{t_{i+1}\wedge t} - \langle M, N\rangle_{t_i\wedge t}\big) = \int_0^t H_s\,\mathrm{d}\langle M, N\rangle_s$$

which gives (5.2) when $H \in \mathscr{E}$. We then observe that the linear mapping $X \mapsto \langle X, N\rangle_\infty$ is continuous from $\mathbb{H}^2$ into L^1. Indeed, by the Kunita–Watanabe inequality,

$$E[|\langle X, N\rangle_\infty|] \le E[\langle X, X\rangle_\infty]^{1/2} E[\langle N, N\rangle_\infty]^{1/2} = \|N\|_{\mathbb{H}^2}\,\|X\|_{\mathbb{H}^2}.$$

If $(H^n)_{n\ge 1}$ is a sequence in $\mathscr{E}$, such that $H^n \to H$ in $L^2(M)$, we have therefore

$$\langle H \cdot M, N\rangle_\infty = \lim_{n\to\infty} \langle H^n \cdot M, N\rangle_\infty = \lim_{n\to\infty} (H^n \cdot \langle M, N\rangle)_\infty = (H \cdot \langle M, N\rangle)_\infty,$$

where the convergences hold in L^1, and the last equality again follows from the Kunita–Watanabe inequality by writing

$$E\left[\left|\int_0^\infty (H_s^n - H_s)\,\mathrm{d}\langle M, N\rangle_s\right|\right] \le E[\langle N, N\rangle_\infty]^{1/2}\,\|H^n - H\|_{L^2(M)}.$$

We have thus obtained the identity $\langle H \cdot M, N \rangle_\infty = (H \cdot \langle M, N \rangle)_\infty$, but replacing N by the stopped martingale N^t in this identity also gives $\langle H \cdot M, N \rangle_t = (H \cdot \langle M, N \rangle)_t$, which completes the proof of (5.2).

It is easy to see that (5.2) characterizes $H \cdot M$ among the martingales of $\mathbb{H}^2$. Indeed, if X is another martingale of $\mathbb{H}^2$ that satisfies the same identity, we get, for every $N \in \mathbb{H}^2$,

$$\langle H \cdot M - X, N \rangle = 0.$$

Taking $N = H \cdot M - X$ and using Proposition 4.12 we obtain that $X = H \cdot M$.

It remains to verify (5.3). Using the properties of the bracket of two continuous local martingales, we observe that, if $N \in \mathbb{H}^2$,

$$\langle (H \cdot M)^T, N \rangle_t = \langle H \cdot M, N \rangle_{t \wedge T} = (H \cdot \langle M, N \rangle)_{t \wedge T} = (\mathbf{1}_{[0,T]} H \cdot \langle M, N \rangle)_t$$

which shows that the stopped martingale $(H \cdot M)^T$ satisfies the characteristic property of the stochastic integral $(\mathbf{1}_{[0,T]} H) \cdot M$. The first equality in (5.3) follows. The second one is proved analogously, writing

$$\langle H \cdot M^T, N \rangle = H \cdot \langle M^T, N \rangle = H \cdot \langle M, N \rangle^T = \mathbf{1}_{[0,T]} H \cdot \langle M, N \rangle.$$

This completes the proof of the theorem.   □

Remark We could have used the relation (5.2) to *define* the stochastic integral $H \cdot M$, observing that the mapping $N \mapsto E[(H \cdot \langle M, N \rangle)_\infty]$ yields a continuous linear form on $\mathbb{H}^2$, and thus there exists a unique martingale $H \cdot M$ in $\mathbb{H}^2$ such that

$$E[(H \cdot \langle M, N \rangle)_\infty] = (H \cdot M, N)_{\mathbb{H}^2} = E[\langle H \cdot M, N \rangle_\infty].$$

Using the notation introduced at the end of Theorem 5.4, we can rewrite (5.2) in the form

$$\langle \int_0^\cdot H_s dM_s, N \rangle_t = \int_0^t H_s \, d\langle M, N \rangle_s.$$

We interpret this by saying that the stochastic integral "commutes" with the bracket. Let us immediately mention a very important consequence. If $M \in \mathbb{H}^2$, and $H \in L^2(M)$, two successive applications of (5.2) give

$$\langle H \cdot M, H \cdot M \rangle = H \cdot (H \cdot \langle M, M \rangle) = H^2 \cdot \langle M, M \rangle,$$

using the "associativity property" (4.1) of integrals with respect to finite variation processes. Put differently, the quadratic variation of the continuous martingale $H \cdot M$ is

$$\langle \int_0^{\cdot} H_s dM_s, \int_0^{\cdot} H_s dM_s \rangle_t = \int_0^t H_s^2 \, d\langle M, M \rangle_s. \tag{5.4}$$

More generally, if N is another martingale of $\mathbb{H}^2$ and $K \in L^2(N)$, the same argument gives

$$\langle \int_0^{\cdot} H_s dM_s, \int_0^{\cdot} K_s dN_s \rangle_t = \int_0^t H_s K_s \, d\langle M, N \rangle_s. \tag{5.5}$$

The following "associativity" property of stochastic integrals, which is analogous to property (4.1) for integrals with respect to finite variation processes, is very useful.

Proposition 5.5 *Let* $H \in L^2(M)$. *If* K *is a progressive process, we have* $KH \in L^2(M)$ *if and only if* $K \in L^2(H \cdot M)$. *If the latter properties hold,*

$$(KH) \cdot M = K \cdot (H \cdot M).$$

Proof Using property (5.4), we have

$$E\left[\int_0^{\infty} K_s^2 H_s^2 \, d\langle M, M \rangle_s \right] = E\left[\int_0^{\infty} K_s^2 \, d\langle H \cdot M, H \cdot M \rangle_s \right],$$

which gives the first assertion. For the second one, we write for $N \in \mathbb{H}^2$,

$$\langle (KH) \cdot M, N \rangle = KH \cdot \langle M, N \rangle = K \cdot (H \cdot \langle M, N \rangle) = K \cdot \langle H \cdot M, N \rangle$$

and, by the uniqueness statement in (5.2), this implies that $(KH) \cdot M = K \cdot (H \cdot M)$. □

Moments of stochastic integrals. Let $M \in \mathbb{H}^2$, $N \in \mathbb{H}^2$, $H \in L^2(M)$ and $K \in L^2(N)$. Since $H \cdot M$ and $K \cdot N$ are martingales in $\mathbb{H}^2$, we have, for every $t \in [0, \infty]$,

$$E\left[\int_0^t H_s \, dM_s \right] = 0 \tag{5.6}$$

$$E\left[\left(\int_0^t H_s dM_s \right) \left(\int_0^t K_s dN_s \right) \right] = E\left[\int_0^t H_s K_s \, d\langle M, N \rangle_s \right], \tag{5.7}$$

using Proposition 4.15 (v) and (5.5) to derive (5.7). In particular,

$$E\left[\left(\int_0^t H_s \, dM_s \right)^2 \right] = E\left[\int_0^t H_s^2 \, d\langle M, M \rangle_s \right]. \tag{5.8}$$

Furthermore, since $H \cdot M$ is a (true) martingale, we also have for every $0 \le s < t \le \infty$,

$$E\left[\int_0^t H_r \, dM_r \,\Big|\, \mathscr{F}_s\right] = \int_0^s H_r \, dM_r, \qquad (5.9)$$

or equivalently

$$E\left[\int_s^t H_r \, dM_r \,\Big|\, \mathscr{F}_s\right] = 0$$

with an obvious notation for $\int_s^t H_r \, dM_r$. It is important to observe that these formulas (and particularly (5.6) and (5.8)) may no longer hold for the extensions of stochastic integrals that we will now describe.

5.1.2 Stochastic Integrals for Local Martingales

We will now use the identities (5.3) to extend the definition of $H \cdot M$ to an arbitrary continuous local martingale. If M is a continuous local martingale, we write $L^2_{\mathrm{loc}}(M)$ (resp. $L^2(M)$) for the set of all progressive processes H such that

$$\int_0^t H_s^2 \, d\langle M, M\rangle_s < \infty, \ \forall t \ge 0, \text{ a.s.} \quad \left(\text{resp. such that } E\left[\int_0^\infty H_s^2 \, d\langle M, M\rangle_s\right] < \infty\right).$$

For future reference, we note that $L^2(M)$ (with the same identifications as in the case where $M \in \mathbb{H}^2$) can again be viewed as an "ordinary" L^2-space and thus has a Hilbert space structure.

Theorem 5.6 *Let M be a continuous local martingale. For every $H \in L^2_{\mathrm{loc}}(M)$, there exists a unique continuous local martingale with initial value 0, which is denoted by $H \cdot M$, such that, for every continuous local martingale N,*

$$\langle H \cdot M, N\rangle = H \cdot \langle M, N\rangle. \qquad (5.10)$$

If T is a stopping time, we have

$$(\mathbf{1}_{[0,T]}H) \cdot M = (H \cdot M)^T = H \cdot M^T. \qquad (5.11)$$

If $H \in L^2_{\mathrm{loc}}(M)$ and K is a progressive process, we have $K \in L^2_{\mathrm{loc}}(H \cdot M)$ if and only if $HK \in L^2_{\mathrm{loc}}(M)$, and then

$$H \cdot (K \cdot M) = HK \cdot M. \qquad (5.12)$$

Finally, if $M \in \mathbb{H}^2$, and $H \in L^2(M)$, the definition of $H \cdot M$ is consistent with that of Theorem 5.4.

Proof We may assume that $M_0 = 0$ (in the general case, we write $M = M_0 + M'$ and we just set $H \cdot M = H \cdot M'$, noting that $\langle M, N \rangle = \langle M', N \rangle$ for every continuous local martingale N). Also we may assume that the property $\int_0^t H_s^2 \, d\langle M, M \rangle_s < \infty$ for every $t \geq 0$ holds for every $\omega \in \Omega$ (on the negligible set where this fails we may replace H by 0).

For every $n \geq 1$, set

$$T_n = \inf\{t \geq 0 : \int_0^t (1 + H_s^2) \, d\langle M, M \rangle_s \geq n\},$$

so that (T_n) is a sequence of stopping times that increase to $+\infty$. Since

$$\langle M^{T_n}, M^{T_n} \rangle_t = \langle M, M \rangle_{t \wedge T_n} \leq n,$$

the stopped martingale M^{T_n} is in $\mathbb{H}^2$ (Theorem 4.13). Furthermore, we also have

$$\int_0^\infty H_s^2 \, d\langle M^{T_n}, M^{T_n} \rangle_s = \int_0^{T_n} H_s^2 \, d\langle M, M \rangle_s \leq n.$$

Hence, $H \in L^2(M^{T_n})$, and the definition of $H \cdot M^{T_n}$ makes sense by Theorem 5.4. Moreover, by property (5.3), we have, if $m > n$,

$$H \cdot M^{T_n} = (H \cdot M^{T_m})^{T_n}.$$

It follows that there exists a unique process denoted by $H \cdot M$ such that, for every n,

$$(H \cdot M)^{T_n} = H \cdot M^{T_n}.$$

Clearly $H \cdot M$ has continuous sample paths and is also adapted since $(H \cdot M)_t = \lim (H \cdot M^{T_n})_t$. Since the processes $(H \cdot M)^{T_n}$ are martingales in $\mathbb{H}^2$, we get that $H \cdot M$ is a continuous local martingale.

Then, to verify (5.10), we may assume that N is a continuous local martingale such that $N_0 = 0$. For every $n \geq 1$, set $T_n' = \inf\{t \geq 0 : |N_t| \geq n\}$, and $S_n = T_n \wedge T_n'$. Then, noting that $N^{T_n'} \in \mathbb{H}^2$, we have

$$\langle H \cdot M, N \rangle^{S_n} = \langle (H \cdot M)^{T_n}, N^{T_n'} \rangle$$
$$= \langle H \cdot M^{T_n}, N^{T_n'} \rangle$$
$$= H \cdot \langle M^{T_n}, N^{T_n'} \rangle$$
$$= H \cdot \langle M, N \rangle^{S_n}$$
$$= (H \cdot \langle M, N \rangle)^{S_n},$$

which gives the equality $\langle H \cdot M, N \rangle = H \cdot \langle M, N \rangle$. since $S_n \uparrow \infty$ as $n \to \infty$. The fact that this equality (written for every continuous local martingale N) characterizes $H \cdot M$ among continuous local martingales with initial value 0 is derived from Proposition 4.12 as in the proof of Theorem 5.4.

The property (5.11) is then obtained by the very same arguments as in the proof of property (5.3) in Theorem 5.4 (these arguments only depended on the characteristic property (5.2) which we have just extended in (5.10)). Similarly, the proof of (5.12) is analogous to the proof of Proposition 5.5.

Finally, if $M \in \mathbb{H}^2$ and $H \in L^2(M)$, the equality $\langle H \cdot M, H \cdot M \rangle = H^2 \cdot \langle M, M \rangle$ follows from (5.10), and implies that $H \cdot M \in \mathbb{H}^2$. Then the characteristic property (5.2) shows that the definitions of Theorems 5.4 and 5.6 are consistent. $\square$

In the setting of Theorem 5.6, we will again write

$$(H \cdot M)_t = \int_0^t H_s \, dM_s.$$

It is worth pointing out that formulas (5.4) and (5.5) **remain valid** when M and N are continuous local martingales and $H \in L^2_{loc}(M)$, $K \in L^2_{loc}(N)$. Indeed, these formulas immediately follow from (5.10).

Connection with the Wiener integral Suppose that B is an $(\mathscr{F}_t)$-Brownian motion, and $h \in L^2(\mathbb{R}_+, \mathscr{B}(\mathbb{R}_+), dt)$ is a deterministic square integrable function. We can then define the Wiener integral $\int_0^t h(s) dB_s = G(f \mathbf{1}_{[0,t]})$, where G is the Gaussian white noise associated with B (see the end of Sect. 2.1). It is easy to verify that this integral coincides with the stochastic integral $(h \cdot B)_t$, which makes sense by viewing h as a (deterministic) progressive process. This is immediate when h is a simple function, and the general case follows from a density argument.

Let us now discuss the extension of the moment formulas that we stated above in the setting of Theorem 5.4. Let M be a continuous local martingale, $H \in L^2_{loc}(M)$ and $t \in [0, \infty]$. Then, **under the condition**

$$E\left[\int_0^t H_s^2 \, d\langle M, M \rangle_s \right] < \infty, \tag{5.13}$$

we can apply Theorem 4.13 to $(H \cdot M)^t$, and get that $(H \cdot M)^t$ is a martingale of $\mathbb{H}^2$. It follows that properties (5.6) and (5.8) still hold:

$$E\left[\int_0^t H_s \, dM_s \right] = 0, \qquad E\left[\left(\int_0^t H_s \, dM_s \right)^2 \right] = E\left[\int_0^t H_s^2 \, d\langle M, M \rangle_s \right],$$

and similarly (5.9) is valid for $0 \le s \le t$. In particular (case $t = \infty$), if $H \in L^2(M)$, the continuous local martingale $H \cdot M$ is in $\mathbb{H}^2$ and its terminal value satisfies

$$E\left[\left(\int_0^\infty H_s \, dM_s \right)^2 \right] = E\left[\int_0^\infty H_s^2 \, d\langle M, M \rangle_s \right].$$

If the condition (5.13) does not hold, the previous formulas may fail. However, we always have the bound

$$E\left[\left(\int_0^t H_s \, dM_s\right)^2\right] \le E\left[\int_0^t H_s^2 \, d\langle M, M\rangle_s\right]. \tag{5.14}$$

Indeed, if the right-hand side is finite, this is an equality by the preceding observations. If the right-hand side is infinite, the bound is trivial.

5.1.3 Stochastic Integrals for Semimartingales

We finally extend the definition of stochastic integrals to continuous semimartingales. We say that a progressive process H is *locally bounded* if

$$\forall t \ge 0, \quad \sup_{s \le t} |H_s| < \infty, \quad \text{a.s.}$$

In particular, any adapted process with continuous sample paths is a locally bounded progressive process. If H is (progressive and) locally bounded, then for every finite variation process V, we have

$$\forall t \ge 0, \quad \int_0^t |H_s| \, |dV_s| < \infty, \quad \text{a.s.}$$

and similarly $H \in L^2_{\text{loc}}(M)$ for every continuous local martingale M.

Definition 5.7 Let X be a continuous semimartingale and let $X = M + V$ be its canonical decomposition. If H is a locally bounded progressive process, the *stochastic integral* $H \cdot X$ is the continuous semimartingale with canonical decomposition

$$H \cdot X = H \cdot M + H \cdot V,$$

and we write

$$(H \cdot X)_t = \int_0^t H_s \, dX_s.$$

Properties

(i) The mapping $(H, X) \mapsto H \cdot X$ is bilinear.

(ii) $H \cdot (K \cdot X) = (HK) \cdot X$, if H and K are progressive and locally bounded.

(iii) For every stopping time T, $(H \cdot X)^T = H\mathbf{1}_{[0,T]} \cdot X = H \cdot X^T$.

(iv) If X is a continuous local martingale, resp. if X is a finite variation process, then the same holds for $H \cdot X$.

(v) If H is of the form $H_s(\omega) = \sum_{i=0}^{p-1} H_{(i)}(\omega) \mathbf{1}_{(t_i,t_{i+1}]}(s)$, where $0 = t_0 < t_1 < \cdots < t_p$, and, for every $i \in \{0, 1, \ldots, p-1\}$, $H_{(i)}$ is $\mathscr{F}_{t_i}$-measurable, then

$$(H \cdot X)_t = \sum_{i=0}^{p-1} H_{(i)} (X_{t_{i+1} \wedge t} - X_{t_i \wedge t}).$$

We can restate the "associativity" property (ii) by saying that, if $Y_t = \int_0^t K_s dX_s$ then

$$\int_0^t H_s \, dY_s = \int_0^t H_s K_s \, dX_s.$$

Properties (i)–(iv) easily follow from the results obtained when X is a continuous local martingale, resp. a finite variation process. As for property (v), we first note that it is enough to consider the case where $X = M$ is a continuous local martingale with $M_0 = 0$, and by stopping M at suitable stopping times (and using (5.11)), we can even assume that M is in $\mathbb{H}^2$. There is a minor difficulty coming from the fact that the variables $H_{(i)}$ are not assumed to be bounded (and therefore we cannot directly use the construction of the integral of elementary processes). To circumvent this difficulty, we set, for every $n \geq 1$,

$$T_n = \inf\{t \geq 0 : |H_t| \geq n\} = \inf\{t_i : |H_{(i)}| \geq n\} \quad (\text{where } \inf \varnothing = \infty).$$

It is easy to verify that T_n is a stopping time, and we have $T_n \uparrow \infty$ as $n \to \infty$. Furthermore, we have for every n,

$$H_s \, \mathbf{1}_{[0,T_n]}(s) = \sum_{i=0}^{p-1} H_{(i)}^n \, \mathbf{1}_{(t_i,t_{i+1}]}(s)$$

where the random variables $H_{(i)}^n = H_{(i)} \mathbf{1}_{\{T_n > t_i\}}$ satisfy the same properties as the $H_{(i)}$'s and additionally are bounded by n. Hence $H \mathbf{1}_{[0,T_n]}$ is an elementary process, and by the very definition of the stochastic integral with respect to a martingale of $\mathbb{H}^2$, we have

$$(H \cdot M)_{t \wedge T_n} = (H \mathbf{1}_{[0,T_n]} \cdot M)_t = \sum_{i=0}^{p-1} H_{(i)}^n (M_{t_{i+1} \wedge t} - M_{t_i \wedge t}).$$

The desired result now follows by letting n tend to infinity.

5.1.4 Convergence of Stochastic Integrals

We start by giving a "dominated convergence theorem" for stochastic integrals.

Proposition 5.8 *Let $X = M + V$ be the canonical decomposition of a continuous semimartingale X, and let $t > 0$. Let $(H^n)_{n \geq 1}$ and H be locally bounded progressive processes, and let K be a nonnegative progressive process. Assume that the following properties hold a.s.:*

(i) $H_s^n \longrightarrow H_s$ *as $n \to \infty$, for every $s \in [0, t]$;*
(ii) $|H_s^n| \leq K_s$, *for every $n \geq 1$ and $s \in [0, t]$;*
(iii) $\int_0^t (K_s)^2 \, d\langle M, M \rangle_s < \infty$ *and $\int_0^t K_s \, |dV_s| < \infty$.*

Then,

$$\int_0^t H_s^n \, dX_s \underset{n \to \infty}{\longrightarrow} \int_0^t H_s \, dX_s$$

in probability.

Remarks

(a) Assertion (iii) holds automatically if K is locally bounded.
(b) Instead of assuming that (i) and (ii) hold for *every* $s \in [0, t]$ (a.s.), it is enough to assume that these conditions hold for $d\langle M, M \rangle_s$-a.e. $s \in [0, t]$ and for $|dV_s|$-a.e. $s \in [0, t]$, a.s. This will be clear from the proof.

Proof The a.s. convergence

$$\int_0^t H_s^n \, dV_s \underset{n \to \infty}{\longrightarrow} \int_0^t H_s \, dV_s$$

follows from the usual dominated convergence theorem. So we just have to verify that $\int_0^t H_s^n \, dM_s$ converges in probability to $\int_0^t H_s \, dM_s$. For every integer $p \geq 1$, consider the stopping time

$$T_p := \inf\{r \in [0, t] : \int_0^r (K_s)^2 d\langle M, M \rangle_s \geq p\} \wedge t,$$

and observe that $T_p = t$ for all large enough p, a.s., by assumption (iii). Then, the bound (5.14) gives

$$E\Big[\Big(\int_0^{T_p} H_s^n \, dM_s - \int_0^{T_p} H_s \, dM_s\Big)^2\Big] \leq E\Big[\int_0^{T_p} (H_s^n - H_s)^2 d\langle M, M \rangle_s\Big],$$

which tends to 0 as $n \to \infty$, by dominated convergence, using assumptions (i) and (ii) and the fact that $\int_0^{T_p} (K_s)^2 \, d\langle M, M \rangle_s \leq p$. Since $P(T_p = t)$ tends to 1 as $p \to \infty$, the desired result follows. □

We apply the preceding proposition to an approximation result in the case of continuous integrands, which will be useful in the next section.

Proposition 5.9 *Let X be a continuous semimartingale, and let H be an adapted process with continuous sample paths. Then, for every $t > 0$, for every sequence $0 = t_0^n < \cdots < t_{p_n}^n = t$ of subdivisions of $[0, t]$ whose mesh tends to 0, we have*

$$\lim_{n \to \infty} \sum_{i=0}^{p_n - 1} H_{t_i^n}(X_{t_{i+1}^n} - X_{t_i^n}) = \int_0^t H_s \, dX_s,$$

in probability.

Proof For every $n \geq 1$, define a process H^n by

$$H_s^n = \begin{cases} H_{t_i^n} & \text{if } t_i^n < s \leq t_{i+1}^n, \text{ for every } i \in \{0, 1, \ldots, p_n - 1\} \\ H_0 & \text{if } s = 0 \\ 0 & \text{if } s > t. \end{cases}$$

Note that H^n is progressive. We then observe that all assumptions of Proposition 5.8 hold if we take

$$K_s = \max_{0 \leq r \leq s} |H_s|,$$

which is a locally bounded progressive process. Hence, we conclude that

$$\int_0^t H_s^n \, dX_s \xrightarrow[n \to \infty]{} \int_0^t H_s \, dX_s$$

in probability. This gives the desired result since, by property (v) in Sect. 5.1.3, we have

$$\int_0^t H_s^n \, dX_s = \sum_{i=0}^{p_n - 1} H_{t_i^n}(M_{t_{i+1}^n} - M_{t_i^n}).$$

$\square$

Remark The preceding proposition can be viewed as a generalization of Lemma 4.3 to stochastic integrals. However, in contrast with that lemma, it is essential in Proposition 5.9 to evaluate H at the **left end** of the interval $(t_i^n, t_{i+1}^n]$: The result will fail if we replace $H_{t_i^n}$ by $H_{t_{i+1}^n}$. Let us give a simple counterexample. We take $H_t = X_t$ and we assume that the sequence of subdivisions $(t_i^n)_{0 \leq i \leq p_n}$ is increasing. By the proposition, we have

$$\lim_{n \to \infty} \sum_{i=0}^{p_n - 1} X_{t_i^n}(X_{t_{i+1}^n} - X_{t_i^n}) = \int_0^t X_s \, dX_s,$$

in probability. On the other hand, writing

$$\sum_{i=0}^{p_n-1} X_{t_{i+1}^n}(X_{t_{i+1}^n} - X_{t_i^n}) = \sum_{i=0}^{p_n-1} X_{t_i^n}(X_{t_{i+1}^n} - X_{t_i^n}) + \sum_{i=0}^{p_n-1}(X_{t_{i+1}^n} - X_{t_i^n})^2,$$

and using Proposition 4.21, we get

$$\lim_{n\to\infty} \sum_{i=0}^{p_n-1} X_{t_{i+1}^n}(X_{t_{i+1}^n} - X_{t_i^n}) = \int_0^t X_s\,dX_s + \langle X, X \rangle_t,$$

in probability. The resulting limit is different from $\int_0^t X_s\,dX_s$ unless the martingale part of X is degenerate. Note that, if we add the previous two convergences, we arrive at the formula

$$(X_t)^2 - (X_0)^2 = 2\int_0^t X_s\,dX_s + \langle X, X \rangle_t$$

which is a special case of Itô's formula of the next section.

5.2 Itô's Formula

Itô's formula is the cornerstone of stochastic calculus. It shows that, if we apply a twice continuously differentiable function to a p-tuple of continuous semimartingales, the resulting process is still a continuous semimartingale, and there is an explicit formula for the canonical decomposition of this semimartingale.

Theorem 5.10 (Itô's formula) *Let $X^1, \ldots, X^p$ be p continuous semimartingales, and let F be a twice continuously differentiable real function on $\mathbb{R}^p$. Then, for every $t \geq 0$,*

$$F(X_t^1, \ldots, X_t^p) = F(X_0^1, \ldots, X_0^p) + \sum_{i=1}^{p} \int_0^t \frac{\partial F}{\partial x^i}(X_s^1, \ldots, X_s^p)\,dX_s^i$$

$$+ \frac{1}{2}\sum_{i,j=1}^{p} \int_0^t \frac{\partial^2 F}{\partial x^i \partial x^j}(X_s^1, \ldots, X_s^p)\,d\langle X^i, X^j \rangle_s.$$

Proof We first deal with the case $p = 1$ and we write $X = X^1$ for simplicity. Fix $t > 0$ and consider an increasing sequence $0 = t_0^n < \cdots < t_{p_n}^n = t$ of subdivisions of $[0, t]$ whose mesh tends to 0. Then, for every n,

$$F(X_t) = F(X_0) + \sum_{i=0}^{p_n-1}(F(X_{t_{i+1}^n}) - F(X_{t_i^n})).$$

For every $i \in \{0, 1, \ldots, p_n - 1\}$, we apply the Taylor–Lagrange formula to the function $[0, 1] \ni \theta \mapsto F(X_{t_i^n} + \theta(X_{t_{i+1}^n} - X_{t_i^n}))$, between $\theta = 0$ and $\theta = 1$, and we get that

$$F(X_{t_{i+1}^n}) - F(X_{t_i^n}) = F'(X_{t_i^n})(X_{t_{i+1}^n} - X_{t_i^n}) + \frac{1}{2} f_{n,i}(X_{t_{i+1}^n} - X_{t_i^n})^2,$$

where the quantity $f_{n,i}$ can be written as $F''(X_{t_i^n} + c(X_{t_{i+1}^n} - X_{t_i^n}))$ for some $c \in [0, 1]$. By Proposition 5.9 with $H_s = F'(X_s)$, we have

$$\lim_{n \to \infty} \sum_{i=0}^{p_n-1} F'(X_{t_i^n})(X_{t_{i+1}^n} - X_{t_i^n}) = \int_0^t F'(X_s)\, dX_s,$$

in probability. To complete the proof of the case $p = 1$ of the theorem, it is therefore enough to verify that

$$\lim_{n \to \infty} \sum_{i=0}^{p_n-1} f_{n,i}(X_{t_{i+1}^n} - X_{t_i^n})^2 = \int_0^t F''(X_s)\, d\langle X, X\rangle_s, \tag{5.15}$$

in probability. We observe that

$$\sup_{0 \le i \le p_n-1} |f_{n,i} - F''(X_{t_i^n})| \le \sup_{0 \le i \le p_n-1} \left(\sup_{x \in [X_{t_i^n} \wedge X_{t_{i+1}^n}, X_{t_i^n} \vee X_{t_{i+1}^n}]} |F''(x) - F''(X_{t_i^n})| \right).$$

The right-hand side of the preceding display tends to 0 a.s. as $n \to \infty$, as a simple consequence of the uniform continuity of F'' (and of the sample paths of X) over a compact interval.

Since $\sum_{i=0}^{p_n-1}(X_{t_{i+1}^n} - X_{t_i^n})^2$ converges in probability (Proposition 4.21), it follows from the last display that

$$\left| \sum_{i=0}^{p_n-1} f_{n,i}(X_{t_{i+1}^n} - X_{t_i^n})^2 - \sum_{i=0}^{p_n-1} F''(X_{t_i^n})(X_{t_{i+1}^n} - X_{t_i^n})^2 \right| \xrightarrow[n \to \infty]{} 0$$

in probability. So the convergence (5.15) will follow if we can verify that

$$\lim_{n \to \infty} \sum_{i=0}^{p_n-1} F''(X_{t_i^n})(X_{t_{i+1}^n} - X_{t_i^n})^2 = \int_0^t F''(X_s)\, d\langle X, X\rangle_s, \tag{5.16}$$

in probability. In fact, we will show that (5.16) holds a.s. along a suitable sequence of values of n (this suffices for our needs, because we can replace the initial sequence of subdivisions by a subsequence). To this end, we note that

$$\sum_{i=0}^{p_n-1} F''(X_{t_i^n})(X_{t_{i+1}^n} - X_{t_i^n})^2 = \int_{[0,t]} F''(X_s)\,\mu_n(ds),$$

where μ_n is the random measure on $[0, t]$ defined by

$$\mu_n(dr) := \sum_{i=0}^{p_n-1} (X_{t_{i+1}^n} - X_{t_i^n})^2\,\delta_{t_i^n}(dr).$$

Write D for the dense subset of $[0, t]$ that consists of all t_i^n for $n \geq 1$ and $0 \leq i \leq p_n$. As a consequence of Proposition 4.21, we get for every $r \in D$,

$$\mu_n([0, r]) \underset{n\to\infty}{\longrightarrow} \langle X, X \rangle_r$$

in probability. Using a diagonal extraction, we can thus find a subsequence of values of n such that, along this subsequence, we have for every $r \in D$,

$$\mu_n([0, r]) \overset{\text{a.s.}}{\underset{n\to\infty}{\longrightarrow}} \langle X, X \rangle_r,$$

which implies that the sequence μ_n converges a.s. to the measure $\mathbf{1}_{[0,t]}(r)\,d\langle X, X \rangle_r$, in the sense of weak convergence of finite measures. We conclude that we have

$$\int_{[0,t]} F''(X_s)\,\mu_n(ds) \overset{\text{a.s.}}{\underset{n\to\infty}{\longrightarrow}} \int_0^t F''(X_s)\,d\langle X, X \rangle_s$$

along the chosen subsequence. This completes the proof of the case $p = 1$.

In the general case, the Taylor–Lagrange formula, applied for every $n \geq 1$ and every $i \in \{0, 1, \ldots, p_n - 1\}$ to the function

$$[0, 1] \ni \theta \mapsto F(X_{t_i^n}^1 + \theta(X_{t_{i+1}^n}^1 - X_{t_i^n}^1), \ldots, X_{t_i^n}^p + \theta(X_{t_{i+1}^n}^p - X_{t_i^n}^p)),$$

gives

$$F(X_{t_{i+1}^n}^1, \ldots, X_{t_{i+1}^n}^p) - F(X_{t_i^n}^1, \ldots, X_{t_i^n}^p) = \sum_{k=1}^p \frac{\partial F}{\partial x^k}(X_{t_i^n}^1, \ldots, X_{t_i^n}^p)\,(X_{t_{i+1}^n}^k - X_{t_i^n}^k)$$

$$+ \sum_{k,l=1}^p \frac{f_{n,i}^{k,l}}{2}(X_{t_{i+1}^n}^k - X_{t_i^n}^k)(X_{t_{i+1}^n}^l - X_{t_i^n}^l)$$

where, for every $k, l \in \{1, \ldots, p\}$,

$$f_{n,i}^{k,l} = \frac{\partial^2 F}{\partial x_k \partial x_l}(X_{t_i^n} + c(X_{t_{i+1}^n} - X_{t_i^n})),$$

for some $c \in [0, 1]$ (here we use the notation $X_t = (X_t^1, \ldots, X_t^p)$).

Proposition 5.9 can again be used to handle the terms involving first derivatives. Moreover, a slight modification of the arguments of the case $p = 1$ shows that, at least along a suitable sequence of values of n, we have for every $k, l \in \{1, \ldots, p\}$,

$$\lim_{n \to \infty} \sum_{i=0}^{p_n - 1} f_{n,i}^{k,l}(X_{t_{i+1}^n}^k - X_{t_i^n}^k)(X_{t_{i+1}^n}^l - X_{t_i^n}^l) = \int_0^t \frac{\partial^2 F}{\partial x_k \partial x_l}(X_s^1, \ldots, X_s^p) \, \mathrm{d}\langle X^k, X^l \rangle_s$$

in probability. This completes the proof of the theorem.   □

An important special case of Itô's formula is the **formula of integration by parts**, which is obtained by taking $p = 2$ and $F(x, y) = xy$: if X and Y are two continuous semimartingales, we have

$$X_t Y_t = X_0 Y_0 + \int_0^t X_s \, \mathrm{d}Y_s + \int_0^t Y_s \, \mathrm{d}X_s + \langle X, Y \rangle_t.$$

In particular, if $Y = X$,

$$X_t^2 = X_0^2 + 2 \int_0^t X_s \, \mathrm{d}X_s + \langle X, X \rangle_t.$$

When $X = M$ is a continuous local martingale, we know from the definition of the quadratic variation that $M^2 - \langle M, M \rangle$ is a continuous local martingale. The previous formula shows that this continuous local martingale is

$$M_0^2 + 2 \int_0^t M_s \, \mathrm{d}M_s.$$

We could have seen this directly from the construction of $\langle M, M \rangle$ in Chap. 4 (this construction involved approximations of the stochastic integral $\int_0^t M_s \mathrm{d}M_s$).

Let B be an $(\mathscr{F}_t)$-real Brownian motion (recall from Definition 3.11 that this means that B is a Brownian motion, which is adapted to the filtration $(\mathscr{F}_t)$ and such that, for every $0 \leq s < t$, the variable $B_t - B_s$ is independent of the σ-field $\mathscr{F}_s$). An $(\mathscr{F}_t)$-Brownian motion is a continuous local martingale (a martingale if $B_0 \in L^1$) and we already noticed that its quadratic variation is $\langle B, B \rangle_t = t$.

In this particular case, Itô's formula reads

$$F(B_t) = F(B_0) + \int_0^t F'(B_s) \, \mathrm{d}B_s + \frac{1}{2} \int_0^t F''(B_s) \mathrm{d}s.$$

Taking $X_t^1 = t$, $X_t^2 = B_t$, we also get for every twice continuously differentiable function $F(t, x)$ on $\mathbb{R}_+ \times \mathbb{R}$,

$$F(t, B_t) = F(0, B_0) + \int_0^t \frac{\partial F}{\partial x}(s, B_s) \, dB_s + \int_0^t \left(\frac{\partial F}{\partial t} + \frac{1}{2}\frac{\partial^2 F}{\partial x^2}\right)(s, B_s) \, ds.$$

Let $B_t = (B_t^1, \ldots, B_t^d)$ be a d-dimensional $(\mathscr{F}_t)$-Brownian motion. Note that the components $B^1, \ldots, B^d$ are $(\mathscr{F}_t)$-Brownian motions. By Proposition 4.16, $\langle B^i, B^j \rangle = 0$ when $i \neq j$ (by subtracting the initial value, which does not change the bracket $\langle B^i, B^j \rangle$, we are reduced to the case where $B^1, \ldots, B^d$ are independent). Itô's formula then shows that, for every twice continuously differentiable function F on $\mathbb{R}^d$,

$$F(B_t^1, \ldots, B_t^d)$$

$$= F(B_0^1, \ldots, B_0^d) + \sum_{i=1}^d \int_0^t \frac{\partial F}{\partial x_i}(B_s^1, \ldots, B_s^d) \, dB_s^i + \frac{1}{2}\int_0^t \Delta F(B_s^1, \ldots, B_s^d) \, ds.$$

The latter formula is often written in the shorter form

$$F(B_t) = F(B_0) + \int_0^t \nabla F(B_s) \cdot dB_s + \frac{1}{2}\int_0^t \Delta F(B_s) \, ds,$$

where ∇F stands for the vector of first partial derivatives of F. There is again an analogous formula for $F(t, B_t)$.

Important remark It frequently occurs that one needs to apply Itô's formula to a function F which is only defined (and twice continuously differentiable) on an open subset U of $\mathbb{R}^p$. In that case, we can argue in the following way. Suppose that there exists another open set V, such that $(X_0^1, \ldots, X_0^p) \in V$ a.s. and $\bar{V} \subset U$ (here $\bar{V}$ denotes the closure of V). Typically V will be the set of all points whose distance from U^c is strictly greater than ε, for some $\varepsilon > 0$. Set $T_V := \inf\{t \geq 0 : (X_t^1, \ldots, X_t^p) \notin V\}$, which is a stopping time by Proposition 3.9. Simple analytic arguments allow us to find a function G which is twice continuously differentiable on $\mathbb{R}^p$ and coincides with F on $\bar{V}$. We can now apply Itô's formula to obtain the canonical decomposition of the semimartingale $G(X_{t \wedge T_V}^1, \ldots, X_{t \wedge T_V}^p) = F(X_{t \wedge T_V}^1, \ldots, X_{t \wedge T_V}^p)$, and this decomposition only involves the first and second derivatives of F on V. If in addition we know that the process $(X_t^1, \ldots, X_t^p)$ a.s. does not exit U, we can let the open set V increase to U, and we get that Itô's formula for $F(X_t^1, \ldots, X_t^p)$ remains valid exactly in the same form as in Theorem 5.10. These considerations can be applied, for instance, to the function $F(x) = \log x$ and to a semimartingale X taking strictly positive values: see the proof of Proposition 5.21 below.

We now use Itô's formula to exhibit a remarkable class of (local) martingales, which extends the exponential martingales associated with processes with independent increments. A random process with values in the complex plane $\mathbb{C}$ is called a complex continuous local martingale if both its real part and its imaginary part are continuous local martingales.

Proposition 5.11 *Let M be a continuous local martingale and, for every $\lambda \in \mathbb{C}$, let*

$$\mathscr{E}(\lambda M)_t = \exp\left(\lambda M_t - \frac{\lambda^2}{2}\langle M, M\rangle_t\right).$$

The process $\mathscr{E}(\lambda M)$ is a complex continuous local martingale, which can be written in the form

$$\mathscr{E}(\lambda M)_t = e^{\lambda M_0} + \lambda \int_0^t \mathscr{E}(\lambda M)_s \, dM_s.$$

Remark The stochastic integral in the right-hand side of the last display is defined by dealing separately with the real and the imaginary part.

Proof If $F(r, x)$ is a twice continuously differentiable function on $\mathbb{R}^2$, Itô's formula gives

$$F(\langle M, M\rangle_t, M_t) = F(0, M_0) + \int_0^t \frac{\partial F}{\partial x}(\langle M, M\rangle_s, M_s) \, dM_s$$

$$+ \int_0^t \left(\frac{\partial F}{\partial r} + \frac{1}{2}\frac{\partial^2 F}{\partial x^2}\right)(\langle M, M\rangle_s, M_s) \, d\langle M, M\rangle_s.$$

Hence, $F(\langle M, M\rangle_t, M_t)$ is a continuous local martingale as soon as F satisfies the equation

$$\frac{\partial F}{\partial r} + \frac{1}{2}\frac{\partial^2 F}{\partial x^2} = 0.$$

This equation holds for $F(r, x) = \exp(\lambda x - \frac{\lambda^2}{2}r)$ (more precisely for both the real and the imaginary part of this function). Moreover, for this choice of F we have $\frac{\partial F}{\partial x} = \lambda F$, which leads to the formula of the statement. $\square$

5.3 A Few Consequences of Itô's Formula

Itô's formula has a huge number of applications. In this section, we derive some of the most important ones.

5.3.1   Lévy's Characterization of Brownian Motion

We start with a striking characterization of real Brownian motion as the unique continuous local martingale M such that $\langle M, M \rangle_t = t$. In fact, we give a multidimensional version of this result, which is known as Lévy's theorem.

Theorem 5.12 *Let $X = (X^1, \ldots, X^d)$ be an adapted process with continuous sample paths. The following are equivalent:*

(i) *X is a d-dimensional $(\mathscr{F}_t)$-Brownian motion.*
(ii) *The processes $X^1, \ldots, X^d$ are continuous local martingales, and $\langle X^i, X^j \rangle_t = \delta_{ij} t$ for every $i, j \in \{1, \ldots, d\}$ (here δ_{ij} is the Kronecker symbol, $\delta_{ij} = \mathbf{1}_{\{i=j\}}$).*

In particular, a continuous local martingale M is an $(\mathscr{F}_t)$-Brownian motion if and only if $\langle M, M \rangle_t = t$, for every $t \geq 0$, or equivalently if and only if $M_t^2 - t$ is a continuous local martingale.

Proof The fact that (i) $\Rightarrow$ (ii) has already been derived. Let us assume that (ii) holds. Let $\xi = (\xi_1, \ldots, \xi_d) \in \mathbb{R}^d$. Then, $\xi \cdot X_t = \sum_{j=1}^d \xi_j X_t^j$ is a continuous local martingale with quadratic variation

$$\sum_{j=1}^d \sum_{k=1}^d \xi_j \xi_k \langle X^j, X^k \rangle_t = |\xi|^2 t.$$

By Proposition 5.11, $\exp(i\xi \cdot X_t + \frac{1}{2}|\xi|^2 t)$ is a complex continuous local martingale. This complex continuous local martingale is bounded on every interval $[0, a]$, $a > 0$, and is therefore a (true) martingale, in the sense that its real and imaginary parts are both martingales. Hence, for every $0 \leq s < t$,

$$E[\exp(i\xi \cdot X_t + \frac{1}{2}|\xi|^2 t) \mid \mathscr{F}_s] = \exp(i\xi \cdot X_s + \frac{1}{2}|\xi|^2 s),$$

and thus

$$E[\exp(i\xi \cdot (X_t - X_s) \mid \mathscr{F}_s] = \exp(-\frac{1}{2}|\xi|^2 (t - s)).$$

It follows that, for every $A \in \mathscr{F}_s$,

$$E[\mathbf{1}_A \exp(i\xi \cdot (X_t - X_s))] = P(A) \exp(-\frac{1}{2}|\xi|^2 (t - s)).$$

Taking $A = \Omega$, we get that $X_t - X_s$ is a centered Gaussian vector with covariance matrix $(t - s)\mathrm{Id}$ (in particular, the components $X_t^j - X_s^j$, $1 \leq j \leq d$ are independent). Furthermore, fix $A \in \mathscr{F}_s$ with $P(A) > 0$, and write P_A for the conditional probability measure $P_A(\cdot) = P(A)^{-1}P(\cdot \cap A)$. We also obtain that

$$P_A[\exp(i\xi \cdot (X_t - X_s))] = \exp(-\frac{1}{2}|\xi|^2(t-s))$$

which means that the law of $X_t - X_s$ under P_A is the same as its law under P. Therefore, for any nonnegative measurable function f on $\mathbb{R}^d$, we have

$$P_A[f(X_t - X_s)] = E[f(X_t - X_s)],$$

or equivalently

$$E[\mathbf{1}_A f(X_t - X_s)] = P(A)\, E[f(X_t - X_s)].$$

This holds for any $A \in \mathscr{F}_s$ (when $P(A) = 0$ the equality is trivial), and thus $X_t - X_s$ is independent of $\mathscr{F}_s$.

It follows that, if $t_0 = 0 < t_1 < \ldots < t_p$, the vectors $X_{t_1} - X_{t_0}, X_{t_2} - X_{t_1}, \ldots, X_{t_p} - X_{t_{p-1}}$ are independent. Since the components of each of these vectors are independent random variables, we obtain that all variables $X_{t_k}^j - X_{t_{k-1}}^j, 1 \leq j \leq d, 1 \leq k \leq p$ are independent, and $X_{t_k}^j - X_{t_{k-1}}^j$ is distributed according to $\mathscr{N}(0, t_k - t_{k-1})$. This implies that $X - X_0$ is a d-dimensional Brownian motion started from 0. Since we also know that $X - X_0$ is independent of X_0 (as an easy consequence of the fact that $X_t - X_s$ is independent of $\mathscr{F}_s$, for every $0 \leq s < t$), we get that X is a d-dimensional Brownian motion. Finally, X is adapted and has independent increments with respect to the filtration $(\mathscr{F}_t)$ so that X is a d-dimensional $(\mathscr{F}_t)$-Brownian motion. $\qquad\square$

5.3.2 Continuous Martingales as Time-Changed Brownian Motions

The next theorem shows that any continuous local martingale M can be written as a "time-changed" Brownian motion (in fact, we prove this only when $\langle M, M \rangle_\infty = \infty$, but see the remarks below). It follows that the sample paths of M are Brownian sample paths run at a different (varying) speed, and certain almost sure properties of sample paths of M can be deduced from the corresponding properties of Brownian sample paths. For instance, under the condition $\langle M, M \rangle_\infty = \infty$, the sample paths of M must oscillate between $+\infty$ and $-\infty$ as $t \to \infty$ (cf. the last assertion of Proposition 2.14).

Theorem 5.13 (Dambis–Dubins–Schwarz) *Let M be a continuous local martingale such that $\langle M, M \rangle_\infty = \infty$ a.s. There exists a Brownian motion $(\beta_s)_{s \geq 0}$ such that*

$$a.s. \; \forall t \geq 0, \quad M_t = \beta_{\langle M, M \rangle_t}.$$

Remarks

(i) One can remove the assumption $\langle M, M \rangle_\infty = \infty$, at the cost of enlarging the underlying probability space, see [70, Chapter V].

(ii) The Brownian motion β is not adapted with respect to the filtration $(\mathscr{F}_t)$, but with respect to a "time-changed" filtration, as the following proof will show.

Proof We first assume that $M_0 = 0$. For every $r \geq 0$, we set

$$\tau_r = \inf\{t \geq 0 : \langle M, M \rangle_t \geq r\}.$$

Note that τ_r is a stopping time by Proposition 3.9. Furthermore, we have $\tau_r < \infty$ for every $r \geq 0$, on the event $\{\langle M, M \rangle_\infty = \infty\}$. It will be convenient to redefine the variables τ_r on the (negligible) event $\mathscr{N} = \{\langle M, M \rangle_\infty < \infty\}$ by taking $\tau_r(\omega) = 0$ for every $r \geq 0$ if $\omega \in \mathscr{N}$. Since the filtration is complete, τ_r remains a stopping time after this modification.

By construction, for every $\omega \in \Omega$, the function $r \mapsto \tau_r(\omega)$ is nondecreasing and left-continuous, and therefore has a right limit at every $r \geq 0$. This right limit is denoted by τ_{r+} and we have

$$\tau_{r+} = \inf\{t \geq 0 : \langle M, M \rangle_t > r\},$$

except of course on the negligible set $\mathscr{N}$, where $\tau_{r+} = 0$.

We set $\beta_r = M_{\tau_r}$ for every $r \geq 0$. By Theorem 3.7, the process $(\beta_r)_{r \geq 0}$ is adapted with respect to the filtration $(\mathscr{G}_r)$ defined by $\mathscr{G}_r = \mathscr{F}_{\tau_r}$ for every $r \geq 0$, and $\mathscr{G}_\infty = \mathscr{F}_\infty$. Note that the filtration $(\mathscr{G}_r)$ is complete since this property holds for $(\mathscr{F}_t)$.

The sample paths $r \mapsto \beta_r(\omega)$ are left-continuous and have right limits given for every $r \geq 0$ by

$$\beta_{r+} = \lim_{s \downarrow \downarrow r} \beta_s = M_{\tau_{r+}}.$$

In fact we have $\beta_{r+} = \beta_r$ for every $r \geq 0$, a.s., as a consequence of the following lemma.

Lemma 5.14 *We have a.s. for every $0 \leq a < b$,*

$$M_t = M_a, \; \forall t \in [a, b] \iff \langle M, M \rangle_b = \langle M, M \rangle_a.$$

Let us postpone the proof of the lemma. Since $\langle M, M \rangle_{\tau_r} = \langle M, M \rangle_{\tau_{r+}}$ for every $r \geq 0$, Lemma 5.14 implies that $M_{\tau_r} = M_{\tau_{r+}}$, for every $r \geq 0$, a.s. Hence the sample paths of β are continuous (to be precise, we should redefine $\beta_r = 0$, for every $r \geq 0$, on the zero probability set where the property of Lemma 5.14 fails).

Let us verify that β_s and $\beta_s^2 - s$ are martingales with respect to the filtration $(\mathscr{G}_s)$. For every integer $n \geq 1$, the stopped continuous local martingales M^{τ_n} and $(M^{\tau_n})^2 - \langle M, M \rangle^{\tau_n}$ are uniformly integrable martingales (by Theorem 4.13, recalling that $M_0 = 0$ and noting that $\langle M^{\tau_n}, M^{\tau_n} \rangle_\infty = \langle M, M \rangle_{\tau_n} = n$ a.s.). The optional stopping theorem (Theorem 3.22) then implies that, for every $0 \leq r \leq s \leq n$,

$$E[\beta_s \mid \mathscr{G}_r] = E[M_{\tau_s}^{\tau_n} \mid \mathscr{F}_{\tau_r}] = M_{\tau_r}^{\tau_n} = \beta_r$$

and similarly

$$E[\beta_s^2 - s \mid \mathscr{G}_r] = E[(M_{\tau_s}^{\tau_n})^2 - \langle M^{\tau_n}, M^{\tau_n} \rangle_{\tau_s} \mid \mathscr{F}_{\tau_r}] = (M_{\tau_r}^{\tau_n})^2 - \langle M^{\tau_n}, M^{\tau_n} \rangle_{\tau_r} = \beta_r^2 - r.$$

Then the case $d = 1$ of Theorem 5.12 shows that β is a $(\mathscr{G}_r)$-Brownian motion. Finally, by the definition of β, we have a.s. for every $t \geq 0$,

$$\beta_{\langle M, M \rangle_t} = M_{\tau_{\langle M, M \rangle_t}}.$$

But since $\tau_{\langle M, M \rangle_t} \leq t \leq \tau_{\langle M, M \rangle_t +}$ and since $\langle M, M \rangle$ takes the same value at $\tau_{\langle M, M \rangle_t}$ and at $\tau_{\langle M, M \rangle_t +}$, Lemma 5.14 shows that $M_t = M_{\tau_{\langle M, M \rangle_t}}$ for every $t \geq 0$, a.s. We conclude that we have $M_t = \beta_{\langle M, M \rangle_t}$ for every $t \geq 0$, a.s. This completes the proof when $M_0 = 0$.

If $M_0 \neq 0$, we write $M_t = M_0 + M_t'$, and we apply the previous argument to M', in order to get a Brownian motion β' with $\beta_0' = 0$, such that $M_t' = \beta'_{\langle M', M' \rangle_t}$ for every $t \geq 0$ a.s. Since β' is a $(\mathscr{G}_r)$-Brownian motion, β' is independent of $\mathscr{G}_0 = \mathscr{F}_0$, hence of M_0. Therefore, $\beta_s = M_0 + \beta_s'$ is also a Brownian motion, and we get the desired representation for M. $\qquad\square$

Proof of Lemma 5.14 Thanks to the continuity of sample paths of M and $\langle M, M \rangle$, it is enough to verify that for any fixed a and b such that $0 \leq a < b$, we have

$$\{M_t = M_a, \ \forall t \in [a, b]\} = \{\langle M, M \rangle_b = \langle M, M \rangle_a\}, \quad \text{a.s.}$$

The fact that the event in the left-hand side is (a.s.) contained in the event in the right-hand side is easy from the approximations of $\langle M, M \rangle$ in Theorem 4.9.

Let us prove the converse. Consider the continuous local martingale $N_t = M_t - M_{t \wedge a}$ and note that

$$\langle N, N \rangle_t = \langle M, M \rangle_t - \langle M, M \rangle_{t \wedge a}.$$

For every $\varepsilon > 0$, introduce the stopping time

$$T_\varepsilon = \inf\{t \geq 0 : \langle N, N \rangle_t \geq \varepsilon\}.$$

Then N^{T_ε} is a martingale in $\mathbb{H}^2$ (since $\langle N^{T_\varepsilon}, N^{T_\varepsilon}\rangle_\infty \le \varepsilon$). Fix $t \in [a, b]$. We have

$$E[N_{t \wedge T_\varepsilon}^2] = E[\langle N, N\rangle_{t \wedge T_\varepsilon}] \le \varepsilon.$$

Hence, considering the event $A := \{\langle M, M\rangle_b = \langle M, M\rangle_a\} \subset \{T_\varepsilon \ge b\}$,

$$E[1_A N_t^2] = E[1_A N_{t \wedge T_\varepsilon}^2] \le E[N_{t \wedge T_\varepsilon}^2] \le \varepsilon.$$

By letting ε go to 0, we get $E[1_A N_t^2] = 0$ and thus $N_t = 0$ a.s. on A, which completes the proof. $\qquad\square$

We can combine the arguments of the proof of Theorem 5.13 with Theorem 5.12 to get the following technical result, which will be useful when we consider the image of planar Brownian motion under holomorphic transformations in Chap. 7.

Proposition 5.15 *Let M and N be two continuous local martingales such that $M_0 = N_0 = 0$. Assume that*

(i) $\langle M, M\rangle_t = \langle N, N\rangle_t$ *for every $t \ge 0$, a.s.*
(ii) *M and N are orthogonal ($\langle M, N\rangle_t = 0$ for every $t \ge 0$, a.s.)*
(iii) $\langle M, M\rangle_\infty = \langle N, N\rangle_\infty = \infty$, *a.s.*

Let $\beta = (\beta_t)_{t \ge 0}$, resp. $\gamma = (\gamma_t)_{t \ge 0}$, be the real Brownian motion such $M_t = \beta_{\langle M, M\rangle_t}$, resp. $N_t = \gamma_{\langle N, N\rangle_t}$, for every $t \ge 0$, a.s. Then β and γ are independent.

Proof We use the notation of the proof of Theorem 5.13 and note that we have $\beta_r = M_{\tau_r}$ and $\gamma_r = N_{\tau_r}$, where

$$\tau_r = \inf\{t \ge 0 : \langle M, M\rangle_t \ge r\} = \inf\{t \ge 0 : \langle N, N\rangle_t \ge r\}.$$

We know that β and γ are $(\mathscr{G}_r)$-Brownian motions. Since M and N are orthogonal martingales, we also know that $M_t N_t$ is a local martingale. As in the proof of Theorem 5.13, and using now Proposition 4.15 (v), we get that, for every $n \ge 1$, $M_t^{\tau_n} N_t^{\tau_n}$ is a uniformly integrable martingale, and by applying the optional stopping theorem, we obtain that for $r \le s \le n$,

$$E[\beta_s \gamma_s \mid \mathscr{G}_r] = E[M_{\tau_s}^{\tau_n} N_{\tau_s}^{\tau_n} \mid \mathscr{F}_{\tau_r}] = M_{\tau_r}^{\tau_n} N_{\tau_r}^{\tau_n} = \beta_s \gamma_r$$

so that $\beta_r \gamma_r$ is a $(\mathscr{G}_r)$-martingale and the bracket $\langle \beta, \gamma\rangle$ (evaluated in the filtration $(\mathscr{G}_r)$) is identically zero. By Theorem 5.12, it follows that (β, γ) is a two-dimensional Brownian motion and, since $\beta_0 = \gamma_0 = 0$, this implies that β and γ are independent. $\qquad\square$

5.3.3   The Burkholder–Davis–Gundy Inequalities

We now state important inequalities connecting a continuous local martingale with its quadratic variation. If M is a continuous local martingale, we set

$$M_t^* = \sup_{s \leq t} |M_s|$$

for every $t \geq 0$. Theorem 5.16 below shows that, under the condition $M_0 = 0$, for every $p > 0$, the p-th moment of M_t^* is bounded above and below (up to universal multiplicative constants) by the p-th moment of $\sqrt{\langle M, M \rangle_t}$. These bounds are very useful because, in particular when M is a stochastic integral, it is often easier to estimate the moments of $\sqrt{\langle M, M \rangle_t}$ than those of M_t^*. Such applications arise, for instance, in the study of stochastic differential equations (see e.g. the proof of Theorem 8.5 below).

Theorem 5.16 (Burkholder–Davis–Gundy inequalities) *For every real $p > 0$, there exist two constants $c_p, C_p > 0$ depending only on p such that, for every continuous local martingale M with $M_0 = 0$, and every stopping time T,*

$$c_p \, E[\langle M, M \rangle_T^{p/2}] \leq E[(M_T^*)^p] \leq C_p \, E[\langle M, M \rangle_T^{p/2}].$$

Remark It may happen that the quantities $E[\langle M, M \rangle_T^{p/2}]$ and $E[(M_T^*)^p]$ are infinite. The theorem says that these quantities are either both finite (then the stated bounds hold) or both infinite.

Proof Replacing M by the stopping martingale M^T, we see that it is enough to treat the special case $T = \infty$. We then observe that it suffices to consider the case when M is bounded: Assuming that the bounded case has been treated, we can replace M by M^{T_n}, where $T_n = \inf\{t \geq 0 : |M_t| = n\}$, and we get the general case by letting n tend to ∞.

The left-hand side inequality, in the case $p \geq 4$, follows from the result of question **4.** in Exercise 4.26. We prove below the right-hand side inequality for all values of p. This is the inequality we will use in the sequel (we refer to [70, Chapter IV] for the remaining case).

We first consider the case $p \geq 2$. We apply Itô's formula to the function $|x|^p$:

$$|M_t|^p = \int_0^t p|M_s|^{p-1}\text{sgn}(M_s)\, dM_s + \frac{1}{2}\int_0^t p(p-1)|M_s|^{p-2}\, d\langle M, M \rangle_s.$$

Since M is bounded, hence in particular $M \in \mathbb{H}^2$, the process

$$\int_0^t p|M_s|^{p-1}\text{sgn}(M_s)\, dM_s$$

is a martingale in $\mathbb{H}^2$. We therefore get

$$
\begin{aligned}
E[|M_t|^p] &= \frac{p(p-1)}{2} E\left[\int_0^t |M_s|^{p-2}\, d\langle M, M\rangle_s\right] \\
&\leq \frac{p(p-1)}{2} E[(M_t^*)^{p-2} \langle M, M\rangle_t] \\
&\leq \frac{p(p-1)}{2} (E[(M_t^*)^p])^{(p-2)/p} (E[\langle M, M\rangle_t^{p/2}])^{2/p},
\end{aligned}
$$

by Hölder's inequality. On the other hand, by Doob's inequality in L^p (Proposition 3.15),

$$
E[(M_t^*)^p] \leq \left(\frac{p}{p-1}\right)^p E[|M_t|^p]
$$

and combining this bound with the previous one, we arrive at

$$
E[(M_t^*)^p] \leq \left(\left(\frac{p}{p-1}\right)^p \frac{p(p-1)}{2}\right)^{p/2} E[\langle M, M\rangle_t^{p/2}].
$$

It now suffices to let t tend to ∞.

Consider then the case $p < 2$. Since $M \in \mathbb{H}^2$, $M^2 - \langle M, M\rangle$ is a uniformly integrable martingale and we have, for every stopping time T,

$$
E[(M_T)^2] = E[\langle M, M\rangle_T].
$$

Let $x > 0$ and consider the stopping time $T_x := \inf\{t \geq 0 : (M_t)^2 \geq x\}$. Then, if T is any bounded stopping time,

$$
\begin{aligned}
P((M_T^*)^2 \geq x) = P(T_x \leq T) = P((M_{T_x \wedge T})^2 \geq x) &\leq \frac{1}{x} E[(M_{T_x \wedge T})^2] \\
&= \frac{1}{x} E[\langle M, M\rangle_{T_x \wedge T}] \\
&\leq \frac{1}{x} E[\langle M, M\rangle_T].
\end{aligned}
$$

Next consider the stopping time $S_x := \inf\{t \geq 0 : \langle M, M\rangle_t \geq x\}$. Observe that, for every $t \geq 0$, we have $\{(M_t^*)^2 \geq x\} \subset (\{(M_{S_x \wedge t}^*)^2 \geq x\} \cup \{S_x \leq t\})$. Using the preceding bound with $T = S_x \wedge t$, we thus get

$$
\begin{aligned}
P((M_t^*)^2 \geq x) &\leq P((M_{S_x \wedge t}^*)^2 \geq x) + P(S_x \leq t) \\
&\leq \frac{1}{x} E[\langle M, M\rangle_{S_x \wedge t}] + P(\langle M, M\rangle_t \geq x)
\end{aligned}
$$

$$= \frac{1}{x}E[\langle M,M \rangle_t \wedge x] + P(\langle M,M \rangle_t \geq x)$$

$$= \frac{1}{x}E[\langle M,M \rangle_t \, \mathbf{1}_{\{\langle M,M \rangle_t < x\}}] + 2 \, P(\langle M,M \rangle_t \geq x).$$

To complete the proof, set $q = p/2 \in (0,1)$ and integrate each side of the last bound with respect to the measure $q \, x^{q-1} \, dx$. We have first

$$\int_0^\infty P((M_t^*)^2 \geq x) \, q \, x^{q-1} \, dx = E\left[\int_0^{(M_t^*)^2} q \, x^{q-1} \, dx \right] = E[(M_t^*)^{2q}],$$

and similarly

$$\int_0^\infty P(\langle M,M \rangle_t \geq x) \, q \, x^{q-1} \, dx = E[\langle M,M \rangle_t^q].$$

Furthermore,

$$\int_0^\infty \frac{1}{x} E[\langle M,M \rangle_t \, \mathbf{1}_{\{\langle M,M \rangle_t < x\}}] \, q \, x^{q-1} \, dx$$

$$= E\left[\langle M,M \rangle_t \int_{\langle M,M \rangle_t}^\infty q \, x^{q-2} \, dx \right] = \frac{q}{1-q} E[\langle M,M \rangle_t^q].$$

Summarizing, we have obtained the bound

$$E[(M_t^*)^{2q}] \leq \left(2 + \frac{q}{1-q} \right) E[\langle M,M \rangle_t^q],$$

and we just have to let $t \to \infty$ to get the desired result.   □

Corollary 5.17 *Let M be a continuous local martingale such that $M_0 = 0$. The condition*

$$E[\langle M,M \rangle_\infty^{1/2}] < \infty$$

implies that M is a uniformly integrable martingale.

Proof By the case $p = 1$ of Theorem 5.16, the condition $E[\langle M,M \rangle_\infty^{1/2}] < \infty$ implies that $E[M_\infty^*] < \infty$. Proposition 4.7 (ii) then shows that the continuous local martingale M, which is dominated by the variable M_∞^*, is a uniformly integrable martingale.   □

The condition $E[\langle M,M \rangle_\infty^{1/2}] < \infty$ is weaker than the condition $E[\langle M,M \rangle_\infty] < \infty$, which ensures that $M \in \mathbb{H}^2$. The corollary can be applied to stochastic integrals. If M is a continuous local martingale and H is a progressive process such that, for every $t \geq 0$,

$$E\left[\left(\int_0^t H_s^2 \, d\langle M, M \rangle_s\right)^{1/2}\right] < \infty,$$

then $\int_0^t H_s dM_s$ is a martingale, and formulas (5.6) and (5.9) for the first moment and the conditional expectations of $\int_0^t H_s dM_s$ hold (of course with $t < \infty$).

5.4 The Representation of Martingales as Stochastic Integrals

In the special setting where the filtration on Ω is the completed canonical filtration of a Brownian motion, we will now show that all martingales can be represented as stochastic integrals with respect to that Brownian motion. For the sake of simplicity, we first consider a one-dimensional Brownian motion, but we will discuss the extension to Brownian motion in higher dimensions at the end of this section.

Theorem 5.18 *Assume that the filtration $(\mathscr{F}_t)$ on Ω is the completed canonical filtration of a real Brownian motion B started from 0. Then, for every random variable $Z \in L^2(\Omega, \mathscr{F}_\infty, P)$, there exists a unique progressive process $h \in L^2(B)$ (i.e. $E[\int_0^\infty h_s^2 ds] < \infty$) such that*

$$Z = E[Z] + \int_0^\infty h_s \, dB_s.$$

Consequently, for every martingale M that is bounded in L^2 (respectively, for every continuous local martingale M), there exists a unique process $h \in L^2(B)$ (resp. $h \in L^2_{loc}(B)$) and a constant $C \in \mathbb{R}$ such that

$$M_t = C + \int_0^t h_s \, dB_s.$$

Remark As the proof will show, the second part of the statement applies to a martingale M that is bounded in L^2, **without any assumption** on the continuity of sample paths of M. This observation will be useful later when we discuss consequences of the representation theorem. Note that continuous local martingales have continuous sample paths by definition.

Lemma 5.19 *Under the assumptions of the theorem, the vector space generated by the random variables*

$$\exp\left(i \sum_{j=1}^n \lambda_j (B_{t_j} - B_{t_{j-1}})\right),$$

for any choice of $0 = t_0 < t_1 < \cdots < t_n$ *and* $\lambda_1, \ldots, \lambda_n \in \mathbb{R}$, *is dense in the space* $L^2_{\mathbb{C}}(\Omega, \mathscr{F}_\infty, P)$ *of all square-integrable complex-valued* $\mathscr{F}_\infty$-*measurable random variables.*

Proof It is enough to prove that, if $Z \in L^2_{\mathbb{C}}(\Omega, \mathscr{F}_\infty, P)$ is such that

$$E\Big[Z \exp\Big(i \sum_{j=1}^{n} \lambda_j (B_{t_j} - B_{t_{j-1}})\Big)\Big] = 0 \qquad (5.17)$$

for any choice of $0 = t_0 < t_1 < \cdots < t_n$ and $\lambda_1, \ldots, \lambda_n \in \mathbb{R}$, then $Z = 0$.

Fix $0 = t_0 < t_1 < \cdots < t_n$, and consider the complex measure μ on $\mathbb{R}^n$ defined by

$$\mu(F) = E\Big[Z\, 1_F(B_{t_1}, B_{t_2} - B_{t_1}, \ldots, B_{t_n} - B_{t_{n-1}})\Big]$$

for any Borel subset F of $\mathbb{R}^n$. Then (5.17) exactly shows that the Fourier transform of μ is identically zero. By the injectivity of the Fourier transform on complex measures on $\mathbb{R}^d$, it follows that $\mu = 0$. We have thus $E[Z\, 1_A] = 0$ for every $A \in \sigma(B_{t_1}, \ldots, B_{t_n})$.

A monotone class argument then shows that the identity $E[Z\, 1_A] = 0$ remains valid for any $A \in \sigma(B_t, t \geq 0)$, and then by completion for any $A \in \mathscr{F}_\infty$. It follows that $Z = 0$. $\square$

Proof of Theorem 5.18 We start with the first assertion. We first observe that the uniqueness of h is easy since, if the representation of a given variable Z holds with two processes h and h' in $L^2(B)$, we have

$$E\Big[\int_0^\infty (h_s - h'_s)^2 ds\Big] = E\Big[\Big(\int_0^\infty h_s\, dB_s - \int_0^\infty h'_s\, dB_s\Big)^2\Big] = 0,$$

hence $h = h'$ in $L^2(B)$.

Let us turn to the existence part. Let $\mathscr{H}$ stand for the vector space of all variables $Z \in L^2(\Omega, \mathscr{F}_\infty, P)$ for which the property of the statement holds. We note that if $Z \in \mathscr{H}$ and h is the associated process in $L^2(B)$, we have

$$E[Z^2] = (E[Z])^2 + E\Big[\int_0^\infty (h_s)^2\, ds\Big].$$

It follows that $\mathscr{H}$ is a closed subspace of $L^2(\Omega, \mathscr{F}_\infty, P)$. Indeed, if (Z_n) is a sequence in $\mathscr{H}$ that converges to Z in $L^2(\Omega, \mathscr{F}_\infty, P)$, the processes $h^{(n)}$ corresponding respectively to the variables Z_n form a Cauchy sequence in $L^2(B)$, hence converge in $L^2(B)$ to a certain process $h \in L^2(B)$ – here we use the Hilbert space structure of $L^2(B)$ – and it immediately follows that $Z = E[Z] + \int_0^\infty h_s\, dB_s$.

Since $\mathscr{H}$ is closed, in order to prove that $\mathscr{H} = L^2(\Omega, \mathscr{F}_\infty, P)$, we just have to verify that $\mathscr{H}$ contains a dense subset of $L^2(\Omega, \mathscr{F}_\infty, P)$. Let $0 = t_0 < t_1 < \cdots < t_n$

and $\lambda_1, \ldots, \lambda_n \in \mathbb{R}$, and set $f(s) = \sum_{j=1}^{n} \lambda_j \mathbf{1}_{(t_{j-1}, t_j]}(s)$. Write $\mathscr{E}_t^f$ for the exponential martingale $\mathscr{E}(i \int_0^{\cdot} f(s) \, dB_s)$ (cf. Proposition 5.11). Proposition 5.11 shows that

$$\exp\left(i \sum_{j=1}^{n} \lambda_j (B_{t_j} - B_{t_{j-1}}) + \frac{1}{2} \sum_{j=1}^{n} \lambda_j^2 (t_j - t_{j-1})\right) = \mathscr{E}_\infty^f = 1 + i \int_0^{\infty} \mathscr{E}_s^f f(s) \, dB_s$$

and it follows that both the real part and the imaginary part of variables of the form $\exp\left(i \sum_{j=1}^{n} \lambda_j (B_{t_j} - B_{t_{j-1}})\right)$ are in $\mathscr{H}$. By Lemma 5.19, linear combinations of such random variables are dense in $L^2(\Omega, \mathscr{F}_\infty, P)$. This completes the proof of the first assertion of the theorem.

Let us turn to the second assertion. If M is a martingale that is bounded in L^2, then $M_\infty \in L^2(\Omega, \mathscr{F}_\infty, P)$, and thus can be written in the form

$$M_\infty = E[M_\infty] + \int_0^{\infty} h_s \, dB_s,$$

where $h \in L^2(B)$. Thanks to (5.9), it follows that

$$M_t = E[M_\infty \mid \mathscr{F}_t] = E[M_\infty] + \int_0^{t} h_s \, dB_s$$

and the uniqueness of h is also immediate from the uniqueness in the first assertion.

Finally, if M is a continuous local martingale, we have first $M_0 = C \in \mathbb{R}$ because the σ-field $\mathscr{F}_0$ contains only events of probability zero or one. If $T_n = \inf\{t \geq 0 : |M_t| \geq n\}$ we can apply the case of martingales bounded in L^2 to M^{T_n} and we get a process $h^{(n)} \in L^2(B)$ such that

$$M_t^{T_n} = C + \int_0^{t} h_s^{(n)} \, dB_s.$$

Using the uniqueness of the progressive process in the representation, we get that $h_s^{(m)} = \mathbf{1}_{[0, T_m]}(s) \, h_s^{(n)}$ if $m < n$, ds a.e., a.s. It is now easy to construct a process $h \in L^2_{loc}(B)$ such that, for every m, $h_s^{(m)} = \mathbf{1}_{[0, T_m]}(s) \, h_s$, ds a.e., a.s. The representation formula of the theorem follows, and the uniqueness of h is also straightforward. $\square$

Consequences Let us give two important consequences of the representation theorem. Under the assumptions of the theorem:

(1) **The filtration $(\mathscr{F}_t)_{t \geq 0}$ is right-continuous.** Indeed, let $t \geq 0$ and let Z be $\mathscr{F}_{t+}$-measurable and bounded. We can find $h \in L^2(B)$ such that

$$Z = E[Z] + \int_0^{\infty} h_s \, dB_s.$$

If $\varepsilon > 0$, Z is $\mathscr{F}_{t+\varepsilon}$-measurable, and thus, using (5.9),

$$Z = E[Z \mid \mathscr{F}_{t+\varepsilon}] = E[Z] + \int_0^{t+\varepsilon} h_s dB_s.$$

When $\varepsilon \to 0$ the right-hand side converges in L^2 to

$$E[Z] + \int_0^t h_s dB_s.$$

Thus Z is equal a.s. to an $\mathscr{F}_t$-measurable random variable, and, since the filtration is complete, Z is $\mathscr{F}_t$-measurable.

A similar argument shows that the filtration $(\mathscr{F}_t)_{t\geq 0}$ is also left-continuous: If, for $t > 0$, we let

$$\mathscr{F}_{t-} = \bigvee_{s \in [0,t)} \mathscr{F}_s$$

be the smallest σ-field that contains all σ-fields $\mathscr{F}_s$ for $s \in [0,t)$, we have $\mathscr{F}_{t-} = \mathscr{F}_t$.

(2) **All martingales of the filtration $(\mathscr{F}_t)_{t\geq 0}$ have a modification with continuous sample paths.** For a martingale that is bounded in L^2, this follows from the representation formula (see the remark after the statement of the theorem). Then consider a uniformly integrable martingale M (if M is not uniformly integrable, we just replace M by $M_{t\wedge a}$ for every $a \geq 0$). In that case, we have, for every $t \geq 0$,

$$M_t = E[M_\infty \mid \mathscr{F}_t].$$

By Theorem 3.18 (whose application is justified as we know that the filtration is right-continuous), the process M_t has a modification with càdlàg sample paths, and we consider this modification. Let $M_\infty^{(n)}$ be a sequence of bounded random variables such that $M_\infty^{(n)} \longrightarrow M_\infty$ in L^1 as $n \to \infty$. Introduce the martingales

$$M_t^{(n)} = E[M_\infty^{(n)} \mid \mathscr{F}_t],$$

which are bounded in L^2. By the beginning of the argument, we can assume that, for every n, the sample paths of $M^{(n)}$ are continuous. On the other hand, Doob's maximal inequality (Proposition 3.15) implies that, for every $\lambda > 0$,

$$P\left[\sup_{t\geq 0} |M_t^{(n)} - M_t| > \lambda \right] \leq \frac{3}{\lambda} E[|M_\infty^{(n)} - M_\infty|].$$

It follows that we can find a sequence $n_k \uparrow \infty$ such that, for every $k \geq 1$,

$$P\left[\sup_{t \geq 0} |M_t^{(n_k)} - M_t| > 2^{-k} \right] \leq 2^{-k}.$$

An application of the Borel–Cantelli lemma now shows that

$$\sup_{t \geq 0} |M_t^{(n_k)} - M_t| \xrightarrow[k \to \infty]{\text{a.s.}} 0$$

and we get that the sample paths of M are continuous as uniform limits of continuous functions.

Multidimensional extension Let us briefly describe the multidimensional extension of the preceding results. We now assume that the filtration $(\mathscr{F}_t)$ on Ω is the completed canonical filtration of a d-dimensional Brownian motion $B = (B^1, \ldots, B^d)$ started from 0. Then, for every random variable $Z \in L^2(\Omega, \mathscr{F}_\infty, P)$, there exists a unique d-tuple $(h^1, \ldots, h^d)$ of progressive processes, satisfying

$$E\left[\int_0^\infty (h_s^i)^2 \, ds \right] < \infty, \qquad \forall i \in \{1, \ldots, d\},$$

such that

$$Z = E[Z] + \sum_{i=1}^d \int_0^\infty h_s^i \, dB_s^i.$$

Similarly, if M is a continuous local martingale, there exist a constant C and a unique d-tuple $(h^1, \ldots, h^d)$ of progressive processes, satisfying

$$\int_0^t (h_s^i)^2 \, ds < \infty, \quad \text{a.s.} \qquad \forall t \geq 0, \ \forall i \in \{1, \ldots, d\},$$

such that

$$M_t = C + \sum_{i=1}^d \int_0^t h_s^i \, dB_s^i.$$

The proofs are exactly the same as in the case $d = 1$ (Theorem 5.18). Consequences (1) and (2) above remain valid.

5.5 Girsanov's Theorem

Throughout this section, we assume that the filtration $(\mathscr{F}_t)$ is both complete and right-continuous. Our goal is to study how the notions of a martingale and of a semimartingale are affected when the underlying probability measure P is replaced by another probability measure Q. Most of the time we will assume that P and Q are mutually absolutely continuous, and then the fact that the filtration $(\mathscr{F}_t)$ is complete with respect to P implies that it is complete with respect to Q. When there is a risk of confusion, we will write E_P for the expectation under the probability measure P, and similarly E_Q for the expectation under Q. Unless otherwise specified, the notions of a (local) martingale or of a semimartingale refer to the underlying probability measure P (when we consider these notions under Q we will say so explicitly). Note that, in contrast with the notion of a martingale, the notion of a finite variation process does not depend on the underlying probability measure.

Proposition 5.20 *Assume that Q is a probability measure on $(\Omega, \mathscr{F})$, which is absolutely continuous with respect to P on the σ-field $\mathscr{F}_\infty$. For every $t \in [0, \infty]$, let*

$$D_t = \frac{dQ}{dP}_{|\mathscr{F}_t}$$

be the Radon–Nikodym derivative of Q with respect to P on the σ-field $\mathscr{F}_t$. The process $(D_t)_{t \geq 0}$ is a uniformly integrable martingale. Consequently $(D_t)_{t \geq 0}$ has a càdlàg modification. Keeping the same notation $(D_t)_{t \geq 0}$ for this modification, we have, for every stopping time T,

$$D_T = \frac{dQ}{dP}_{|\mathscr{F}_T}.$$

Finally, if we assume furthermore that P and Q are mutually absolutely continuous on $\mathscr{F}_\infty$, we have

$$\inf_{t \geq 0} D_t > 0, \quad P \text{ a.s.}$$

Proof If $A \in \mathscr{F}_t$, we have

$$Q(A) = E_Q[1_A] = E_P[1_A D_\infty] = E_P[1_A E_P[D_\infty \mid \mathscr{F}_t]]$$

and, by the uniqueness of the Radon–Nikodym derivative on $\mathscr{F}_t$, it follows that

$$D_t = E_P[D_\infty \mid \mathscr{F}_t], \qquad \text{a.s.}$$

Hence D is a uniformly integrable martingale, which is closed by D_∞. Theorem 3.18 (using the fact that $(\mathscr{F}_t)$ is both complete and right-continuous) then allows us to find a càdlàg modification of $(D_t)_{t \geq 0}$, which we consider from now on.

Then, if T is a stopping time, the optional stopping theorem (Theorem 3.22) gives for every $A \in \mathscr{F}_T$,

$$Q(A) = E_Q[1_A] = E_P[1_A D_\infty] = E_P[1_A E_P[D_\infty \mid \mathscr{F}_T]] = E_P[1_A D_T],$$

and, since D_T is $\mathscr{F}_T$-measurable, it follows that

$$D_T = \frac{dQ}{dP}\Big|_{\mathscr{F}_T}.$$

Let us prove the last assertion. For every $\varepsilon > 0$, set

$$T_\varepsilon = \inf\{t \geq 0 : D_t < \varepsilon\}$$

and note that T_ε is a stopping time as the first hitting time of an open set by a càdlàg process (recall Proposition 3.9 and the fact that the filtration is right-continuous). Then, noting that the event $\{T_\varepsilon < \infty\}$ is $\mathscr{F}_{T_\varepsilon}$-measurable,

$$Q(T_\varepsilon < \infty) = E_P[1_{\{T_\varepsilon < \infty\}} D_{T_\varepsilon}] \leq \varepsilon$$

since $D_{T_\varepsilon} \leq \varepsilon$ on $\{T_\varepsilon < \infty\}$ by the right-continuity of sample paths. It immediately follows that

$$Q\left(\bigcap_{n=1}^{\infty}\{T_{1/n} < \infty\}\right) = 0$$

and since P is absolutely continuous with respect to Q we have also

$$P\left(\bigcap_{n=1}^{\infty}\{T_{1/n} < \infty\}\right) = 0.$$

But this exactly means that, P a.s., there exists an integer $n \geq 1$ such that $T_{1/n} = \infty$, giving the last assertion of the proposition. $\qquad\square$

Proposition 5.21 *Let D be a continuous local martingale taking (strictly) positive values. There exists a unique continuous local martingale L such that*

$$D_t = \exp\left(L_t - \frac{1}{2}\langle L, L\rangle_t\right) = \mathscr{E}(L)_t.$$

Moreover, L is given by the formula

$$L_t = \log D_0 + \int_0^t D_s^{-1}\, dD_s.$$

Proof Uniqueness is an easy consequence of Theorem 4.8. Then, since D takes positive values, we can apply Itô's formula to $\log D_t$ (see the remark before Proposition 5.11), and we get

$$\log D_t = \log D_0 + \int_0^t \frac{dD_s}{D_s} - \frac{1}{2}\int_0^t \frac{d\langle D, D\rangle_s}{D_s^2} = L_t - \frac{1}{2}\langle L, L\rangle_t,$$

where L is as in the statement.   □

We now state the main theorem of this section, which explains the relation between continuous local martingales under P and continuous local martingales under Q.

Theorem 5.22 (Girsanov) *Assume that the probability measures P and Q are mutually absolutely continuous on $\mathscr{F}_\infty$. Let $(D_t)_{t\geq 0}$ be the martingale with càdlàg sample paths such that, for every $t \geq 0$,*

$$D_t = \frac{dQ}{dP}_{|\mathscr{F}_t}.$$

Assume that D has continuous sample paths, and let L be the unique continuous local martingale such that $D_t = \mathscr{E}(L)_t$. Then, if M is a continuous local martingale under P, the process

$$\tilde{M} = M - \langle M, L\rangle$$

is a continuous local martingale under Q.

Remark By consequences of the martingale representation theorem explained at the end of the previous section, the continuity assumption for the sample paths of D always holds when $(\mathscr{F}_t)$ is the (completed) canonical filtration of a Brownian motion. In applications of Theorem 5.22, one often starts from the martingale (D_t) to define the probability measure Q, so that the continuity assumption is satisfied by construction (see the examples in the next section).

Proof The fact that D_t can be written in the form $D_t = \mathscr{E}(L)_t$ follows from Proposition 5.21 (we are assuming that D has continuous sample paths, and we also know from Proposition 5.20 that D takes positive values). Then, let T be a stopping time and let X be an adapted process with continuous sample paths. We claim that, if $(XD)^T$ is a martingale under P, then X^T is a martingale under Q. Let us verify the claim. By Proposition 5.20, $E_Q[|X_{T\wedge t}|] = E_P[|X_{T\wedge t}D_{T\wedge t}|] < \infty$, and it follows that $X_t^T \in L^1(Q)$. Then, let $A \in \mathscr{F}_s$ and $s < t$. Since $A \cap \{T > s\} \in \mathscr{F}_s$, we have, using the fact that $(XD)^T$ is a martingale under P,

$$E_P[\mathbf{1}_{A\cap\{T>s\}}X_{T\wedge t}D_{T\wedge t}] = E_P[\mathbf{1}_{A\cap\{T>s\}}X_{T\wedge s}D_{T\wedge s}].$$

By Proposition 5.20,

$$D_{T \wedge t} = \frac{dQ}{dP}\Big|_{\mathscr{F}_{T \wedge t}}, \quad D_{T \wedge s} = \frac{dQ}{dP}\Big|_{\mathscr{F}_{T \wedge s}},$$

and thus, since $A \cap \{T > s\} \in \mathscr{F}_{T \wedge s} \subset \mathscr{F}_{T \wedge t}$, it follows that

$$E_Q[1_{A \cap \{T > s\}} X_{T \wedge t}] = E_Q[1_{A \cap \{T > s\}} X_{T \wedge s}].$$

On the other hand, it is immediate that

$$E_Q[1_{A \cap \{T \le s\}} X_{T \wedge t}] = E_Q[1_{A \cap \{T \le s\}} X_{T \wedge s}].$$

By combining with the previous display, we have $E_Q[1_A X_{T \wedge t}] = E_Q[1_A X_{T \wedge s}]$, giving our claim. As a consequence of the claim, we get that, if XD is a continuous local martingale under P, then X is a continuous local martingale under Q.

Next let M be a continuous local martingale under P, and let $\tilde{M}$ be as in the statement of the theorem. We apply the preceding observation to $X = \tilde{M}$, noting that, by Itô's formula,

$$\tilde{M}_t D_t = M_0 D_0 + \int_0^t \tilde{M}_s \, dD_s + \int_0^t D_s \, dM_s - \int_0^t D_s \, d\langle M, L \rangle_s + \langle M, D \rangle_t$$

$$= M_0 D_0 + \int_0^t \tilde{M}_s \, dD_s + \int_0^t D_s \, dM_s$$

since $d\langle M, L \rangle_s = D_s^{-1} d\langle M, D \rangle_s$ by Proposition 5.21. We get that $\tilde{M}D$ is a continuous local martingale under P, and thus $\tilde{M}$ is a continuous local martingale under Q. $\square$

Consequences

(a) A process M which is a continuous local martingale under P remains a semimartingale under Q, and its canonical decomposition under Q is $M = \tilde{M} + \langle M, L \rangle$ (recall that the notion of a finite variation process does not depend on the underlying probability measure). It follows that the class of semimartingales under P is contained in the class of semimartingales under Q.

In fact these two classes are equal. Indeed, under the assumptions of Theorem 5.22, P and Q play symmetric roles, since the Radon–Nikodym derivative of P with respect to Q on the σ-field $\mathscr{F}_t$ is D_t^{-1}, which has continuous sample paths if D does.

We may furthermore notice that

$$D_t^{-1} = \exp\left(-L_t + \langle L, L \rangle_t - \frac{1}{2}\langle L, L \rangle_t\right) = \exp\left(-\tilde{L}_t - \frac{1}{2}\langle \tilde{L}, \tilde{L} \rangle_t\right) = \mathscr{E}(-\tilde{L})_t,$$

where $\tilde{L} = L - \langle L, L \rangle$ is a continuous local martingale under Q, and $\langle \tilde{L}, \tilde{L} \rangle = \langle L, L \rangle$. So, under the assumptions of Theorem 5.22, the roles of P and Q can be interchanged provided D is replaced by D^{-1} and L is replaced by $-\tilde{L}$.

(b) Let X and Y be two semimartingales (under P or under Q). The bracket $\langle X, Y \rangle$ is the same under P and under Q. In fact this bracket is given in both cases by the approximation of Proposition 4.21 (this observation was used implicitly in (a) above).

Similarly, if H is a locally bounded progressive process, the stochastic integral $H \cdot X$ is the same under P and under Q. To see this it is enough to consider the case when $X = M$ is a continuous local martingale (under P). Write $(H \cdot M)_P$ for the stochastic integral under P and $(H \cdot M)_Q$ for the one under Q. By linearity,

$$(H \cdot \tilde{M})_P = (H \cdot M)_P - H \cdot \langle M, L \rangle = (H \cdot M)_P - \langle (H \cdot M)_P, L \rangle,$$

and Theorem 5.22 shows that $(H \cdot \tilde{M})_P$ is a continuous local martingale under Q. Furthermore the bracket of this continuous local martingale with any continuous local martingale N under Q is equal to $H \cdot \langle M, N \rangle = H \cdot \langle \tilde{M}, N \rangle$, and it follows from Theorem 5.6 that $(H \cdot \tilde{M})_P = (H \cdot \tilde{M})_Q$ hence also $(H \cdot M)_P = (H \cdot M)_Q$.

With the notation of Theorem 5.22, set $\tilde{M} = \mathscr{G}_Q^P(M)$. Then $\mathscr{G}_Q^P$ maps the set of all P-continuous local martingales onto the set of all Q-continuous local martingales. One easily verifies, using the remarks in (a) above, that $\mathscr{G}_P^Q \circ \mathscr{G}_Q^P = \mathrm{Id}$. Furthermore, the mapping $\mathscr{G}_Q^P$ commutes with the stochastic integral: if H is a locally bounded progressive process, $H \cdot \mathscr{G}_Q^P(M) = \mathscr{G}_Q^P(H \cdot M)$.

(c) Suppose that $M = B$ is an $(\mathscr{F}_t)$-Brownian motion under P, then $\tilde{B} = B - \langle B, L \rangle$ is a continuous local martingale under Q, with quadratic variation $\langle \tilde{B}, \tilde{B} \rangle_t = \langle B, B \rangle_t = t$. By Theorem 5.12, it follows that $\tilde{B}$ is an $(\mathscr{F}_t)$-Brownian motion under Q.

In most applications of Girsanov's theorem, one constructs the probability measure Q in the following way. Start from a continuous local martingale L such that $L_0 = 0$ and $\langle L, L \rangle_\infty < \infty$ a.s. The latter condition implies that the limit $L_\infty := \lim_{t \to \infty} L_t$ exists a.s. (see Exercise 4.24). Then $\mathscr{E}(L)_t$ is a nonnegative continuous local martingale hence a supermartingale (Proposition 4.7), which converges a.s. to $\mathscr{E}(L)_\infty = \exp(L_\infty - \frac{1}{2}\langle L, L \rangle_\infty)$, and $E[\mathscr{E}(L)_\infty] \leq 1$ by Fatou's lemma. If the property

$$E[\mathscr{E}(L)_\infty] = 1 \tag{5.18}$$

holds, then $\mathscr{E}(L)$ is a uniformly integrable martingale (by Fatou's lemma again, one has $\mathscr{E}(L)_t \geq E[\mathscr{E}(L)_\infty \mid \mathscr{F}_t]$, but (5.18) implies that $E[\mathscr{E}(L)_\infty] = E[\mathscr{E}(L)_0] = E[\mathscr{E}(L)_t]$ for every $t \geq 0$). If we let Q be the probability measure with density $\mathscr{E}(L)_\infty$ with respect to P, we are in the setting of Theorem 5.22, with $D_t = \mathscr{E}(L)_t$. It is therefore very important to give conditions that ensure that (5.18) holds.

Theorem 5.23 *Let L be a continuous local martingale such that $L_0 = 0$. Consider the following properties:*

(i) $E[\exp \frac{1}{2}\langle L, L \rangle_\infty] < \infty$ *(Novikov's criterion);*

(ii) *L is a uniformly integrable martingale, and* $E[\exp \frac{1}{2}L_\infty] < \infty$ *(Kazamaki's criterion);*

(iii) $\mathscr{E}(L)$ *is a uniformly integrable martingale.*

Then, (i) $\Rightarrow$ (ii) $\Rightarrow$ (iii).

Proof (i) $\Rightarrow$ (ii) Property (i) implies that $E[\langle L, L \rangle_\infty] < \infty$ hence also that L is a continuous martingale bounded in L^2 (Theorem 4.13). Then,

$$\exp \frac{1}{2}L_\infty = (\mathscr{E}(L)_\infty)^{1/2} (\exp(\frac{1}{2}\langle L, L \rangle_\infty))^{1/2}$$

so that, by the Cauchy–Schwarz inequality,

$$E[\exp \frac{1}{2}L_\infty] \leq (E[\mathscr{E}(L)_\infty])^{1/2}(E[\exp(\frac{1}{2}\langle L, L \rangle_\infty)])^{1/2}$$

$$\leq (E[\exp(\frac{1}{2}\langle L, L \rangle_\infty)])^{1/2} < \infty.$$

(ii) $\Rightarrow$ (iii) Since L is a uniformly integrable martingale, Theorem 3.22 shows that, for any stopping time T, we have $L_T = E[L_\infty \mid \mathscr{F}_T]$. Jensen's inequality then gives

$$\exp \frac{1}{2}L_T \leq E[\exp \frac{1}{2}L_\infty \mid \mathscr{F}_T].$$

By assumption, $E[\exp \frac{1}{2}L_\infty] < \infty$, which implies that the collection of all variables of the form $E[\exp \frac{1}{2}L_\infty \mid \mathscr{F}_T]$, for any stopping time T, is uniformly integrable. The preceding bound then shows that the collection of all variables $\exp \frac{1}{2}L_T$, for any stopping time T, is also uniformly integrable.

For $0 < a < 1$, set $Z_t^{(a)} = \exp(\frac{aL_t}{1+a})$. Then, one easily verifies that

$$\mathscr{E}(aL)_t = (\mathscr{E}(L)_t)^{a^2} (Z_t^{(a)})^{1-a^2}.$$

If $\Gamma \in \mathscr{F}$ and T is a stopping time, Hölder's inequality gives

$$E[1_\Gamma \mathscr{E}(aL)_T] \leq E[\mathscr{E}(L)_T]^{a^2} E[1_\Gamma Z_T^{(a)}]^{1-a^2} \leq E[1_\Gamma Z_T^{(a)}]^{1-a^2} \leq E[1_\Gamma \exp \frac{1}{2}L_T]^{2a(1-a)}.$$

In the second inequality, we used the property $E[\mathscr{E}(L)_T] \leq 1$, which holds by Proposition 3.25 because $\mathscr{E}(L)$ is a nonnegative supermartingale and $\mathscr{E}(L)_0 = 1$. In the third inequality, we use Jensen's inequality, noting that $\frac{1+a}{2a} > 1$. Since the collection of all variables of the form $\exp \frac{1}{2} L_T$, for any stopping time T, is uniformly integrable, the preceding display shows that so is the collection of all variables $\mathscr{E}(aL)_T$ for any stopping time T. By the definition of a continuous local martingale, there is an increasing sequence $T_n \uparrow \infty$ of stopping times, such that, for every n, $\mathscr{E}(aL)_{t \wedge T_n}$ is a martingale. If $0 \le s \le t$, we can use uniform integrability to pass to the limit $n \to \infty$ in the equality $E[\mathscr{E}(aL)_{t \wedge T_n} \mid \mathscr{F}_s] = \mathscr{E}(aL)_{s \wedge T_n}$ and we get that $\mathscr{E}(aL)$ is a uniformly integrable martingale. It follows that

$$1 = E[\mathscr{E}(aL)_\infty] \le E[\mathscr{E}(L)_\infty]^{a^2} E[Z_\infty^{(a)}]^{1-a^2} \le E[\mathscr{E}(L)_\infty]^{a^2} E[\exp \frac{1}{2} L_\infty]^{2a(1-a)},$$

using again Jensen's inequality as above. When $a \to 1$, this gives $E[\mathscr{E}(L)_\infty] \ge 1$ hence $E[\mathscr{E}(L)_\infty] = 1$.   □

5.6   A Few Applications of Girsanov's Theorem

In this section, we describe a few applications of Girsanov's theorem, which illustrate the strength of the previous results.

Constructing solutions of stochastic differential equations Let b be a bounded measurable function on $\mathbb{R}_+ \times \mathbb{R}$. We assume that there exists a function $g \in L^2(\mathbb{R}_+, \mathscr{B}(\mathbb{R}_+), dt)$ such that $|b(t, x)| \le g(t)$ for every $(t, x) \in \mathbb{R}_+ \times \mathbb{R}$. This holds in particular if there exists an $A > 0$ such that $|b|$ is bounded on $[0, A] \times \mathbb{R}_+$ and vanishes on $(A, \infty) \times \mathbb{R}_+$.

Let B be an $(\mathscr{F}_t)$-Brownian motion. Consider the continuous local martingale

$$L_t = \int_0^t b(s, B_s) \, dB_s$$

and the associated exponential martingale

$$D_t = \mathscr{E}(L)_t = \exp \left(\int_0^t b(s, B_s) \, dB_s - \frac{1}{2} \int_0^t b(s, B_s)^2 ds \right).$$

Our assumption on b ensures that condition (i) of Theorem 5.23 holds, and thus D is a uniformly integrable martingale. We set $Q = D_\infty \cdot P$. Girsanov's theorem, and remark (c) following the statement of this theorem, show that the process

$$\beta_t := B_t - \int_0^t b(s, B_s) \, ds$$

is an $(\mathscr{F}_t)$-Brownian motion under Q.

We can restate the latter property by saying that, under the probability measure Q, there exists an $(\mathscr{F}_t)$-Brownian motion β such that the process $X = B$ solves the stochastic differential equation

$$dX_t = d\beta_t + b(t, X_t)\, dt.$$

This equation is of the type that will be considered in Chap. 7 below, but in contrast with the statements of this chapter, we are not making any regularity assumption on the function b. It is remarkable that Girsanov's theorem allows one to construct solutions of stochastic differential equations without regularity conditions on the coefficients.

The Cameron–Martin formula We now specialize the preceding discussion to the case where $b(t, x)$ does not depend on x. We assume that $b(t, x) = g(t)$, where $g \in L^2(\mathbb{R}_+, \mathscr{B}(\mathbb{R}_+), dt)$, and we also set, for every $t \geq 0$,

$$h(t) = \int_0^t g(s)\, ds.$$

The set $\mathscr{H}$ of all functions h that can be written in this form is called the Cameron–Martin space. If $h \in \mathscr{H}$, we sometimes write $\dot{h} = g$ for the associated function in $L^2(\mathbb{R}_+, \mathscr{B}(\mathbb{R}_+), dt)$ (this is the derivative of h in the sense of distributions).

As a special case of the previous discussion, under the probability measure

$$Q := D_\infty \cdot P = \exp\left(\int_0^\infty g(s)\, dB_s - \frac{1}{2}\int_0^\infty g(s)^2 ds\right) \cdot P,$$

the process $\beta_t := B_t - h(t)$ is a Brownian motion. Hence, for every nonnegative measurable function Φ on $C(\mathbb{R}_+, \mathbb{R})$,

$$E_P[D_\infty \, \Phi((B_t)_{t \geq 0})] = E_Q[\Phi((B_t)_{t \geq 0})] = E_Q[\Phi((\beta_t + h(t))_{t \geq 0})]$$

$$= E_P[\Phi((B_t + h(t))_{t \geq 0})].$$

The equality between the two ends of the last display is the Cameron–Martin formula. In the next proposition, we write this formula in the special case of the canonical construction of Brownian motion on the Wiener space (see the end of Sect. 2.2).

Proposition 5.24 (Cameron–Martin formula) *Let $W(dw)$ be the Wiener measure on $C(\mathbb{R}_+, \mathbb{R})$, and let h be a function in the Cameron–Martin space $\mathscr{H}$. Then, for every nonnegative measurable function Φ on $C(\mathbb{R}_+, \mathbb{R})$,*

$$\int W(dw)\, \Phi(w + h) = \int W(dw) \exp\left(\int_0^\infty \dot{h}(s)\, dw(s) - \frac{1}{2}\int_0^\infty \dot{h}(s)^2\, ds\right)\Phi(w).$$

Remark The integral $\int_0^\infty \dot{h}(s)\,dw(s)$ is a stochastic integral with respect to $w(s)$ (which is a Brownian motion under $W(dw)$), but it can also be viewed as a Wiener integral since the function $\dot{h}(s)$ is deterministic. The Cameron–Martin formula can be established by Gaussian calculations that do not involve stochastic integrals or Girsanov's theorem (see e.g. Chapter 1 of [62]). Still it is instructive to derive this formula as a special case of Girsanov's theorem.

The Cameron–Martin formula gives a "quasi-invariance" property of Wiener measure under the translations by functions of the Cameron–Martin space: The image of Wiener measure $W(dw)$ under the mapping $w \mapsto w + h$ has a density with respect to $W(dw)$ and this density is the terminal value of the exponential martingale associated with the martingale $\int_0^t \dot{h}(s)dw(s)$.

Law of hitting times for Brownian motion with drift Let B be a real Brownian motion with $B_0 = 0$, and for every $a > 0$, let $T_a := \inf\{t \geq 0 : B_t = a\}$. If $c \in \mathbb{R}$ is given, we aim at computing the law of the stopping time

$$U_a := \inf\{t \geq 0 : B_t + ct = a\}.$$

Of course, if $c = 0$, we have $U_a = T_a$, and the desired distribution is given by Corollary 2.22. Girsanov's theorem (or rather the Cameron–Martin formula) will allow us to derive the case where c is arbitrary from the special case $c = 0$.

Fix $t > 0$ and apply the Cameron–Martin formula with

$$\dot{h}(s) = c\,\mathbf{1}_{\{s \leq t\}} \quad , \quad h(s) = c(s \wedge t)\,,$$

and, for every $w \in C(\mathbb{R}_+, \mathbb{R})$,

$$\Phi(w) = \mathbf{1}_{\{\max_{[0,t]} w(s) \geq a\}}.$$

It follows that

$$\begin{aligned}
P(U_a \leq t) &= E[\Phi(B + h)] \\
&= E\left[\Phi(B) \exp\left(\int_0^\infty \dot{h}(s)\,dB_s - \frac{1}{2}\int_0^\infty \dot{h}(s)^2\,ds\right)\right] \\
&= E[\mathbf{1}_{\{T_a \leq t\}} \exp(cB_t - \frac{c^2}{2}t))] \\
&= E[\mathbf{1}_{\{T_a \leq t\}} \exp(cB_{t \wedge T_a} - \frac{c^2}{2}(t \wedge T_a))] \\
&= E[\mathbf{1}_{\{T_a \leq t\}} \exp(ca - \frac{c^2}{2}T_a)]
\end{aligned}$$

$$= \int_0^t ds\, \frac{a}{\sqrt{2\pi s^3}} e^{-\frac{a^2}{2s}} e^{ca-\frac{c^2}{2}s}$$

$$= \int_0^t ds\, \frac{a}{\sqrt{2\pi s^3}} e^{-\frac{1}{2s}(a-cs)^2},$$

where, in the fourth equality, we used the optional stopping theorem (Corollary 3.23) to write

$$E[\exp(cB_t - \frac{c^2}{2}t) \mid \mathscr{F}_{t\wedge T_a}] = \exp(cB_{t\wedge T_a} - \frac{c^2}{2}(t \wedge T_a)),$$

and we also made use of the explicit density of T_a given in Corollary 2.22. This calculation shows that the variable U_a has a density on $\mathbb{R}_+$ given by

$$\psi(s) = \frac{a}{\sqrt{2\pi s^3}} e^{-\frac{1}{2s}(a-cs)^2}.$$

By integrating this density, we can verify that

$$P(U_a < \infty) = \begin{cases} 1 & \text{if } c \geq 0, \\ e^{2ca} & \text{if } c \leq 0, \end{cases}$$

which may also be checked more easily by applying the optional stopping theorem to the continuous martingale $\exp(-2c(B_t + ct))$.

Exercises

In the following exercises, processes are defined on a probability space $(\Omega, \mathscr{F}, P)$ equipped with a complete filtration $(\mathscr{F}_t)_{t\in[0,\infty]}$.

Exercise 5.25 Let B be an $(\mathscr{F}_t)$-Brownian motion with $B_0 = 0$, and let H be an adapted process with continuous sample paths. Show that $\frac{1}{B_t} \int_0^t H_s dB_s$ converges in probability when $t \to 0$ and determine the limit.

Exercise 5.26

1. Let B be a one-dimensional $(\mathscr{F}_t)$-Brownian motion with $B_0 = 0$. Let f be a twice continuously differentiable function on $\mathbb{R}$, and let g be a continuous function on $\mathbb{R}$. Verify that the process

$$X_t = f(B_t) \exp\left(-\int_0^t g(B_s)\, ds\right)$$

is a semimartingale, and give its decomposition as the sum of a continuous local martingale and a finite variation process.

2. Prove that X is a continuous local martingale if and only if the function f satisfies the differential equation

$$f'' = 2gf.$$

3. From now on, we suppose in addition that g is nonnegative and vanishes outside a compact subinterval of $(0, \infty)$. Justify the existence and uniqueness of a solution f_1 of the equation $f'' = 2gf$ such that $f_1(0) = 1$ and $f_1'(0) = 0$. Let $a > 0$ and $T_a = \inf\{t \geq 0 : B_t = a\}$. Prove that

$$E\left[\exp\left(- \int_0^{T_a} g(B_s) \, ds \right) \right] = \frac{1}{f_1(a)}.$$

Exercise 5.27 (*Stochastic calculus with the supremum*) *Preliminary question.* Let $m : \mathbb{R}_+ \longrightarrow \mathbb{R}$ be a continuous function such that $m(0) = 0$, and let $s : \mathbb{R}_+ \longrightarrow \mathbb{R}$ be the monotone increasing function defined by

$$s(t) = \sup_{0 \leq r \leq t} m(r).$$

Show that, for every bounded Borel function h on $\mathbb{R}$ and every $t > 0$,

$$\int_0^t (s(r) - m(r)) \, h(r) \, ds(r) = 0.$$

(One may first observe that $\int \mathbf{1}_I(r) \, ds(r) = 0$ for every open interval I that does not intersect $\{r \geq 0 : s(r) = m(r)\}$.)

1. Let M be a continuous local martingale such that $M_0 = 0$, and for every $t \geq 0$, let

$$S_t = \sup_{0 \leq r \leq t} M_r.$$

Let $\varphi : \mathbb{R}_+ \longrightarrow \mathbb{R}$ be a twice continuously differentiable function. Justify the equality

$$\varphi(S_t) = \varphi(0) + \int_0^t \varphi'(S_s) \, dS_s.$$

2. Show that

$$(S_t - M_t)\,\varphi(S_t) = \Phi(S_t) - \int_0^t \varphi(S_s)\,dM_s$$

where $\Phi(x) = \int_0^x \varphi(y)\,dy$ for every $x \in \mathbb{R}$.

3. Infer that, for every $\lambda > 0$,

$$e^{-\lambda S_t} + \lambda(S_t - M_t)e^{-\lambda S_t}$$

is a continuous local martingale.

4. Let $a > 0$ and $T = \inf\{t \geq 0 : S_t - M_t = a\}$. We assume that $\langle M, M \rangle_\infty = \infty$ a.s. Show that $T < \infty$ a.s. and S_T is exponentially distributed with parameter $1/a$.

Exercise 5.28 Let B be an $(\mathscr{F}_t)$-Brownian motion started from 1. We fix $\varepsilon \in (0, 1)$ and set $T_\varepsilon = \inf\{t \geq 0 : B_t = \varepsilon\}$. We also let $\lambda > 0$ and $\alpha \in \mathbb{R}\backslash\{0\}$.

1. Show that $Z_t = (B_{t \wedge T_\varepsilon})^\alpha$ is a semimartingale and give its canonical decomposition as the sum of a continuous local martingale and a finite variation process.

2. Show that the process

$$Z_t = (B_{t \wedge T_\varepsilon})^\alpha \, \exp\left(-\lambda \int_0^{t \wedge T_\varepsilon} \frac{ds}{B_s^2}\right)$$

is a continuous local martingale if α and λ satisfy a polynomial equation to be determined.

3. Compute

$$E\left[\exp\left(-\lambda \int_0^{T_\varepsilon} \frac{ds}{B_s^2}\right)\right].$$

Exercise 5.29 Let $(X_t)_{t \geq 0}$ be a semimartingale. We assume that there exists an $(\mathscr{F}_t)$-Brownian motion $(B_t)_{t \geq 0}$ started from 0 and a continuous function $b : \mathbb{R} \longrightarrow \mathbb{R}$, such that

$$X_t = B_t + \int_0^t b(X_s)\,ds.$$

1. Let $F : \mathbb{R} \longrightarrow \mathbb{R}$ be a twice continuously differentiable function on $\mathbb{R}$. Show that, for $F(X_t)$ to be a continuous local martingale, it suffices that F satisfies a second-order differential equation to be determined.

2. Give the solution of this differential equation which is such that $F(0) = 0$ and $F'(0) = 1$. In what follows, F stands for this particular solution, which can be written in the form $F(x) = \int_0^x \exp(-2\beta(y))\,dy$, with a function β that will be determined in terms of b.

3. In this question only, we assume that b is integrable, i.e. $\int_{\mathbb{R}} |b(x)|\,dx < \infty$.

(a) Show that the continuous local martingale $M_t = F(X_t)$ is a martingale.
(b) Show that $\langle M, M \rangle_\infty = \infty$ a.s.
(c) Infer that

$$\limsup_{t\to\infty} X_t = +\infty \,, \ \liminf_{t\to\infty} X_t = -\infty \,, \ \text{a.s.}$$

4. We come back to the general case. Let $c < 0$ and $d > 0$, and

$$T_c = \inf\{t \geq 0 : X_t \leq c\} \,, \ T_d = \inf\{t \geq 0 : X_t \geq d\} \,.$$

Show that, on the event $\{T_c \wedge T_d = \infty\}$, the random variables $|B_{n+1} - B_n|$, for integers $n \geq 0$, are bounded above by a (deterministic) constant which does not depend on n. Infer that $P(T_c \wedge T_d = \infty) = 0$.
5. Compute $P(T_c < T_d)$ in terms of $F(c)$ of $F(d)$.
6. We assume that b vanishes on $(-\infty, 0]$ and that there exists a constant $\alpha > 1/2$ such that $b(x) \geq \alpha/x$ for every $x \geq 1$. Show that, for every $\varepsilon > 0$, one can choose $c < 0$ such that

$$P(T_n < T_c, \ \text{for every } n \geq 1) \geq 1 - \varepsilon.$$

Infer that $X_t \longrightarrow +\infty$ as $t \to \infty$, a.s. (*Hint:* Observe that the continuous local martingale $M_{t \wedge T_c}$ is bounded.)
7. Suppose now $b(x) = 1/(2x)$ for every $x \geq 1$. Show that

$$\liminf_{t\to\infty} X_t = -\infty \,, \ \text{a.s.}$$

Exercise 5.30 (*Lévy area*) Let $(X_t, Y_t)_{t \geq 0}$ be a two-dimensional $(\mathscr{F}_t)$-Brownian motion started from 0. We set, for every $t \geq 0$:

$$\mathscr{A}_t = \int_0^t X_s \, dY_s - \int_0^t Y_s \, dX_s \qquad (\textit{Lévy's area}).$$

1. Compute $\langle \mathscr{A}, \mathscr{A} \rangle_t$ and infer that $(\mathscr{A}_t)_{t \geq 0}$ is a square-integrable (true) martingale.
2. Let $\lambda > 0$. Justify the equality

$$E[e^{i\lambda \mathscr{A}_t}] = E[\cos(\lambda \mathscr{A}_t)].$$

3. Let f be a twice continuously differentiable function on $\mathbb{R}_+$. Give the canonical decomposition of the semimartingales

$$Z_t = \cos(\lambda \mathscr{A}_t),$$

$$W_t = -\frac{f'(t)}{2}(X_t^2 + Y_t^2) + f(t).$$

Verify that $\langle Z, W \rangle_t = 0$.

4. Show that, for the process $Z_t e^{W_t}$ to be a continuous local martingale, it suffices that f solves the differential equation

$$f''(t) = f'(t)^2 - \lambda^2 .$$

5. Let $r > 0$. Verify that the function

$$f(t) = -\log \cosh(\lambda(r - t))$$

solves the differential equation of question **4.** and derive the formula

$$E[e^{i\lambda \mathscr{A}_r}] = \frac{1}{\cosh(\lambda r)}.$$

Exercise 5.31 (*Squared Bessel processes*)   Let B be an $(\mathscr{F}_t)$-Brownian motion started from 0, and let X be a continuous semimartingale. We assume that X takes values in $\mathbb{R}_+$, and is such that, for every $t \geq 0$,

$$X_t = x + 2 \int_0^t \sqrt{X_s}\, dB_s + \alpha t$$

where x and α are nonnegative real numbers.

1. Let $f : \mathbb{R}_+ \longrightarrow \mathbb{R}_+$ be a continuous function, and let φ be a twice continuously differentiable function on $\mathbb{R}_+$, taking **strictly positive** values, which solves the differential equation

$$\varphi'' = 2f\varphi$$

and satisfies $\varphi(0) = 1$ and $\varphi'(1) = 0$. Observe that the function φ must then be decreasing over the interval $[0, 1]$.

   We set

$$u(t) = \frac{\varphi'(t)}{2\varphi(t)}.$$

for every $t \geq 0$. Verify that we have, for every $t \geq 0$,

$$u'(t) + 2u(t)^2 = f(t),$$

then show that, for every $t \geq 0$,

$$u(t)X_t - \int_0^t f(s)X_s\,ds = u(0)x + \int_0^t u(s)\,dX_s - 2\int_0^t u(s)^2 X_s\,ds.$$

We set

$$Y_t = u(t)X_t - \int_0^t f(s)X_s\,ds.$$

2. Show that, for every $t \geq 0$,

$$\varphi(t)^{-\alpha/2}\, e^{Y_t} = \mathscr{E}(N)_t$$

where $\mathscr{E}(N)_t = \exp(N_t - \frac{1}{2}\langle N, N\rangle_t)$ denotes the exponential martingale associated with the continuous local martingale

$$N_t = u(0)x + 2\int_0^t u(s)\sqrt{X_s}\,dB_s.$$

3. Infer from the previous question that

$$E\left[\exp\left(-\int_0^1 f(s)X_s\,ds\right)\right] = \varphi(1)^{\alpha/2}\,\exp(\frac{x}{2}\varphi'(0)).$$

4. Let $\lambda > 0$. Show that

$$E\left[\exp\left(-\lambda\int_0^1 X_s\,ds\right)\right] = (\cosh(\sqrt{2\lambda}))^{-\alpha/2}\exp(-\frac{x}{2}\sqrt{2\lambda}\,\tanh(\sqrt{2\lambda})).$$

5. Show that, if $\beta = (\beta_t)_{t\geq 0}$ is a real Brownian motion started from y, one has, for every $\lambda > 0$,

$$E\left[\exp\left(-\lambda\int_0^1 \beta_s^2\,ds\right)\right] = (\cosh(\sqrt{2\lambda}))^{-1/2}\exp(-\frac{y^2}{2}\sqrt{2\lambda}\,\tanh(\sqrt{2\lambda})).$$

Exercise 5.32 (*Tanaka's formula and local time*) Let B be an $(\mathscr{F}_t)$-Brownian motion started from 0. For every $\varepsilon > 0$, we define a function $g_\varepsilon : \mathbb{R} \longrightarrow \mathbb{R}$ by setting $g_\varepsilon(x) = \sqrt{\varepsilon + x^2}$.

1. Show that

$$g_\varepsilon(B_t) = g_\varepsilon(0) + M_t^\varepsilon + A_t^\varepsilon$$

where M^ε is a square integrable continuous martingale that will be identified in the form of a stochastic integral, and A^ε is an increasing process.

2. We set $\mathrm{sgn}(x) = \mathbf{1}_{\{x>0\}} - \mathbf{1}_{\{x<0\}}$ for every $x \in \mathbb{R}$. Show that, for every $t \geq 0$,

$$M_t^\varepsilon \xrightarrow[\varepsilon \to 0]{L^2} \int_0^t \mathrm{sgn}(B_s)\, dB_s.$$

Infer that there exists an increasing process L such that, for every $t \geq 0$,

$$|B_t| = \int_0^t \mathrm{sgn}(B_s)\, dB_s + L_t .$$

3. Observing that $A_t^\varepsilon \longrightarrow L_t$ when $\varepsilon \to 0$, show that, for every $\delta > 0$, for every choice of $0 < u < v$, the condition ($|B_t| \geq \delta$ for every $t \in [u, v]$) a.s. implies that $L_v = L_u$. Infer that the function $t \mapsto L_t$ is a.s. constant on every connected component of the open set $\{t \geq 0 : B_t \neq 0\}$.

4. We set $\beta_t = \int_0^t \mathrm{sgn}(B_s)\, dB_s$ for every $t \geq 0$. Show that $(\beta_t)_{t\geq 0}$ is an $(\mathscr{F}_t)$-Brownian motion started from 0.

5. Show that $L_t = \sup_{s \leq t}(-\beta_s)$, a.s. (In order to derive the bound $L_t \leq \sup_{s \leq t}(-\beta_s)$, one may consider the last zero of B before time t, and use question **3.**) Give the law of L_t.

6. For every $\varepsilon > 0$, we define two sequences of stopping times $(S_n^\varepsilon)_{n\geq 1}$ and $(T_n^\varepsilon)_{n\geq 1}$, by setting

$$S_1^\varepsilon = 0, \quad T_1^\varepsilon = \inf\{t \geq 0 : |B_t| = \varepsilon\}$$

and then, by induction,

$$S_{n+1}^\varepsilon = \inf\{t \geq T_n^\varepsilon : B_t = 0\}, \quad T_{n+1}^\varepsilon = \inf\{t \geq S_{n+1}^\varepsilon : |B_t| = \varepsilon\}.$$

For every $t \geq 0$, we set $N_t^\varepsilon = \sup\{n \geq 1 : T_n^\varepsilon \leq t\}$, where $\sup \varnothing = 0$. Show that

$$\varepsilon N_t^\varepsilon \xrightarrow[\varepsilon \to 0]{L^2} L_t.$$

(One may observe that

$$L_t + \int_0^t \left(\sum_{n=1}^\infty \mathbf{1}_{[S_n^\varepsilon, T_n^\varepsilon]}(s) \right) \mathrm{sgn}(B_s)\, dB_s = \varepsilon N_t^\varepsilon + r_t^\varepsilon$$

where the "remainder" r_t^ε satisfies $|r_t^\varepsilon| \leq \varepsilon$.)

7. Show that $N_t^1/\sqrt{t}$ converges in law as $t \to \infty$ to $|U|$, where U is $\mathcal{N}(0,1)$-distributed.

(Many results of Exercise 5.32 are reproved and generalized in Chap. 8.)

Exercise 5.33 (*Study of multidimensional Brownian motion*) Let $B_t = (B_t^1, B_t^2, \ldots, B_t^N)$ be an N-dimensional $(\mathcal{F}_t)$-Brownian motion started from $x = (x_1, \ldots, x_N) \in \mathbb{R}^N$. We suppose that $N \geq 2$.

1. Verify that $|B_t|^2$ is a continuous semimartingale, and that the martingale part of $|B_t|^2$ is a true martingale.
2. We set

$$\beta_t = \sum_{i=1}^{N} \int_0^t \frac{B_s^i}{|B_s|}\, dB_s^i$$

with the convention that $\frac{B_s^i}{|B_s|} = 0$ if $|B_s| = 0$. Justify the definition of the stochastic integrals appearing in the definition of β_t, then show that the process $(\beta_t)_{t \geq 0}$ is an $(\mathcal{F}_t)$-Brownian motion started from 0.
3. Show that

$$|B_t|^2 = |x|^2 + 2\int_0^t |B_s|\, d\beta_s + N\,t.$$

4. From now on, we assume that $x \neq 0$. Let $\varepsilon \in (0, |x|)$ and $T_\varepsilon = \inf\{t \geq 0 : |B_t| \leq \varepsilon\}$. We set $f(a) = \log a$ if $N = 2$, and $f(a) = a^{2-N}$ if $N \geq 3$, for every $a > 0$. Verify that $f(|B_{t \wedge T_\varepsilon}|)$ is a continuous local martingale.
5. Let $R > |x|$ and $S_R = \inf\{t \geq 0 : |B_t| \geq R\}$. Show that

$$P(T_\varepsilon < S_R) = \frac{f(R) - f(|x|)}{f(R) - f(\varepsilon)}.$$

Observing that $P(T_\varepsilon < S_R) \longrightarrow 0$ when $\varepsilon \to 0$, show that $B_t \neq 0$ for every $t \geq 0$, a.s.
6. Show that, a.s., for every $t \geq 0$,

$$|B_t| = |x| + \beta_t + \frac{N-1}{2} \int_0^t \frac{ds}{|B_s|}.$$

7. We assume that $N \geq 3$. Show that $|B_t| \longrightarrow \infty$ when $t \to \infty$, a.s. (*Hint*: Observe that $|B_t|^{2-N}$ is a nonnegative supermartingale.)
8. We assume $N = 3$. Using the form of the Gaussian density, verify that the collection of random variables $(|B_t|^{-1})_{t \geq 0}$ is bounded in L^2. Show that $(|B_t|^{-1})_{t \geq 0}$ is a continuous local martingale but is not a (true) martingale.

(*Chapter 7 presents a slightly different approach to the results of this exercise, see in particular Proposition 7.16.*)

Exercise 5.34 (*Application of the Cameron–Martin formula*) Let B be an $(\mathscr{F}_t)$-Brownian motion started from 0. We set $B_t^* = \sup\{|B_s| : s \leq t\}$ for every $t \geq 0$.

1. Set $U_1 = \inf\{t \geq 0 : |B_t| = 1\}$ and $V_1 = \inf\{t \geq U_1 : B_t = 0\}$. Justify the equality $P(B_{V_1}^* < 2) = 1/2$, and then show that one can find two constants $\alpha > 0$ and $\gamma > 0$ such that

$$P(V_1 \geq \alpha, \, B_{V_1}^* < 2) = \gamma > 0.$$

2. Show that, for every integer $n \geq 1$, $P(B_{n\alpha}^* < 2) \geq \gamma^n$. *Hint*: Construct a suitable sequence $V_1, V_2, \ldots$ of stopping times such that, for every $n \geq 2$,

$$P(V_n \geq n\alpha, \, B_{V_n}^* < 2) \geq \gamma \, P(V_{n-1} \geq (n-1)\alpha, \, B_{V_{n-1}}^* < 2).$$

   Conclude that, for every $\varepsilon > 0$ and $t \geq 0$, $P(B_t^* \leq \varepsilon) > 0$.
3. Let h be a twice continuously differentiable function on $\mathbb{R}_+$ such that $h(0) = 0$, and let $K > 0$. Via a suitable application of Itô's formula, show that there exists a constant A such that, for every $\varepsilon > 0$,

$$\left| \int_0^K h'(s) \, dB_s \right| \leq A\varepsilon \quad \text{a.s. on the event } \{B_K^* \leq \varepsilon\}.$$

4. We set $X_t = B_t - h(t)$ and $X_t^* = \sup\{|X_s| : s \leq t\}$. Infer from question **3.** that

$$\lim_{\varepsilon \downarrow 0} \frac{P(X_K^* \leq \varepsilon)}{P(B_K^* \leq \varepsilon)} = \exp\left(-\frac{1}{2} \int_0^K h'(s)^2 \, ds \right).$$

Notes and Comments

The reader who wishes to learn more about the topics of this chapter is strongly advised to look at the excellent books by Karatzas and Shreve [49], Revuz and Yor [70] and Rogers and Williams [72]. A more concise introduction to stochastic integration can also be found in Chung and Williams [10].

Stochastic integrals with respect to Brownian motion were introduced by Itô [36] in 1944. His motivation was to give a rigorous approach to the stochastic differential equations that govern diffusion processes. Doob [15] suggested to study stochastic integrals as martingales. Several authors then contributed to the theory, including Kunita and Watanabe [50] and Meyer [60]. We have chosen to restrict our attention to stochastic integration with respect to continuous semimartingales. The reader interested in the more general case of semimartingales with jumps can consult the treatise of Dellacherie and Meyer [14] and the more recent books of Protter [63] or Jacod and Shiryaev [44]. Itô's formula was derived in [40] for processes that are stochastic integrals with respect to Brownian motions, and in our general context it appeared in the work of Kunita and Watanabe [50]. Theorem 5.12, at least in the case $d = 1$, is usually attributed to Lévy, although it seems difficult to find this statement in Lévy's work (see however [54, Chapitre III]). Theorem 5.13 showing that any continuous martingale can be written as a time-changed Brownian motion is due to Dambis [11] and Dubins–Schwarz [17]. The Burkholder–Davis–Gundy inequalities appear in [7], see also the expository article of Burkholder [6] for the history of these famous inequalities. Theorem 5.18 goes back to Itô [39] – in the different form of the chaos decomposition of Wiener functionals – and was a great success of the theory of stochastic integration. This theorem and its numerous extensions have found many applications in the area of mathematical finance. Girsanov's theorem appears in [29] in 1960, whereas the Cameron–Martin formula goes back to [8] in 1944. Applications of Girsanov's theorem to stochastic differential equations are developed in the book [77] of Stroock and Varadhan. Exercise 5.30 is concerned with the so-called Lévy area of planar Brownian motion, which was studied by Lévy [53, 54] with a different definition. Exercise 5.31 is inspired by Pitman and Yor [67].

Chapter 6
General Theory of Markov Processes

Our goal in this chapter is to give a concise introduction to the main ideas of the theory of continuous time Markov processes. Markov processes form a fundamental class of stochastic processes, with many applications in real life problems outside mathematics. The reason why Markov processes are so important comes from the so-called Markov property, which enables many explicit calculations that would be intractable for more general random processes. Although the theory of Markov processes is by no means the central topic of this book, it will play a significant role in the next chapters, in particular in our discussion of stochastic differential equations. In fact the whole invention of Itô's stochastic calculus was motivated by the study of the Markov processes obtained as solutions of stochastic differential equations, which are also called diffusion processes.

This chapter is mostly independent of the previous ones, even though Brownian motion is used as a basic example, and the martingale theory developed in Chap. 3 plays an important role. After a section dealing with the general definitions and the problem of existence, we focus on the particular case of Feller processes, and in that framework we introduce the key notion of the generator. We establish regularity properties of Feller processes as consequences of the analogous results for supermartingales. We then discuss the strong Markov property, and we conclude the chapter by presenting three important classes of Markov processes.

6.1 General Definitions and the Problem of Existence

Let $(E, \mathscr{E})$ be a measurable space. A *Markovian transition kernel* from E into E is a mapping $Q : E \times \mathscr{E} \longrightarrow [0, 1]$ satisfying the following two properties:

(i) For every $x \in E$, the mapping $\mathscr{E} \ni A \mapsto Q(x, A)$ is a probability measure on $(E, \mathscr{E})$.
(ii) For every $A \in \mathscr{E}$, the mapping $E \ni x \mapsto Q(x, A)$ is $\mathscr{E}$-measurable.

© Springer International Publishing Switzerland 2016
J.-F. Le Gall, *Brownian Motion, Martingales, and Stochastic Calculus*,
Graduate Texts in Mathematics 274, DOI 10.1007/978-3-319-31089-3_6

In what follows we say transition kernel instead of Markovian transition kernel.

Remark In the case where E is finite or countable (and equipped with the σ-field of all subsets of E), Q is characterized by the "matrix" $(Q(x, \{y\}))_{x,y \in E}$.

If $f : E \longrightarrow \mathbb{R}$ is bounded and measurable (resp. nonnegative and measurable), the function Qf defined by

$$Qf(x) = \int Q(x, dy) f(y)$$

is also bounded and measurable (resp. nonnegative and measurable) on E. Indeed, if f is an indicator function, the measurability of Qf is just property (ii) and the general case follows from standard arguments.

Definition 6.1 A collection $(Q_t)_{t \geq 0}$ of transition kernels on E is called a *transition semigroup* if the following three properties hold.

(i) For every $x \in E$, $Q_0(x, dy) = \delta_x(dy)$.
(ii) For every $s, t \geq 0$ and $A \in \mathscr{E}$,

$$Q_{t+s}(x, A) = \int_E Q_t(x, dy) Q_s(y, A)$$

  (Chapman–Kolmogorov identity).
(iii) For every $A \in \mathscr{E}$, the function $(t, x) \mapsto Q_t(x, A)$ is measurable with respect to the σ-field $\mathscr{B}(\mathbb{R}_+) \otimes \mathscr{E}$.

Let $B(E)$ be the vector space of all bounded measurable real functions on E, which is equipped with the norm $\|f\| = \sup\{|f(x)| : x \in E\}$. Then the linear mapping $B(E) \ni f \mapsto Q_t f$ is a contraction of $B(E)$. From this point of view, the Chapman–Kolmogorov identity is equivalent to the relation

$$Q_{t+s} = Q_t Q_s$$

for every $s, t \geq 0$. This allows one to view $(Q_t)_{t \geq 0}$ as a semigroup of contractions of $B(E)$.

We now consider a filtered probability space $(\Omega, \mathscr{F}, (\mathscr{F}_t)_{t \in [0, \infty]}, P)$.

Definition 6.2 Let $(Q_t)_{t \geq 0}$ be a transition semigroup on E. A *Markov process* (with respect to the filtration $(\mathscr{F}_t)$) with transition semigroup $(Q_t)_{t \geq 0}$ is an $(\mathscr{F}_t)$-adapted process $(X_t)_{t \geq 0}$ with values in E such that, for every $s, t \geq 0$ and $f \in B(E)$,

$$E[f(X_{s+t}) \mid \mathscr{F}_s] = Q_t f(X_s).$$

Remark When we speak about a Markov process X without specifying the filtration, we implicitly mean that the property of the definition holds with the canonical filtration $\mathscr{F}_t^X = \sigma(X_r, 0 \leq r \leq t)$. We may also notice that, if X is a

Markov process with respect to a filtration $(\mathscr{F}_t)$, it is automatically also a Markov process (with the same semigroup) with respect to $(\mathscr{F}_t^X)$.

The definition of a Markov process can be interpreted as follows. Taking $f = \mathbf{1}_A$, we have

$$P[X_{s+t} \in A \mid \mathscr{F}_s] = Q_t(X_s, A)$$

and in particular

$$P[X_{s+t} \in A \mid X_r, 0 \le r \le s] = Q_t(X_s, A).$$

Hence the conditional distribution of X_{s+t} knowing the "past" $(X_r, 0 \le r \le s)$ before time s is given by $Q_t(X_s, \cdot)$, and this conditional distribution only depends on the "present" state X_s. This is the *Markov property* (informally, if one wants to predict the future after time s, the past up to time s does not give more information than just the present at time s).

Consequences of the definition Let $\gamma(dx)$ be the law of X_0. Then if $0 < t_1 < t_2 < \cdots < t_p$ and $A_0, A_1, \ldots, A_p \in \mathscr{E}$,

$$P(X_0 \in A_0, X_{t_1} \in A_1, X_{t_2} \in A_2, \ldots, X_{t_p} \in A_p)$$

$$= \int_{A_0} \gamma(dx_0) \int_{A_1} Q_{t_1}(x_0, dx_1) \int_{A_2} Q_{t_2-t_1}(x_1, dx_2) \cdots \int_{A_p} Q_{t_p-t_{p-1}}(x_{p-1}, dx_p).$$

More generally, if $f_0, f_1, \ldots, f_p \in B(E)$,

$$E[f_0(X_0)f_1(X_{t_1}) \cdots f_p(X_{t_p})] = \int \gamma(dx_0)f_0(x_0) \int Q_{t_1}(x_0, dx_1)f_1(x_1)$$

$$\times \int Q_{t_2-t_1}(x_1, dx_2)f_2(x_2) \cdots \int Q_{t_p-t_{p-1}}(x_{p-1}, dx_p)f_p(x_p).$$

This last formula is derived from the definition by induction on p. Note that, conversely, if the latter formula holds for any choice of $0 < t_1 < t_2 < \cdots < t_p$ and $f_0, f_1, \ldots, f_p \in B(E)$, then $(X_t)_{t\ge0}$ is a Markov process of semigroup $(Q_t)_{t\ge0}$, with respect to its canonical filtration $\mathscr{F}_t^X = \sigma(X_r, 0 \le r \le t)$ (use a monotone class argument to see that the property of the definition holds with $\mathscr{F}_t = \mathscr{F}_t^X$, see Appendix A1).

From the preceding formulas, we see that the finite-dimensional marginals of the process X are completely determined by the semigroup $(Q_t)_{t\ge0}$ and the law of X_0 (initial distribution).

Example If $E = \mathbb{R}^d$, we can take, for every $t > 0$ and $x \in \mathbb{R}^d$,

$$Q_t(x, dy) = p_t(y - x)\, dy$$

where, for $z \in \mathbb{R}^d$,

$$p_t(z) = (2\pi t)^{-d/2} \exp -\frac{|z|^2}{2t},$$

is the density of the Gaussian vector in $\mathbb{R}^d$ with covariance matrix $t \,\mathrm{Id}$. It is straightforward to verify that this defines a transition semigroup on $\mathbb{R}^d$, and the associated Markov process is d-dimensional Brownian motion (it would be more accurate to say pre-Brownian motion since we have not yet said anything about sample paths). In the case $d = 1$, compare with Corollary 2.4.

We now address the problem of the existence of a Markov process with a given semigroup. To this end, we will need a general theorem of construction of random processes, namely the Kolmogorov extension theorem. We give without proof the special case of this theorem that is of interest to us (a proof in a more general setting can be found in [64, Chapter III], see also [47, Chapter VII], and [49, Section 2.2] for the special case $E = \mathbb{R}$).

Let $\Omega^* = E^{\mathbb{R}_+}$ be the space of all mappings $\omega : \mathbb{R}_+ \longrightarrow E$. We equip Ω^* with the σ-field $\mathscr{F}^*$ generated by the coordinate mappings $\omega \mapsto \omega(t)$ for $t \in \mathbb{R}_+$. Let $F(\mathbb{R}_+)$ be the collection of all finite subsets of $\mathbb{R}_+$, and, for every $U \in F(\mathbb{R}_+)$, let $\pi_U : \Omega^* \longrightarrow E^U$ be the mapping which associates with every $\omega : \mathbb{R}_+ \longrightarrow E$ its restriction to U. If $U, V \in F(\mathbb{R}_+)$ and $U \subset V$, we similarly write $\pi_U^V : E^V \longrightarrow E^U$ for the obvious restriction mapping.

We recall that a topological space is Polish if its topology is separable (there exists a dense sequence) and can be defined by a complete metric.

Theorem 6.3 *Assume that E is a Polish space equipped with its Borel σ-field $\mathscr{E}$. For every $U \in F(\mathbb{R}_+)$, let μ_U be a probability measure on E^U. Assume that the collection $(\mu_U, U \in F(\mathbb{R}_+))$ is consistent in the following sense: If $U \subset V$, μ_U is the image of μ_V under π_U^V. Then there exists a unique probability measure μ on $(\Omega^*, \mathscr{F}^*)$ such that $\pi_U(\mu) = \mu_U$ for every $U \in F(\mathbb{R}_+)$.*

Remark The uniqueness of μ is an immediate consequence of the monotone class lemma (cf. Appendix A1).

The Kolmogorov extension theorem allows one to construct random processes having prescribed finite-dimensional marginals. To see this, let $(X_t)_{t \geq 0}$ be the canonical process on Ω^* :

$$X_t(\omega) = \omega(t), \qquad t \geq 0.$$

If μ is a probability measure on Ω^* and $U = \{t_1, \ldots, t_p\} \in F(\mathbb{R}_+)$, with $t_1 < t_2 < \cdots < t_p$, then $(X_{t_1}, \ldots, X_{t_p})$ can be viewed as a random variable with values in E^U, provided we identify E^U with E^p via the mapping $\omega \longrightarrow (\omega(t_1), \ldots, \omega(t_p))$. Furthermore, the distribution of $(X_{t_1}, \ldots, X_{t_p})$ under μ is $\pi_U(\mu)$. The Kolmogorov theorem can thus be rephrased by saying that given a collection $(\mu_U, U \in F(\mathbb{R}_+))$ of finite-dimensional marginal distributions, which satisfies the consistency condition (this condition is clearly necessary for the desired conclusion), one can construct a probability measure μ on the space Ω^*, under which the finite-dimensional marginals of the canonical process X are the measures $\mu_U, U \in F(\mathbb{R}_+)$.

Corollary 6.4 *We assume that E satisfies the assumption of the previous theorem and that $(Q_t)_{t\geq0}$ is a transition semigroup on E. Let γ be a probability measure on E. Then there exists a (unique) probability measure P on Ω^* under which the canonical process $(X_t)_{t\geq0}$ is a Markov process with transition semigroup $(Q_t)_{t\geq0}$ and the law of X_0 is γ.*

Proof Let $U = \{t_1, \ldots, t_p\} \in F(\mathbb{R}_+)$, with $0 \leq t_1 < \cdots < t_p$. We define a probability measure P^U on E^U (identified with E^p as explained above) by setting

$$\int P^U(\mathrm{d}x_1 \ldots \mathrm{d}x_p)\, 1_A(x_1, \ldots, x_p)$$

$$= \int \gamma(\mathrm{d}x_0) \int Q_{t_1}(x_0, \mathrm{d}x_1) \int Q_{t_2-t_1}(x_1, \mathrm{d}x_2) \cdots \int Q_{t_p-t_{p-1}}(x_{p-1}, \mathrm{d}x_p)\, 1_A(x_1, \ldots, x_p)$$

for any measurable subset A of E^U.

Using the Chapman–Kolmogorov relation, one verifies that the measures P^U satisfy the consistency condition. The Kolmogorov theorem then gives the existence (and uniqueness) of a probability measure P on Ω^* whose finite-dimensional marginals are the measures $P_U, U \in F(\mathbb{R}_+)$. By a previous observation, this implies that $(X_t)_{t\geq0}$ is under P a Markov process with semigroup $(Q_t)_{t\geq0}$, with respect to the canonical filtration. $\qquad\square$

For $x \in E$, let P_x be the measure given by the preceding corollary when $\gamma = \delta_x$. Then, the mapping $x \mapsto P_x$ is measurable in the sense that $x \mapsto P_x(A)$ is measurable, for every $A \in \mathscr{F}^*$. In fact, the latter property holds when A only depends on a finite number of coordinates (in that case, there is an explicit formula for $P_x(A)$) and a monotone class argument gives the general case. Moreover, for any probability measure γ on E, the measure defined by

$$P_{(\gamma)}(A) = \int \gamma(\mathrm{d}x)\, P_x(A)$$

is the unique probability measure on Ω^* under which the canonical process $(X_t)_{t\geq0}$ is a Markov process with semigroup $(Q_t)_{t\geq0}$ and the law of X_0 is γ.

Summarizing, the preceding corollary allows one to construct (under a topological assumption on E) a Markov process $(X_t)_{t\geq0}$ with semigroup $(Q_t)_{t\geq0}$, which starts with a given initial distribution. More precisely, we get a measurable collection of probability measures $(P_x)_{x\in E}$ such that the Markov process X starts from x under P_x. However, a drawback of the method that we used is the fact that it does not give any information on the regularity properties of sample paths of X – at present we cannot even assert that these sample paths are measurable. We will come back to this question later, but this will require additional assumptions on the semigroup $(Q_t)_{t\geq 0}$.

We end this section by introducing the important notion of the resolvent.

Definition 6.5 Let $\lambda > 0$. The λ-*resolvent* of the transition semigroup $(Q_t)_{t\geq 0}$ is the linear operator $R_\lambda : B(E) \longrightarrow B(E)$ defined by

$$R_\lambda f(x) = \int_0^\infty e^{-\lambda t} Q_t f(x)\, dt$$

for $f \in B(E)$ and $x \in E$.

Remark Property (iii) of the definition of a transition semigroup is used here to get the measurability of the mapping $t \mapsto Q_t f(x)$, which is required to make sense of the definition of $R_\lambda f(x)$.

Properties of the resolvent.

(i) $\|R_\lambda f\| \leq \frac{1}{\lambda} \|f\|$.
(ii) If $0 \leq f \leq 1$, then $0 \leq \lambda R_\lambda f \leq 1$.
(iii) If $\lambda, \mu > 0$, we have $R_\lambda - R_\mu + (\lambda - \mu) R_\lambda R_\mu = 0$ (*resolvent equation*).

Proof Properties (i) and (ii) are very easy. Let us only prove (iii). We may assume that $\lambda \neq \mu$. Then,

$$
\begin{aligned}
R_\lambda (R_\mu f)(x) &= \int_0^\infty e^{-\lambda s} Q_s \left(\int_0^\infty e^{-\mu t} Q_t f\, dt \right)(x)\, ds \\
&= \int_0^\infty e^{-\lambda s} \left(\int Q_s(x, dy) \int_0^\infty e^{-\mu t} Q_t f(y)\, dt \right) ds \\
&= \int_0^\infty e^{-\lambda s} \left(\int_0^\infty e^{-\mu t} Q_{s+t} f(y)\, dt \right) ds \\
&= \int_0^\infty e^{-(\lambda - \mu)s} \left(\int_0^\infty e^{-\mu(s+t)} Q_{s+t} f(y)\, dt \right) ds \\
&= \int_0^\infty e^{-(\lambda - \mu)s} \left(\int_s^\infty e^{-\mu r} Q_r f(y)\, dr \right) ds \\
&= \int_0^\infty Q_r f(y)\, e^{-\mu r} \left(\int_0^r e^{-(\lambda - \mu)s}\, ds \right) dr \\
&= \int_0^\infty Q_r f(y) \left(\frac{e^{-\mu r} - e^{-\lambda r}}{\lambda - \mu} \right) dr
\end{aligned}
$$

giving the desired result. $\square$

Example In the case of real Brownian motion, one verifies that

$$R_\lambda f(x) = \int r_\lambda(y-x)f(y)\,dy$$

where

$$r_\lambda(y-x) = \int_0^\infty (2\pi t)^{-1/2}\exp\left(-\frac{|y-x|^2}{2t} - \lambda t\right)dt = \frac{1}{\sqrt{2\lambda}}\exp(-|y-x|\sqrt{2\lambda}).$$

A neat way of getting the last equality is to use the formula $E[e^{-\lambda T_a}] = e^{-a\sqrt{2\lambda}}$ for the Laplace transform of the hitting time $a > 0$ by a real Brownian motion started from 0 (see formula (3.7)). By differentiating with respect to λ, we get $E[T_a\,e^{-\lambda T_a}] = (a/\sqrt{2\lambda})e^{-a\sqrt{2\lambda}}$ and using the density of T_a (Corollary 2.22) to rewrite $E[T_a\,e^{-\lambda T_a}]$, we exactly find the integral that comes up in the calculation of $r_\lambda(y-x)$.

A key motivation of the introduction of the resolvent is the fact that it allows one to construct certain supermartingales associated with a Markov process.

Lemma 6.6 *Let X be a Markov process with semigroup $(Q_t)_{t\geq0}$ with respect to the filtration $(\mathscr{F}_t)$. Let $h \in B(E)$ be nonnegative and let $\lambda > 0$. Then the process*

$$e^{-\lambda t}R_\lambda h(X_t)$$

is an $(\mathscr{F}_t)$-supermartingale.

Proof The random variables $e^{-\lambda t}R_\lambda h(X_t)$ are bounded and thus in L^1. Then, for every $s \geq 0$,

$$Q_s R_\lambda h = \int_0^\infty e^{-\lambda t}Q_{s+t}h\,dt$$

and it follows that

$$e^{-\lambda s}Q_s R_\lambda h = \int_0^\infty e^{-\lambda(s+t)}Q_{s+t}h\,dt = \int_s^\infty e^{-\lambda t}Q_t h\,dt \leq R_\lambda h.$$

Hence, for every $s, t \geq 0$,

$$E[e^{-\lambda(t+s)}R_\lambda h(X_{t+s}) \mid \mathscr{F}_t] = e^{-\lambda(t+s)}Q_s R_\lambda h(X_t) \leq e^{-\lambda t}R_\lambda h(X_t),$$

giving the desired supermartingale property.   □

6.2   Feller Semigroups

From now on, we assume that E is a metrizable locally compact topological space. We also assume that E is countable at infinity, meaning that E is a countable union of compact sets. The space E is equipped with its Borel σ-field.

Under the previous assumptions, it is well known that the space E is Polish. Moreover, one can find an increasing sequence $(K_n)_{n\geq 1}$ of compact subsets of E, such that any compact set of E is contained in K_n for some n. A function $f : E \longrightarrow \mathbb{R}$ tends to 0 at infinity if, for every $\varepsilon > 0$, there exists a compact subset K of E such that $|f(x)| \leq \varepsilon$ for every $x \in E\backslash K$. This is equivalent to requiring that

$$\sup_{x\in E\backslash K_n} |f(x)| \xrightarrow[n\to\infty]{} 0.$$

We let $C_0(E)$ stand for the set of all continuous real functions on E that tend to 0 at infinity. The space $C_0(E)$ is a Banach space for the supremum norm

$$\|f\| = \sup_{x\in E} |f(x)|.$$

Definition 6.7 Let $(Q_t)_{t\geq 0}$ be a transition semigroup on E. We say that $(Q_t)_{t\geq 0}$ is a *Feller semigroup* if:

(i) $\forall f \in C_0(E), Q_t f \in C_0(E)$;
(ii) $\forall f \in C_0(E), \|Q_t f - f\| \longrightarrow 0$ as $t \to 0$.

A Markov process with values in E is a *Feller process* if its semigroup if Feller.

Remark One can prove (see for instance [70, Proposition III.2.4]) that condition (ii) can be replaced by the seemingly weaker property

$$\forall f \in C_0(E), \ \forall x \in E, \ Q_t f(x) \xrightarrow[t\to 0]{} f(x).$$

We will not use this, except in one particular example at the end of this chapter.

Condition (ii) implies that, for every $s \geq 0$,

$$\lim_{t\downarrow 0} \|Q_{s+t}f - Q_s f\| = \lim_{t\downarrow 0} \|Q_s(Q_t f - f)\| = 0$$

since Q_s is a contraction of $C_0(E)$. We note that the convergence is uniform when s varies over $\mathbb{R}_+$, which ensures that the mapping $t \mapsto Q_t f$ is uniformly continuous from $\mathbb{R}_+$ into $C_0(E)$, for any fixed $f \in C_0(E)$.

In what follows, we fix a Feller semigroup $(Q_t)_{t\geq 0}$ on E. Using property (i) of the definition and the dominated convergence theorem, one easily verifies that $R_\lambda f \in C_0(E)$ for every $f \in C_0(E)$ and $\lambda > 0$.

Proposition 6.8 *Let $\lambda > 0$, and set $\mathscr{R} = \{R_\lambda f : f \in C_0(E)\}$. Then $\mathscr{R}$ does not depend on the choice $\lambda > 0$. Furthermore, $\mathscr{R}$ is a dense subspace of $C_0(E)$.*

Proof If $\lambda \neq \mu$, the resolvent equation gives

$$R_\lambda f = R_\mu(f + (\mu - \lambda)R_\lambda f).$$

Hence any function of the form $R_\lambda f$ with $f \in C_0(E)$ is also of the form $R_\mu g$ for some $g \in C_0(E)$. This gives the first assertion.

Clearly, $\mathscr{R}$ is a linear subspace of $C_0(E)$. To see that it is dense, we simply note that, for every $f \in C_0(E)$,

$$\lambda R_\lambda f = \lambda \int_0^\infty e^{-\lambda t} Q_t f \, dt = \int_0^\infty e^{-t} Q_{t/\lambda} f \, dt \xrightarrow[\lambda \to \infty]{} f \quad \text{in } C_0(E),$$

by property (ii) of the definition of a Feller semigroup and dominated convergence. $\qquad \square$

Definition 6.9 We set

$$D(L) = \{f \in C_0(E) : \frac{Q_t f - f}{t} \text{ converges in } C_0(E) \text{ when } t \downarrow 0\}$$

and, for every $f \in D(L)$,

$$Lf = \lim_{t \downarrow 0} \frac{Q_t f - f}{t}.$$

Then $D(L)$ is a linear subspace of $C_0(E)$ and $L : D(L) \longrightarrow C_0(E)$ is a linear operator called the *generator* of the semigroup $(Q_t)_{t \geq 0}$. The subspace $D(L)$ is called the *domain* of L.

Let us start with two simple properties of the generator.

Proposition 6.10 *Let $f \in D(L)$ and $s > 0$. Then $Q_s f \in D(L)$ and $L(Q_s f) = Q_s(Lf)$.*

Proof Writing

$$\frac{Q_t(Q_s f) - Q_s f}{t} = Q_s \left(\frac{Q_t f - f}{t} \right)$$

and using the fact that Q_s is a contraction of $C_0(E)$, we get that $t^{-1}(Q_t(Q_s f) - Q_s f)$ converges to $Q_s(Lf)$, which gives the desired result. $\qquad \square$

Proposition 6.11 *If $f \in D(L)$, we have, for every $t \geq 0$,*

$$Q_t f = f + \int_0^t Q_s(Lf) \, ds = f + \int_0^t L(Q_s f) \, ds.$$

Proof Let $f \in D(L)$. For every $t \geq 0$,

$$\varepsilon^{-1}(Q_{t+\varepsilon}f - Q_t f) = Q_t(\varepsilon^{-1}(Q_\varepsilon f - f)) \xrightarrow[\varepsilon \downarrow 0]{} Q_t(Lf).$$

Moreover, the preceding convergence is uniform when t varies over $\mathbb{R}_+$. This implies that, for every $x \in E$, the function $t \mapsto Q_t f(x)$ is differentiable on $\mathbb{R}_+$ and its derivative is $Q_t(Lf)(x)$, which is a continuous function of t. The formula of the proposition follows, also using the preceding proposition. $\qquad\square$

The next proposition identifies the domain $D(L)$ in terms of the resolvent operators R_λ.

Proposition 6.12 *Let* $\lambda > 0$.

(i) *For every* $g \in C_0(E)$, $R_\lambda g \in D(L)$ *and* $(\lambda - L)R_\lambda g = g$.
(ii) *If* $f \in D(L)$, $R_\lambda(\lambda - L)f = f$.

Consequently, $D(L) = \mathscr{R}$ *and the operators* $R_\lambda : C_0(E) \to \mathscr{R}$ *and* $\lambda - L : D(L) \to C_0(E)$ *are the inverse of each other.*

Proof

(i) If $g \in C_0(E)$, we have for every $\varepsilon > 0$,

$$
\begin{aligned}
\varepsilon^{-1}(Q_\varepsilon R_\lambda g - R_\lambda g) &= \varepsilon^{-1}\left(\int_0^\infty e^{-\lambda t} Q_{\varepsilon + t} g \, dt - \int_0^\infty e^{-\lambda t} Q_t g \, dt \right) \\
&= \varepsilon^{-1}\left((1 - e^{-\lambda \varepsilon}) \int_0^\infty e^{-\lambda t} Q_{\varepsilon + t} g \, dt - \int_0^\varepsilon e^{-\lambda t} Q_t g \, dt \right) \\
&\xrightarrow[\varepsilon \to 0]{} \lambda R_\lambda g - g
\end{aligned}
$$

using property (ii) of the definition of a Feller semigroup (and the fact that this property implies the continuity of the mapping $t \mapsto Q_t g$ from $\mathbb{R}_+$ into $C_0(E)$). The preceding calculation shows that $R_\lambda g \in D(L)$ and $L(R_\lambda g) = \lambda R_\lambda g - g$.

(ii) Let $f \in D(L)$. By Proposition 6.11, $Q_t f = f + \int_0^t Q_s(Lf) \, ds$, hence

$$
\begin{aligned}
\int_0^\infty e^{-\lambda t} Q_t f(x) \, dt &= \frac{f(x)}{\lambda} + \int_0^\infty e^{-\lambda t} \left(\int_0^t Q_s(Lf)(x) \, ds \right) dt \\
&= \frac{f(x)}{\lambda} + \int_0^\infty \frac{e^{-\lambda s}}{\lambda} Q_s(Lf)(x) \, ds.
\end{aligned}
$$

We have thus obtained the equality

$$\lambda R_\lambda f = f + R_\lambda Lf$$

giving the result in (ii).

The last assertions of the proposition follow from (i) and (ii): (i) shows that $\mathcal{R} \subset D(L)$ and (ii) gives the reverse inclusion, then the identities in (i) and (ii) show that R_λ and $\lambda - L$ are inverse of each other. $\qquad \square$

Corollary 6.13 *The semigroup $(Q_t)_{t \geq 0}$ is determined by the generator L (including also the domain $D(L)$).*

Proof Let f be a nonnegative function in $C_0(E)$. Then $R_\lambda f$ is the unique element of $D(L)$ such that $(\lambda - L)R_\lambda f = f$. On the other hand, knowing $R_\lambda f(x) = \int_0^\infty e^{-\lambda t} Q_t f(x) dt$ for every $\lambda > 0$ determines the continuous function $t \mapsto Q_t f(x)$. To complete the argument, note that Q_t is characterized by the values of $Q_t f$ for every nonnegative function f in $C_0(E)$. $\qquad \square$

Example It is easy to verify that the semigroup $(Q_t)_{t \geq 0}$ of real Brownian motion is Feller. Let us compute its generator L. We saw that, for every $\lambda > 0$ and $f \in C_0(\mathbb{R})$,

$$R_\lambda f(x) = \int \frac{1}{\sqrt{2\lambda}} \exp(-\sqrt{2\lambda}|y - x|) f(y) \, dy.$$

If $h \in D(L)$, we know that there exists an $f \in C_0(\mathbb{R})$ such that $h = R_\lambda f$. Taking $\lambda = \frac{1}{2}$, we have

$$h(x) = \int \exp(-|y - x|) f(y) \, dy.$$

By differentiating under the integral sign (we leave the justification to the reader), we get that h is differentiable on $\mathbb{R}$, and

$$h'(x) = \int \operatorname{sgn}(y - x) \exp(-|y - x|) f(y) \, dy$$

with the notation $\operatorname{sgn}(z) = \mathbf{1}_{\{z > 0\}} - \mathbf{1}_{\{z \leq 0\}}$ (the value of $\operatorname{sgn}(0)$ is unimportant). Let us also show that h' is differentiable on $\mathbb{R}$. Let $x_0 \in \mathbb{R}$. Then, for $x > x_0$,

$$h'(x) - h'(x_0) = \int \Big(\operatorname{sgn}(y - x) \exp(-|y - x|) - \operatorname{sgn}(y - x_0) \exp(-|y - x_0|) \Big) f(y) dy$$

$$= \int_{x_0}^x \Big(-\exp(-|y - x|) - \exp(-|y - x_0|) \Big) f(y) \, dy$$

$$+ \int_{\mathbb{R} \setminus [x_0, x]} \operatorname{sgn}(y - x_0) \Big(\exp(-|y - x|) - \exp(-|y - x_0|) \Big) f(y) \, dy.$$

It follows that

$$\frac{h'(x) - h'(x_0)}{x - x_0} \xrightarrow[x \downarrow x_0]{} -2f(x_0) + h(x_0).$$

We get the same limit when $x \uparrow x_0$, and we thus obtain that h is twice differentiable, and $h'' = -2f + h$.

On the other hand, since $h = R_{1/2} f$, Proposition 6.12 shows that

$$(\frac{1}{2} - L) h = f$$

hence $Lh = -f + \frac{1}{2}h = \frac{1}{2}h''$.

Summarizing, we have obtained that

$$D(L) \subset \{h \in C^2(\mathbb{R}) : h \text{ and } h'' \in C_0(\mathbb{R})\}$$

and that, if $h \in D(L)$, we have $Lh = \frac{1}{2}h''$.

In fact, the preceding inclusion is an equality. To see this, we may argue in the following way. If g is a twice differentiable function such that g and g'' are in $C_0(\mathbb{R})$, we set $f = \frac{1}{2}(g - g'') \in C_0(\mathbb{R})$, then $h = R_{1/2} f \in D(L)$. By the preceding argument, h is twice differentiable and $h'' = -2f + h$. It follows that $(h - g)'' = h - g$. Since the function $h - g$ belongs to $C_0(\mathbb{R})$, it must vanish identically and we get $g = h \in D(L)$.

Remark In general, it is very difficult to determine the exact domain of the generator. The following theorem often allows one to identify elements of this domain using martingales associated with the Markov process with semigroup $(Q_t)_{t \geq 0}$.

We consider again a general Feller semigroup $(Q_t)_{t \geq 0}$. We assume that on some probability space, we are given, for every $x \in E$, a process $(X_t^x)_{t \geq 0}$ which is Markov with semigroup $(Q_t)_{t \geq 0}$, with respect to a filtration $(\mathscr{F}_t)_{t \geq 0}$, and such that $P(X_0^x = x) = 1$. To make sense of the integrals that will appear below, we also assume that the sample paths of $(X_t^x)_{t \geq 0}$ are càdlàg (we will see in the next section that this assumption is not restrictive).

Theorem 6.14 *Let $h, g \in C_0(E)$. The following two conditions are equivalent:*

(i) *$h \in D(L)$ and $Lh = g$.*
(ii) *For every $x \in E$, the process*

$$h(X_t^x) - \int_0^t g(X_s^x) \, ds$$

is a martingale, with respect to the filtration $(\mathscr{F}_t)$.

Proof We first prove that (i) $\Rightarrow$ (ii). Let $h \in D(L)$ and $g = Lh$. By Proposition 6.11, we have then, for every $s \geq 0$,

$$Q_s h = h + \int_0^s Q_r g \, dr.$$

It follows that, for $t \geq 0$ and $s \geq 0$,

$$E[h(X^x_{t+s}) \mid \mathcal{F}_t] = Q_s h(X^x_t) = h(X^x_t) + \int_0^s Q_r g(X^x_t)\, dr.$$

On the other hand,

$$E\left[\int_t^{t+s} g(X^x_r)\, dr \;\middle|\; \mathcal{F}_t \right] = \int_t^{t+s} E[g(X^x_r) \mid \mathcal{F}_t]\, dr = \int_t^{t+s} Q_{r-t} g(X^x_t)\, dr$$

$$= \int_0^s Q_r g(X^x_t)\, dr.$$

The fact that the conditional expectation and the integral can be interchanged (in the first equality of the last display) is easy to justify using the characteristic property of conditional expectations. It follows from the last two displays that

$$E\left[h(X^x_{t+s}) - \int_0^{t+s} g(X^x_r)\, dr \;\middle|\; \mathcal{F}_t \right] = h(X^x_t) - \int_0^t g(X^x_r)\, dr$$

giving property (ii).

Conversely, suppose that (ii) holds. Then, for every $t \geq 0$,

$$E\left[h(X^x_t) - \int_0^t g(X^x_r)\, dr \right] = h(x)$$

and on the other hand, from the definition of a Markov process,

$$E\left[h(X^x_t) - \int_0^t g(X^x_r)\, dr \right] = Q_t h(x) - \int_0^t Q_r g(x)\, dr.$$

Consequently,

$$\frac{Q_t h - h}{t} = \frac{1}{t} \int_0^t Q_r g\, dr \xrightarrow[t \downarrow 0]{} g$$

in $C_0(E)$, by property (ii) of the definition of a Feller semigroup. We conclude that $h \in D(L)$ and $Lh = g$. $\square$

Example In the case of d-dimensional Brownian motion, Itô's formula shows that, if $h \in C^2(\mathbb{R}^d)$,

$$h(X_t) - \frac{1}{2} \int_0^t \Delta h(X_s)\, ds$$

is a continuous local martingale. This continuous local martingale is a martingale if we furthermore assume that h and Δh are in $C_0(\mathbb{R}^d)$ (hence bounded). It then follows from Theorem 6.14 that $h \in D(L)$ and $Lh = \frac{1}{2}\Delta h$. Recall that we already obtained this result by a direct computation of L when $d = 1$ (in fact in a more precise form since here we can only assert that $D(L) \supset \{h \in C^2(\mathbb{R}^d) : h \text{ and } \Delta h \in C_0(\mathbb{R}^d)\}$, whereas equality holds if $d = 1$).

6.3   The Regularity of Sample Paths

Our aim in this section is to show that one construct Feller processes in such a way that they have càdlàg sample paths. We consider a Feller semigroup $(Q_t)_{t\geq 0}$ on a topological space E (assumed to be metrizable, locally compact and countable at infinity as above).

Theorem 6.15 *Let $(X_t)_{t\geq 0}$ be a Markov process with semigroup $(Q_t)_{t\geq 0}$, with respect to the filtration $(\mathscr{F}_t)_{t\in[0,\infty]}$. Set $\widetilde{\mathscr{F}}_\infty = \mathscr{F}_\infty$ and, for every $t \geq 0$,*

$$\widetilde{\mathscr{F}}_t = \mathscr{F}_{t+} \vee \sigma(\mathscr{N}),$$

where $\mathscr{N}$ denotes the class of all $\mathscr{F}_\infty$-measurable sets that have zero probability.

Then, the process $(X_t)_{t\geq 0}$ has a càdlàg modification $(\tilde{X}_t)_{t\geq 0}$, which is adapted to the filtration $(\widetilde{\mathscr{F}}_t)$. Moreover, $(\tilde{X}_t)_{t\geq 0}$ is a Markov process with semigroup $(Q_t)_{t\geq 0}$, with respect to the filtration $(\widetilde{\mathscr{F}}_t)_{t\in[0,\infty]}$.

Remark The filtration $(\widetilde{\mathscr{F}}_t)$ is right-continuous because so is the filtration $(\mathscr{F}_{t+})$ and the right-continuity property is preserved when adding the class of negligible sets $\mathscr{N}$.

Proof Let $E_\Delta = E \cup \{\Delta\}$ be the Alexandroff compactification of E, which is obtained by adding the point at infinity Δ to E (and by definition the neighborhoods of Δ are the complements of compact subsets of E). We agree that every function $f \in C_0(E)$ is extended to a continuous function on E_Δ by setting $f(\Delta) = 0$.

Write $C_0^+(E)$ for the set of all nonnegative functions in $C_0(E)$. We can find a sequence $(f_n)_{n\geq 0}$ in $C_0^+(E)$ which separates the points of E_Δ, in the sense that, for every $x, y \in E_\Delta$ with $x \neq y$, there is an integer n such that $f_n(x) \neq f_n(y)$. Then,

$$\mathscr{H} = \{R_p f_n : p \geq 1, n \geq 0\}$$

is also a countable subset of $C_0^+(E)$ which separates the points of E_Δ (use the fact that $\|pR_p f - f\| \longrightarrow 0$ when $p \to \infty$).

If $h \in \mathscr{H}$, Lemma 6.6 shows that there exists an integer $p \geq 1$ such that $e^{-pt}h(X_t)$ is a supermartingale. Let D be a countable dense subset of $\mathbb{R}_+$. Then Theorem 3.17 (i) shows that the limits

$$\lim_{D \ni s \downarrow \downarrow t} h(X_s), \quad \lim_{D \ni s \uparrow \uparrow t} h(X_s)$$

exist simultaneously for every $t \in \mathbb{R}_+$ (the second one only for $t > 0$) outside an $\mathscr{F}_\infty$-measurable event N_h of zero probability. Indeed, as in the proof of Theorem 3.17, we may define the complementary event N_h^c as the set of all $\omega \in \Omega$ for which the function $D \ni s \mapsto e^{-ps} h(X_s)$ makes a finite number of upcrossings along any interval $[a, b]$ $(a, b \in \mathbb{Q}, a < b)$ on every finite time interval. We then set

$$N = \bigcup_{h \in \mathscr{H}} N_h$$

in such a way that we still have $N \in \mathscr{N}$. Then if $\omega \notin N$, the limits

$$\lim_{D \ni s \downarrow \downarrow t} X_s(\omega) , \quad \lim_{D \ni s \uparrow \uparrow t} X_s(\omega)$$

exist for every $t \geq 0$ (the second one only for $t > 0$) in E_Δ. In fact, if we assume that $X_s(\omega)$ has two distinct accumulation points in E_Δ when $D \ni s \downarrow\downarrow t$, we get a contradiction by considering a function $h \in \mathscr{H}$ that separates these two points. We can then set, for every $\omega \in \Omega \backslash N$ and every $t \geq 0$,

$$\tilde{X}_t(\omega) = \lim_{D \ni s \downarrow \downarrow t} X_s(\omega).$$

If $\omega \in N$, we set $\tilde{X}_t(\omega) = x_0$ for every $t \geq 0$, where x_0 is a fixed point in E. Then, for every $t \geq 0$, $\tilde{X}_t$ is an $\widetilde{\mathscr{F}}_t$-measurable random variable with values in E_Δ. Furthermore, for every $\omega \in \Omega$, $t \mapsto \tilde{X}_t(\omega)$, viewed as a mapping with values in E_Δ, is càdlàg by Lemma 3.16 (this lemma shows that the functions $t \mapsto h(\tilde{X}_t(\omega))$, for $h \in \mathscr{H}$, are càdlàg, and this suffices since $\mathscr{H}$ separates points of E).

Let us now show that $P(X_t = \tilde{X}_t) = 1$, for every fixed $t \geq 0$. Let $f, g \in C_0(E)$ and let (t_n) be a sequence in D that decreases (strictly) to t. Then,

$$\begin{aligned}
E[f(X_t)g(\tilde{X}_t)] &= \lim_{n \to \infty} E[f(X_t)g(X_{t_n})] \\
&= \lim_{n \to \infty} E[f(X_t)Q_{t_n-t}g(X_t)] \\
&= E[f(X_t)g(X_t)]
\end{aligned}$$

since $Q_{t_n-t}g \longrightarrow g$ by the definition of a Feller semigroup. The preceding equality entails that the two pairs $(X_t, \tilde{X}_t)$ and (X_t, X_t) have the same distribution and thus $P(X_t = \tilde{X}_t) = 1$.

Let us then verify that $(\tilde{X}_t)_{t \geq 0}$ is a Markov process with semigroup $(Q_t)_{t \geq 0}$ with respect to the filtration $(\widetilde{\mathscr{F}}_t)$. It is enough to prove that, for every $s \geq 0, t > 0$ and $A \in \widetilde{\mathscr{F}}_s, f \in C_0(E)$, we have

$$E[\mathbf{1}_A f(\tilde{X}_{s+t})] = E[\mathbf{1}_A Q_t f(\tilde{X}_s)].$$

Since $\tilde{X}_s = X_s$ a.s. and $\tilde{X}_{s+t} = X_{s+t}$ a.s., this is equivalent to proving that

$$E[1_A f(X_{s+t})] = E[1_A Q_t f(X_s)].$$

Because A is equal a.s. to an element of $\mathscr{F}_{s+}$, we may assume that $A \in \mathscr{F}_{s+}$. Let (s_n) be a sequence in D that decreases to s, so that $A \in \mathscr{F}_{s_n}$ for every n. Then, as soon as $s_n \le s + t$, we have

$$E[1_A f(X_{s+t})] = E[1_A E[f(X_{s+t}) \mid \mathscr{F}_{s_n}]] = E[1_A Q_{s+t-s_n} f(X_{s_n})].$$

But $Q_{s+t-s_n} f$ converges (uniformly) to $Q_t f$ by properties of Feller semigroups, and since $X_{s_n} = \tilde{X}_{s_n}$ a.s. we also know that X_{s_n} converges a.s. to $\tilde{X}_s = X_s$ a.s. We thus obtain the desired result by letting n tend to ∞.

It remains to verify that the sample paths $t \mapsto \tilde{X}_t(\omega)$ are càdlàg as E-valued mappings, and not only as E_Δ-valued mappings (we already know that, for every fixed $t \ge 0$, $\tilde{X}_t(\omega) = X_t(\omega)$ a.s. is in E with probability one, but this does not imply that the sample paths, and their left-limits, remain in E). Fix a function $g \in C_0^+(E)$ such that $g(x) > 0$ for every $x \in E$. The function $h = R_1 g$ then satisfies the same property. Set, for every $t \ge 0$,

$$Y_t = e^{-t} h(\tilde{X}_t).$$

Then Lemma 6.6 shows that $(Y_t)_{t \ge 0}$ is a nonnegative supermartingale with respect to the filtration $(\widetilde{\mathscr{F}}_t)$. Additionally, we know that the sample paths of $(Y_t)_{t \ge 0}$ are càdlàg (recall that $h(\Delta) = 0$ by convention).

For every integer $n \ge 1$, set

$$T_{(n)} = \inf\{t \ge 0 : Y_t < \frac{1}{n}\}.$$

Then $T_{(n)}$ is a stopping time of the filtration $(\widetilde{\mathscr{F}}_t)$ (we can apply Proposition 3.9, because $T_{(n)}$ is the first hitting time of an open set by an adapted process with càdlàg sample paths, and the filtration $(\widetilde{\mathscr{F}}_t)$ is right-continuous). Consequently,

$$T = \lim_{n \to \infty} \uparrow T_{(n)}$$

is a stopping time. The desired result will follow if we can verify that $P(T < \infty) = 0$. Indeed, it is clear that, for every $t \in [0, T_{(n)})$, $\tilde{X}_t \in E$ and $\tilde{X}_{t-} \in E$, and we may redefine $\tilde{X}_t(\omega) = x_0$ (for every $t \ge 0$) for all ω belonging to the event $\{T < \infty\}$ $\in \mathcal{N}$.

To verify that $P(T < \infty) = 0$, we apply Theorem 3.25 and the subsequent remark to $Z = Y$ and $U = T_{(n)}$, $V = T + q$, where $q > 0$ is a rational number. We get

$$E[Y_{T+q} \mathbf{1}_{\{T < \infty\}}] \leq E[Y_{T_{(n)}} \mathbf{1}_{\{T_{(n)} < \infty\}}] \leq \frac{1}{n}.$$

By letting n tend to ∞, we thus have

$$E[Y_{T+q} \mathbf{1}_{\{T < \infty\}}] = 0,$$

hence $Y_{T+q} = 0$ a.s. on $\{T < \infty\}$. By the right-continuity of sample paths of Y, we conclude that $Y_t = 0$, for every $t \in [T, \infty)$, a.s. on $\{T < \infty\}$. But we also know that, for every integer $k \geq 0$, $Y_k = e^{-k} h(\tilde{X}_k) > 0$ a.s., since $\tilde{X}_k \in E$ a.s. This suffices to get $P(T < \infty) = 0$. $\qquad\square$

Remark The previous proof applies with minor modifications to the different setting where we are given the process $(X_t)_{t \geq 0}$ together with a collection $(P_x)_{x \in E}$ of probability measures such that, under P_x, $(X_t)_{t \geq 0}$ is a Markov process with semigroup $(Q_t)_{t \geq 0}$, with respect to a filtration $(\mathcal{F}_t)_{t \in [0, \infty]}$, and $P_x(X_0 = x) = 1$ (in the first section above, we saw that these properties will hold for the canonical process $(X_t)_{t \geq 0}$ on the space $\Omega^* = E^{\mathbb{R}_+}$ if the measures P_x are constructed from the semigroup $(Q_t)_{t \geq 0}$ using the Kolmogorov extension theorem). In that setting, we can define the filtration $(\widetilde{\mathcal{F}}_t)_{t \in [0, \infty]}$ by

$$\widetilde{\mathcal{F}}_t = \mathcal{F}_{t+} \vee \sigma(\mathcal{N}'),$$

where $\mathcal{N}'$ denotes the class of all $\mathcal{F}_\infty$-measurable sets that have zero P_x-probability for every $x \in E$. By the same arguments as in the preceding proof, we can then construct an $(\widetilde{\mathcal{F}}_t)$-adapted process $(\tilde{X}_t)_{t \geq 0}$ with càdlàg sample paths, such that, for every $x \in E$,

$$P_x(\tilde{X}_t = X_t) = 1, \quad \forall t \geq 0,$$

and $(\tilde{X}_t)_{t \geq 0}$ is under P_x a Markov process with semigroup $(Q_t)_{t \geq 0}$, with respect to the filtration $(\widetilde{\mathcal{F}}_t)_{t \in [0, \infty]}$, such that $P_x(\tilde{X}_0 = x) = 1$.

6.4 The Strong Markov Property

In the first part of this section, we come back to the general setting of Sect. 6.1 above, where $(Q_t)_{t \geq 0}$ is a (not necessarily Feller) transition semigroup on E. We assume here that E is a metric space (equipped with its Borel σ-field), and moreover that, for every $x \in E$, one can construct a Markov process $(X_t^x)_{t \geq 0}$ with semigroup

$(Q_t)_{t\geq 0}$ such that $X_0^x = x$ a.s. and the sample paths of X are càdlàg. In the case of a Feller semigroup, the existence of such a process follows from Corollary 6.4 and Theorem 6.15.

The space of all càdlàg paths $f : \mathbb{R}_+ \longrightarrow E$ is denoted by $\mathbb{D}(E)$. This space is equipped with the σ-field $\mathscr{D}$ generated by the coordinate mappings $f \mapsto f(t)$. For every $x \in E$, we write $\mathbb{P}_x$ for the probability measure on $\mathbb{D}(E)$ which is the law of the random path $(X_t^x)_{t\geq 0}$. Notice that $\mathbb{P}_x$ does not depend on the choice of X^x, nor of the probability space where X^x is defined: This follows from the fact that the finite-dimensional marginals of a Markov process are determined by its semigroup and initial distribution.

We first give a version of the (simple) Markov property, which is a simple extension of the definition of a Markov process. We use the notation $\mathbb{E}_x$ for the expectation under $\mathbb{P}_x$.

Theorem 6.16 (Simple Markov property) *Let $(Y_t)_{t\geq 0}$ be a Markov process with semigroup $(Q_t)_{t\geq 0}$, with respect to the filtration $(\mathscr{F}_t)_{t\geq 0}$. We assume that the sample paths of Y are càdlàg. Let $s \geq 0$ and let $\Phi : \mathbb{D}(E) \longrightarrow \mathbb{R}_+$ be a measurable function. Then,*

$$E[\Phi((Y_{s+t})_{t\geq 0}) \mid \mathscr{F}_s] = \mathbb{E}_{Y_s}[\Phi].$$

Remark The right-hand side of the last display is the composition of Y_s and of the mapping $y \mapsto \mathbb{E}_y[\Phi]$. To see that the latter mapping is measurable, it is enough to consider the case where $\Phi = 1_A, A \in \mathscr{D}$. When A only depends on a finite number of coordinates, there is an explicit formula, and an application of the monotone class lemma completes the argument.

Proof As in the preceding remark, it suffices to consider the case where $\Phi = 1_A$ and

$$A = \{f \in \mathbb{D}(E) : f(t_1) \in B_1, \ldots, f(t_p) \in B_p\},$$

where $0 \leq t_1 < t_2 < \cdots < t_p$ and $B_1, \ldots, B_p$ are measurable subsets of E. In that case, we need to verify that

$$P(Y_{s+t_1} \in B_1, \ldots, Y_{s+t_p} \in B_p \mid \mathscr{F}_s)$$
$$= \int_{B_1} Q_{t_1}(Y_s, dx_1) \int_{B_2} Q_{t_2-t_1}(x_1, dx_2) \cdots \int_{B_p} Q_{t_p-t_{p-1}}(x_{p-1}, dx_p).$$

We in fact prove more generally that, if $\varphi_1, \ldots, \varphi_p \in B(E)$,

$$E[\varphi_1(Y_{s+t_1}) \cdots \varphi_p(Y_{s+t_p}) \mid \mathscr{F}_s]$$
$$= \int Q_{t_1}(Y_s, dx_1)\varphi_1(x_1) \int Q_{t_2-t_1}(x_1, dx_2)\varphi_2(x_2) \cdots \int Q_{t_p-t_{p-1}}(x_{p-1}, dx_p)\varphi_p(x_p).$$

If $p = 1$ this is the definition of a Markov process. We then argue by induction, writing

$$E[\varphi_1(Y_{s+t_1}) \cdots \varphi_p(Y_{s+t_p}) \mid \mathscr{F}_s]$$
$$= E[\varphi_1(Y_{s+t_1}) \cdots \varphi_{p-1}(Y_{s+t_{p-1}}) E[\varphi_p(Y_{s+t_p}) \mid \mathscr{F}_{s+t_{p-1}}] \mid \mathscr{F}_s]$$
$$= E[\varphi_1(Y_{s+t_1}) \cdots \varphi_{p-1}(Y_{s+t_{p-1}}) Q_{t_p - t_{p-1}} \varphi_p(Y_{s+t_{p-1}}) \mid \mathscr{F}_s]$$

and the desired result easily follows.   □

We now turn to the strong Markov property.

Theorem 6.17 (Strong Markov property) *Retain the assumptions of the previous theorem, and suppose in addition that $(Q_t)_{t\geq0}$ is a Feller semigroup (in particular, E is assumed to be metrizable locally compact and countable at infinity). Let T be a stopping time of the filtration $(\mathscr{F}_{t+})$, and let $\Phi : \mathbb{D}(E) \longrightarrow \mathbb{R}_+$ be a measurable function. Then, for every $x \in E$,*

$$E[\mathbf{1}_{\{T<\infty\}} \Phi((Y_{T+t})_{t\geq0}) \mid \mathscr{F}_T] = \mathbf{1}_{\{T<\infty\}} \mathbb{E}_{Y_T}[\Phi].$$

Remark We allow T to be a stopping time of $(\mathscr{F}_{t+})$, which is slightly more general than saying that T is a stopping time of the filtration $(\mathscr{F}_t)$.

Proof We first observe that the right-hand side of the last display is $\mathscr{F}_T$-measurable, because $\{T < \infty\} \ni \omega \mapsto Y_T(\omega)$ is $\mathscr{F}_T$-measurable (Theorem 3.7) and the function $y \mapsto \mathbb{E}_y[\Phi]$ is measurable. It is then enough to show that, for $A \in \mathscr{F}_T$ fixed,

$$E[\mathbf{1}_{A\cap\{T<\infty\}} \Phi((Y_{T+t})_{t\geq0})] = E[\mathbf{1}_{A\cap\{T<\infty\}} \mathbb{E}_{Y_T}[\Phi]].$$

As above, we can restrict our attention to the case where

$$\Phi(f) = \varphi_1(f(t_1)) \cdots \varphi_p(f(t_p))$$

where $0 \leq t_1 < t_2 < \cdots < t_p$ and $\varphi_1, \ldots, \varphi_p \in B(E)$. It is in fact enough to take $p = 1$: If the desired result holds in that case, we can argue by induction, writing

$$E[\mathbf{1}_{A\cap\{T<\infty\}}\varphi_1(Y_{T+t_1}) \cdots \varphi_p(Y_{T+t_p})]$$
$$= E[\mathbf{1}_{A\cap\{T<\infty\}}\varphi_1(Y_{T+t_1}) \cdots \varphi_{p-1}(Y_{T+t_{p-1}}) E[\varphi_p(Y_{T+t_p}) \mid \mathscr{F}_{T+t_{p-1}}]]$$
$$= E[\mathbf{1}_{A\cap\{T<\infty\}}\varphi_1(Y_{T+t_1}) \cdots \varphi_{p-1}(Y_{T+t_{p-1}})Q_{t_p-t_{p-1}}\varphi_p(Y_{T+t_{p-1}})].$$

We thus fix $t \geq 0$ and $\varphi \in B(E)$ and we aim to prove that

$$E[\mathbf{1}_{A\cap\{T<\infty\}}\varphi(Y_{T+t})] = E[\mathbf{1}_{A\cap\{T<\infty\}}Q_t\varphi(Y_T)].$$

We may assume that $\varphi \in C_0(E)$ (a finite measure on E is determined by its values against functions of $C_0(E)$).

On the event $T < \infty$, write $[T]_n$ for the smallest real number of the form $i2^{-n}$, $i \in \mathbb{Z}_+$, which is strictly greater than T. Then,

$$
\begin{aligned}
E[\mathbf{1}_{A \cap \{T < \infty\}} \varphi(Y_{T+t})] &= \lim_{n \to \infty} E[\mathbf{1}_{A \cap \{T < \infty\}} \varphi(Y_{[T]_n + t})] \\
&= \lim_{n \to \infty} \sum_{i=1}^{\infty} E[\mathbf{1}_{A \cap \{(i-1)2^{-n} \leq T < i2^{-n}\}} \varphi(Y_{i2^{-n} + t})] \\
&= \lim_{n \to \infty} \sum_{i=1}^{\infty} E[\mathbf{1}_{A \cap \{(i-1)2^{-n} \leq T < i2^{-n}\}} Q_t \varphi(Y_{i2^{-n}})] \\
&= \lim_{n \to \infty} E[\mathbf{1}_{A \cap \{T < \infty\}} Q_t \varphi(Y_{[T]_n})] \\
&= E[\mathbf{1}_{A \cap \{T < \infty\}} Q_t \varphi(Y_T)]
\end{aligned}
$$

giving the desired result. In the first (and in the last) equality, we use the right continuity of sample paths. In the third equality, we observe that the event

$$
A \cap \{(i-1)2^{-n} \leq T < i2^{-n}\}
$$

belongs to $\mathscr{F}_{i2^{-n}}$ because $A \in \mathscr{F}_T$ and T is a stopping time of the filtration $(\mathscr{F}_{t+})$ (use Proposition 3.6). Finally, and this is the key point, in the last equality we also use the fact that $Q_t \varphi$ is continuous, since $\varphi \in C_0(E)$ and the semigroup is Feller. $\square$

Remark In the special case of (linear) Brownian motion, the result of Theorem 6.17 is essentially equivalent to Theorem 2.20 stated in Chap. 2. The reason why the formulation in Theorem 2.20 looks different comes from the property of stationarity and independence of the increments of Brownian motion, which of course does not subsist in our general setting. Even for Brownian motion, the formulation of Theorem 6.17 turns out to be more appropriate in a number of situations: A convincing illustration is the proof of Proposition 7.7 (ii) in the next chapter.

6.5   Three Important Classes of Feller Processes

6.5.1   *Jump Processes on a Finite State Space*

In this subsection, we assume that the state space E is finite (and equipped with the discrete topology). Note that any càdlàg function $f \in \mathbb{D}(E)$ must be of the following type: There exists a real $t_1 \in (0, \infty]$ such that $f(t) = f(0)$ for every $t \in [0, t_1)$, then, if $t_1 < \infty$, there exists a real $t_2 \in (t_1, \infty]$ such that $f(t) = f(t_1) \neq f(0)$ for every $t \in [t_1, t_2)$, and so on.

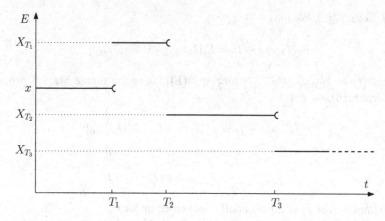

Fig. 6.1 A sample path of the jump process X under P_x

Consider a Feller semigroup $(Q_t)_{t\geq 0}$ on E. By the remark of the end of Sect. 6.3, we can construct, on a probability space Ω equipped with a right-continuous filtration $(\mathscr{F}_t)_{t\in[0,\infty]}$, a collection $(P_x)_{x\in E}$ of probability measures and a process $(X_t)_{t\geq 0}$ with càdlàg sample paths such that, under P_x, X is Markov with semigroup $(Q_t)_{t\geq 0}$ with respect to the filtration $(\mathscr{F}_t)$, and $P_x(X_0 = x) = 1$. As previously, E_x stands for the expectation under P_x. Since the sample paths of X are càdlàg, we know that, for every $\omega \in \Omega$, there exists a sequence

$$T_0(\omega) = 0 < T_1(\omega) \leq T_2(\omega) \leq T_3(\omega) \leq \cdots \leq \infty,$$

such that $X_t(\omega) = X_0(\omega)$ for every $t \in [0, T_1(\omega))$ and, for every integer $i \geq 1$, the condition $T_i(\omega) < \infty$ implies $T_i(\omega) < T_{i+1}(\omega)$, $X_{T_i(\omega)}(\omega) \neq X_{T_{i-1}(\omega)}(\omega)$ and $X_t(\omega) = X_{T_i(\omega)}(\omega)$ for every $t \in [T_i(\omega), T_{i+1}(\omega))$. Moreover, $T_n(\omega) \uparrow \infty$ as $n \to \infty$. See Fig. 6.1.

It is not hard to verify that $T_0, T_1, T_2, \ldots$ are stopping times. For instance,

$$\{T_1 < t\} = \bigcup_{q\in[0,t)\cap\mathbb{Q}} \{X_q \neq X_0\} \in \mathscr{F}_t.$$

Recall that, for $\lambda > 0$, a positive random variable U is exponentially distributed with parameter λ if $P(U > r) = e^{-\lambda r}$ for every $r \geq 0$. In the following lemma, we make the convention that an exponential variable with parameter 0 is equal to ∞ a.s.

Lemma 6.18 *Let $x \in E$. There exists a real number $q(x) \geq 0$ such that the random variable T_1 is exponentially distributed with parameter $q(x)$ under P_x. Furthermore, if $q(x) > 0$, T_1 and X_{T_1} are independent under P_x.*

Proof Let $s, t \geq 0$. We have

$$P_x(T_1 > s + t) = E_x[\mathbf{1}_{\{T_1 > s\}} \Phi((X_{s+r})_{r \geq 0})],$$

where $\Phi(f) = \mathbf{1}_{\{f(r) = f(0), \forall r \in [0,t]\}}$ for $f \in \mathbb{D}(E)$. Using the simple Markov property (Theorem 6.16), we get

$$\begin{aligned}
P_x(T_1 > s + t) &= E_x[\mathbf{1}_{\{T_1 > s\}} E_{X_s}[\Phi((X_r)_{r \geq 0})]] \\
&= E_x[\mathbf{1}_{\{T_1 > s\}} P_x(T_1 > t)] \\
&= P_x(T_1 > s) P_x(T_1 > t),
\end{aligned}$$

which implies that T_1 is exponentially distributed under P_x.

Assume that $q(x) > 0$, so that $T_1 < \infty$, P_x a.s. Then, for every $t \geq 0$ and $y \in E$,

$$P_x(T_1 > t, X_{T_1} = y) = E_x[\mathbf{1}_{\{T_1 > t\}} \Psi((X_{t+r})_{r \geq 0})],$$

where for $f \in \mathbb{D}(E)$, $\Psi(f) = 0$ is f is constant, and otherwise $\Psi(f) = \mathbf{1}_{\{\gamma_1(f) = y\}}$, if $\gamma_1(f)$ is the value of f after its first jump. We thus get

$$\begin{aligned}
P_x(T_1 > t, X_{T_1} = y) &= E_x[\mathbf{1}_{\{T_1 > t\}} E_{X_t}[\Psi((X_r)_{r \geq 0})]] \\
&= E_x[\mathbf{1}_{\{T_1 > t\}} P_x(X_{T_1} = y)] \\
&= P_x(T_1 > t) P_x(X_{T_1} = y),
\end{aligned}$$

which gives the desired independence.   □

Points x such that $q(x) = 0$ are absorbing states for the Markov process, in the sense that $P_x(X_t = x, \forall t \geq 0) = 1$.

For every $x \in E$ such that $q(x) > 0$, and every $y \in E$, we set

$$\Pi(x, y) = P_x(X_{T_1} = y).$$

Note that $\Pi(x, \cdot)$ is a probability measure on E, and $\Pi(x, x) = 0$.

Proposition 6.19 *Let L denote the generator of $(Q_t)_{t \geq 0}$. Then $D(L) = C_0(E) = B(E)$, and, for every $\varphi \in B(E)$, for every $x \in E$:*

- *if $q(x) = 0$, $L\varphi(x) = 0$;*
- *if $q(x) > 0$,*

$$L\varphi(x) = q(x) \sum_{y \in E, y \neq x} \Pi(x, y)(\varphi(y) - \varphi(x)) = \sum_{y \in E} L(x, y)\,\varphi(y),$$

where

$$L(x, y) = \begin{cases} q(x)\Pi(x, y) & \text{if } y \neq x, \\ -q(x) & \text{if } y = x. \end{cases}$$

Proof Let $\varphi \in B(E)$ and $x \in E$. If $q(x) = 0$, it is trivial that $Q_t\varphi(x) = \varphi(x)$ and so

$$\lim_{t\downarrow 0} \frac{Q_t\varphi(x) - \varphi(x)}{t} = 0.$$

Suppose then that $q(x) > 0$. We first observe that

$$P_x(T_2 \leq t) = O(t^2) \tag{6.1}$$

as $t \to 0$. Indeed, using the strong Markov property at T_1,

$$P_x(T_2 \leq t) \leq P_x(T_1 \leq t,\, T_2 \leq T_1 + t) = E_x[\mathbf{1}_{\{T_1 \leq t\}} P_{X_{T_1}}(T_1 \leq t)],$$

and we can bound

$$P_{X_{T_1}}(T_1 \leq t) \leq \sup_{y\in E} P_y(T_1 \leq t) \leq t \sup_{y\in E} q(y),$$

giving the desired result since we have also $P_x(T_1 \leq t) \leq q(x)t$.

It follows from (6.1) that

$$Q_t\varphi(x) = E_x[\varphi(X_t)] = E_x[\varphi(X_t)\,\mathbf{1}_{\{t<T_1\}}] + E_x[\varphi(X_{T_1})\,\mathbf{1}_{\{T_1\leq t\}}] + O(t^2)$$

$$= \varphi(x)\,e^{-q(x)t} + (1 - e^{-q(x)t}) \sum_{y\in E, y\neq x} \Pi(x, y)\,\varphi(y) + O(t^2),$$

using the independence of T_1 and X_{T_1} and the definition of $\Pi(x, y)$. We conclude that

$$\frac{Q_t\varphi(x) - \varphi(x)}{t} \xrightarrow[t\to 0]{} -q(x)\varphi(x) + q(x) \sum_{y\in E, y\neq x} \Pi(x, y)\,\varphi(y),$$

and this completes the proof.   □

In particular, taking $\varphi(y) = \mathbf{1}_{\{y\}}$, we have if $y \neq x$,

$$L(x, y) = \frac{d}{dt} P_x(X_t = y)_{|t=0},$$

so that $L(x, y)$ can be interpreted as the instantaneous rate of transition from x to y.

The next proposition provides a complete description of the sample paths of X under P_x. For the sake of simplicity, we assume that there are no absorbing states, but the reader will easily extend the statement to the general case.

Proposition 6.20 *We assume that $q(y) > 0$ for every $y \in E$. Let $x \in E$. Then, P_x a.s., the jump times $T_1 < T_2 < T_3 < \cdots$ are all finite and the sequence $X_0, X_{T_1}, X_{T_2}, \ldots$ is under P_x a discrete Markov chain with transition kernel Π started from x. Furthermore, conditionally on $(X_0, X_{T_1}, X_{T_2}, \ldots)$, the random variables $T_1 - T_0, T_2 - T_1, \ldots$ are independent and, for every integer $i \geq 0$, the conditional distribution of $T_{i+1} - T_i$ is exponential with parameter $q(X_{T_i})$.*

Proof An application of the strong Markov property shows that all stopping times $T_1, T_2, \ldots$ are finite P_x a.s. Then, let $y, z \in E$, and $f_1, f_2 \in B(\mathbb{R}_+)$. By the strong Markov property at T_1,

$$E_x[\mathbf{1}_{\{X_{T_1}=y\}} f_1(T_1)\, \mathbf{1}_{\{X_{T_2}=z\}} f_2(T_2 - T_1)]$$

$$= E_x[\mathbf{1}_{\{X_{T_1}=y\}} f_1(T_1)\, E_{X_{T_1}}[\mathbf{1}_{\{X_{T_1}=z\}} f_2(T_1)]]$$

$$= \Pi(x, y) \Pi(y, z) \int_0^\infty ds_1\, e^{-q(x)s_1} f_1(s_1) \int_0^\infty ds_2 e^{-q(y)s_2} f_2(s_2).$$

Arguing by induction, we get for every $y_1, \ldots, y_p \in E$ and $f_1, \ldots, f_p \in B(\mathbb{R}_+)$,

$$E_x[\mathbf{1}_{\{X_{T_1}=y_1\}} \mathbf{1}_{\{X_{T_2}=y_2\}} \cdots \mathbf{1}_{\{X_{T_p}=y_p\}} f_1(T_1) f_2(T_2 - T_1) \cdots f_p(T_p - T_{p-1})]$$

$$= \Pi(x, y_1) \Pi(y_1, y_2) \cdots \Pi(y_{p-1}, y_p) \prod_{i=1}^{p} \left(\int_0^\infty ds\, e^{-q(y_{i-1})s} f_i(s) \right),$$

where $y_0 = x$ by convention. The various assertions of the proposition follow. $\square$

Jump processes play an important role in various models of applied probability, in particular in reliability and in queueing theory. In such applications, one usually starts from the transition rates of the process. It is thus important to know whether, given a collection $(q(x))_{x \in E}$ of nonnegative real numbers and, for every x such that $q(x) > 0$, a probability measure $\Pi(x, \cdot)$ on E such that $\Pi(x, x) = 0$, there exists a corresponding Feller semigroup $(Q_t)_{t \geq 0}$ and therefore an associated Markov process. The answer to this question is yes, and one can give two different arguments:

- *Probabilistic method.* Use the description of Proposition 6.20 (or its extension to the case where there are absorbing states) to construct the process $(X_t)_{t \geq 0}$ starting from any $x \in E$, and thus the semigroup $(Q_t)_{t \geq 0}$ via the formula $Q_t \varphi(x) = E_x[\varphi(X_t)]$.

- *Analytic method.* Define the generator L via the formulas of Proposition 6.19, and observe that the semigroup $(Q_t)_{t\geq 0}$, if it exists, must solve the differential equation

$$\frac{d}{dt}Q_t(x,y) = Q_t L(x,y)$$

by Proposition 6.11. This leads to

$$Q_t = \exp(t L),$$

in the sense of the exponential of matrices. Since $\lambda \mathrm{Id} + L$ has nonnegative entries if $\lambda > 0$ is large enough, one immediately gets that Q_t has nonnegative entries. Writing $\mathbf{1}$ for the vector $(1,1,\dots,1)$, the property $L\mathbf{1} = 0$ gives $Q_t \mathbf{1} = \mathbf{1}$, so that $(Q_t(x,\cdot))_{x\in E}$ defines a transition kernel. Finally, the property $\exp((s+t)L) = \exp(sL)\exp(tL)$ gives the Chapman–Kolmogorov property, and we get that $(Q_t)_{t\geq 0}$ is a transition semigroup on E, whose Feller property is also immediate.

Many of the preceding results can be extended to Feller Markov processes on a *countable* state space E. Note, however, that certain difficulties arise in the question of the existence of a process with given transition rates. In fact, starting from the probabilistic description of Proposition 6.20, one needs to avoid the possibility of an accumulation of jumps in a finite time interval, which may occur if the rates $(q(y), y \in E)$ are unbounded – of course this problem does not occur when E is finite.

6.5.2 Lévy Processes

Consider a real process $(Y_t)_{t\geq 0}$ which satisfies the following three assumptions:

- (i) $Y_0 = 0$ a.s.
- (ii) For every $0 \leq s \leq t$, the variable $Y_t - Y_s$ is independent of $(Y_r, 0 \leq r \leq s)$ and has the same law as Y_{t-s}.
- (iii) Y_t converges in probability to 0 when $t \downarrow 0$.

Two special cases are real Brownian motion (started from 0) and the process $(T_a)_{a\geq 0}$ of hitting times of a real Brownian motion (cf. Exercise 2.26).

Notice that we do not assume that sample paths of Y are càdlàg, but only the weaker regularity assumption (iii). The preceding theory will allow us to find a modification of Y with càdlàg sample paths.

For every $t \geq 0$, we denote the law of Y_t by $Q_t(0, dy)$, and, for every $x \in \mathbb{R}$, we let $Q_t(x, dy)$ be the image of $Q_t(0, dy)$ under the translation $y \mapsto x + y$.

Proposition 6.21 *The collection $(Q_t)_{t\geq 0}$ is a Feller semigroup on $\mathbb{R}$. Furthermore, $(Y_t)_{t\geq 0}$ is a Markov process with semigroup $(Q_t)_{t\geq 0}$.*

Proof Let us show that $(Q_t)_{t\geq 0}$ is a transition semigroup. Let $\varphi \in B(\mathbb{R})$, $s, t \geq 0$ and $x \in \mathbb{R}$. Property (ii) shows that the law of $(Y_t, Y_{t+s} - Y_t)$ is the product probability measure $Q_t(0, \cdot) \otimes Q_s(0, \cdot)$. Hence,

$$\int Q_t(x, dy) \int Q_s(y, dz)\varphi(z) = \int Q_t(0, dy) \int Q_s(0, dz)\varphi(x + y + z)$$
$$= E[\varphi(x + Y_t + (Y_{t+s} - Y_t))]$$
$$= E[\varphi(x + Y_{t+s})]$$
$$= \int Q_{t+s}(x, dz)\varphi(z)$$

giving the Chapman–Kolmogorov relation. We should also verify the measurability of the mapping $(t, x) \mapsto Q_t(x, A)$, but this will follow from the stronger continuity properties that we will establish in order to verify the Feller property.

Let us start with the first property of the definition of a Feller semigroup. If $\varphi \in C_0(\mathbb{R})$, the mapping

$$x \mapsto Q_t\varphi(x) = E[\varphi(x + Y_t)]$$

is continuous by dominated convergence, and, again by dominated convergence, we have

$$E[\varphi(x + Y_t)] \xrightarrow[x\to\infty]{} 0$$

showing that $Q_t\varphi \in C_0(\mathbb{R})$. Then,

$$Q_t\varphi(x) = E[\varphi(x + Y_t)] \xrightarrow[t\to 0]{} \varphi(x)$$

thanks to property (iii). The uniform continuity of φ even shows that the latter convergence is uniform in x. This completes the proof of the first assertion of the proposition. To get the second one, we write, for every $s, t \geq 0$ and every $\varphi \in B(\mathbb{R})$,

$$E[\varphi(Y_{s+t}) \mid Y_r, 0 \leq r \leq s] = E[\varphi(Y_s + (Y_{s+t} - Y_s)) \mid Y_r, 0 \leq r \leq s]$$
$$= \int \varphi(Y_s + y) Q_t(0, dy)$$
$$= \int \varphi(y) Q_t(Y_s, dy),$$

using property (ii) and the definition of $Q_t(0, \cdot)$ in the second equality. $\qquad\square$

It then follows from Theorem 6.15 that there exists a modification of $(Y_t)_{t\geq0}$ with càdlàg sample paths. Obviously this modification still satisfies (i) and (ii).

A *Lévy process* is a process satisfying properties (i) and (ii) above, and having càdlàg sample paths (which implies (iii)). We refer to [3] for a thorough account of the theory of Lévy processes.

6.5.3 Continuous-State Branching Processes

A Markov process $(X_t)_{t\geq0}$ with values in $E = \mathbb{R}_+$ is called a *continuous-state branching process* if its semigroup $(Q_t)_{t\geq0}$ satisfies the following property: for every $x, y \in \mathbb{R}_+$ and $t \geq 0$,

$$Q_t(x, \cdot) * Q_t(y, \cdot) = Q_t(x + y, \cdot),$$

where $\mu * \nu$ denotes the convolution of the probability measures μ and ν on $\mathbb{R}_+$. Note that this implies $Q_t(0, \cdot) = \delta_0$ for every $t \geq 0$.

Exercise Verify that, if X and X' are two *independent* continuous-state branching processes with the same semigroup $(Q_t)_{t\geq0}$, then $(X_t + X'_t)_{t\geq0}$ is also a Markov process with semigroup $(Q_t)_{t\geq0}$. This is the so-called branching property: compare with discrete time Galton–Watson processes.

Let us fix the semigroup $(Q_t)_{t\geq0}$ of a continuous-state branching process, and assume that:

(i) $Q_t(x, \{0\}) < 1$ for every $x > 0$ and $t > 0$;
(ii) $Q_t(x, \cdot) \longrightarrow \delta_x(\cdot)$ when $t \to 0$, in the sense of weak convergence of probability measures.

Proposition 6.22 *Under the preceding assumptions, the semigroup $(Q_t)_{t\geq0}$ is Feller. Furthermore, for every $\lambda > 0$, and every $x \geq 0$,*

$$\int Q_t(x, dy)\, e^{-\lambda y} = e^{-x\psi_t(\lambda)}$$

where the functions $\psi_t : (0, \infty) \longrightarrow (0, \infty)$ satisfy $\psi_t \circ \psi_s = \psi_{t+s}$ for every $s, t \geq 0$.

Proof Let us start with the second assertion. If $x, y > 0$, the equality $Q_t(x, \cdot) * Q_t(y, \cdot) = Q_t(x + y, \cdot)$ implies that

$$\left(\int Q_t(x, dz)\, e^{-\lambda z}\right)\left(\int Q_t(y, dz)\, e^{-\lambda z}\right) = \int Q_t(x + y, dz)\, e^{-\lambda z}.$$

Thus the function

$$x \mapsto -\log \left(\int Q_t(x, dz)\, e^{-\lambda z} \right)$$

is nondecreasing and linear on $\mathbb{R}_+$, hence of the form $x\psi_t(\lambda)$ for some constant $\psi_t(\lambda) > 0$ (the case $\psi_t(\lambda) = 0$ is excluded by assumption (i)). To obtain the identity $\psi_t \circ \psi_s = \psi_{t+s}$, we write

$$\int Q_{t+s}(x, dz)\, e^{-\lambda z} = \int Q_t(x, dy) \int Q_s(y, dz)\, e^{-\lambda z}$$

$$= \int Q_t(x, dy)\, e^{-y\psi_s(\lambda)}$$

$$= e^{-x\psi_t(\psi_s(\lambda))}.$$

We still have to prove that the semigroup is Feller. For every $\lambda > 0$, set $\varphi_\lambda(x) = e^{-\lambda x}$. Then,

$$Q_t\varphi_\lambda = \varphi_{\psi_t(\lambda)} \in C_0(\mathbb{R}_+).$$

Furthermore, an application of the Stone–Weierstrass theorem shows that the vector space generated by the functions φ_λ, $\lambda > 0$, is dense in $C_0(\mathbb{R}_+)$. It easily follows that $Q_t\varphi \in C_0(\mathbb{R}_+)$ for every $\varphi \in C_0(\mathbb{R}_+)$.

Finally, if $\varphi \in C_0(\mathbb{R}_+)$, for every $x \geq 0$,

$$Q_t\varphi(x) = \int Q_t(x, dy)\, \varphi(y) \xrightarrow[t \to 0]{} \varphi(x)$$

by assumption (ii). Using a remark following the definition of Feller semigroups, this suffices to show that $\|Q_t\varphi - \varphi\| \longrightarrow 0$ when $t \to 0$, which completes the proof.
□

Example For every $t > 0$ and every $x \geq 0$, define $Q_t(x, dy)$ as the law of $\mathbf{e}_1 + \mathbf{e}_2 + \cdots + \mathbf{e}_N$, where $\mathbf{e}_1, \mathbf{e}_2, \ldots$ are independent random variables with exponential distribution of parameter $1/t$, and N is Poisson with parameter x/t, and is independent of the sequence $(\mathbf{e}_i)$. Then a simple calculation shows that

$$\int Q_t(x, dy)\, e^{-\lambda y} = e^{-x\psi_t(\lambda)}$$

where

$$\psi_t(\lambda) = \frac{\lambda}{1 + \lambda t}.$$

Noting that $\psi_t \circ \psi_s = \psi_{t+s}$, we obtain that $(Q_t)_{t\geq0}$ satisfies the Chapman–Kolmogorov identity, and then that $(Q_t)_{t\geq0}$ is the transition semigroup of a continuous-state branching process. Furthermore, $(Q_t)_{t\geq0}$ satisfies assumptions (i) and (ii) above. In particular, the semigroup $(Q_t)_{t\geq0}$ is Feller, and one can construct an associated Markov process $(X_t)_{t\geq0}$ with càdlàg sample paths. One can in fact prove that the sample paths of $(X_t)_{t\geq0}$ are continuous, and this process is called Feller's branching diffusion, see Sect. 8.4.3 below.

Exercises

Exercise 6.23 (*Reflected Brownian motion*)   We consider a probability space equipped with a filtration $(\mathscr{F}_t)_{t\in[0,\infty]}$. Let $a \geq 0$ and let $B = (B_t)_{t\geq0}$ be an $(\mathscr{F}_t)$-Brownian motion such that $B_0 = a$. For every $t > 0$ and every $z \in \mathbb{R}$, we set

$$p_t(z) = \frac{1}{\sqrt{2\pi t}}\,\exp(-\frac{z^2}{2t}).$$

1. We set $X_t = |B_t|$ for every $t \geq 0$. Verify that, for every $s \geq 0$ and $t \geq 0$, for every bounded measurable function $f : \mathbb{R}_+ \longrightarrow \mathbb{R}$,

$$E[f(X_{s+t})\,|\,\mathscr{F}_s] = Q_t f(X_s),$$

where $Q_0 f = f$ and, for every $t > 0$, for every $x \geq 0$,

$$Q_t f(x) = \int_0^\infty \big(p_t(y-x) + p_t(y+x)\big) f(y)\,\mathrm{d}y.$$

2. Infer that $(Q_t)_{t\geq0}$ is a transition semigroup, then that $(X_t)_{t\geq0}$ is a Markov process with values in $E = \mathbb{R}_+$, with respect to the filtration $(\mathscr{F}_t)$, with semigroup $(Q_t)_{t\geq0}$.
3. Verify that $(Q_t)_{t\geq0}$ is a Feller semigroup. We denote its generator by L.
4. Let f be a twice continuously differentiable function on $\mathbb{R}_+$, such that f and f'' belong to $C_0(\mathbb{R}_+)$. Show that, if $f'(0) = 0$, f belongs to the domain of L, and $Lf = \frac{1}{2}f''$. (*Hint:* One may observe that the function $g : \mathbb{R} \to \mathbb{R}$ defined by $g(y) = f(|y|)$ is then twice continuously differentiable on $\mathbb{R}$.) Show that, conversely, if $f'(0) \neq 0$, f does not belong to the domain of L.

Exercise 6.24 Let $(Q_t)_{t\geq0}$ be a transition semigroup on a measurable space E. Let π be a measurable mapping from E onto another measurable space F. We assume that, for any measurable subset A of F, for every $x, y \in E$ such that $\pi(x) = \pi(y)$, we have

$$Q_t(x, \pi^{-1}(A)) = Q_t(y, \pi^{-1}(A)),$$

for every $t > 0$. We then set, for every $z \in F$ and every measurable subset A of F, for every $t > 0$,

$$Q'_t(z, A) = Q_t(x, \pi^{-1}(A))$$

where x is an arbitrary point of E such that $\pi(x) = z$. We also set $Q'_0(z, A) = \mathbf{1}_A(z)$. We assume that the mapping $(t, z) \mapsto Q'_t(z, A)$ is measurable on $\mathbb{R}_+ \times F$, for every fixed A.

1. Verify that $(Q'_t)_{t \geq 0}$ forms a transition semigroup on F.
2. Let $(X_t)_{t \geq 0}$ be a Markov process in E with transition semigroup $(Q_t)_{t \geq 0}$ with respect to the filtration $(\mathscr{F}_t)_{t \geq 0}$. Set $Y_t = \pi(X_t)$ for every $t \geq 0$. Verify that $(Y_t)_{t \geq 0}$ is a Markov process in F with transition semigroup $(Q'_t)_{t \geq 0}$ with respect to the filtration $(\mathscr{F}_t)_{t \geq 0}$.
3. Let $(B_t)_{t \geq 0}$ be a d-dimensional Brownian motion, and set $R_t = |B_t|$ for every $t \geq 0$. Verify that $(R_t)_{t \geq 0}$ is a Markov process and give a formula for its transition semigroup. (The case $d = 1$ was treated via a different approach in Exercise 6.23.)

In the remaining exercises, we use the following notation. (E, d) is a locally compact metric space, which is countable at infinity, and $(Q_t)_{t \geq 0}$ is a Feller semigroup on E. We consider an E-valued process $(X_t)_{t \geq 0}$ with càdlàg sample paths, and a collection $(P_x)_{x \in E}$ of probability measures on E, such that, under P_x, $(X_t)_{t \geq 0}$ is a Markov process with semigroup $(Q_t)_{t \geq 0}$, with respect to the filtration $(\mathscr{F}_t)$, and $P_x(X_0 = x) = 1$. We write L for the generator of the semigroup $(Q_t)_{t \geq 0}$, $D(L)$ for the domain of L and R_λ for the λ-resolvent, for every $\lambda > 0$.

Exercise 6.25 (*Scale function*) In this exercise, we assume that $E = \mathbb{R}_+$ and that the sample paths of X are continuous. For every $x \in \mathbb{R}_+$, we set

$$T_x := \inf\{t \geq 0 : X_t = x\}$$

and

$$\varphi(x) := P_x(T_0 < \infty).$$

1. Show that, if $0 \leq x \leq y$,

$$\varphi(y) = \varphi(x) P_y(T_x < \infty).$$

2. We assume that $\varphi(x) < 1$ and $P_x(\sup_{t \geq 0} X_t = +\infty) = 1$, for every $x > 0$. Show that, if $0 < x \leq y$,

$$P_x(T_0 < T_y) = \frac{\varphi(x) - \varphi(y)}{1 - \varphi(y)}.$$

Exercise 6.26 (*Feynman–Kac formula*) Let v be a nonnegative function in $C_0(E)$. For every $x \in E$ and every $t \geq 0$, we set, for every $\varphi \in B(E)$,

$$Q_t^* \varphi(x) = E_x\left[\varphi(X_t) \exp\left(-\int_0^t v(X_s)\, ds\right)\right].$$

1. Show that, for every $\varphi \in B(E)$ and $s, t \geq 0$, $Q_{t+s}^* \varphi = Q_t^*(Q_s^* \varphi)$.
2. After observing that

$$1 - \exp\left(-\int_0^t v(X_s)\, ds\right) = \int_0^t v(X_s) \exp\left(-\int_s^t v(X_r)\, dr\right) ds$$

show that, for every $\varphi \in B(E)$,

$$Q_t \varphi - Q_t^* \varphi = \int_0^t Q_s(v\, Q_{t-s}^* \varphi)\, ds.$$

3. Assume that $\varphi \in D(L)$. Show that

$$\frac{d}{dt} Q_t^* \varphi_{|t=0} = L\varphi - v\varphi.$$

Exercise 6.27 (*Quasi left-continuity*) Throughout the exercise we fix the starting point $x \in E$. For every $t > 0$, we write $X_{t-}(\omega)$ for the left-limit of the sample path $s \mapsto X_s(\omega)$ at t.

Let $(T_n)_{n \geq 1}$ be a strictly increasing sequence of stopping times, and $T = \lim \uparrow T_n$. We assume that there exists a constant $C < \infty$ such that $T \leq C$. The goal of the exercise is to verify that $X_{T-} = X_T$, P_x a.s.

1. Let $f \in D(L)$ and $h = Lf$. Show that, for every $n \geq 1$,

$$E_x[f(X_T) \mid \mathscr{F}_{T_n}] = f(X_{T_n}) + E_x\left[\int_{T_n}^T h(X_s)\, ds \,\Big|\, \mathscr{F}_{T_n}\right].$$

2. We recall from the theory of discrete time martingales that

$$E_x[f(X_T) \mid \mathscr{F}_{T_n}] \xrightarrow[n \to \infty]{\text{a.s.}, L^1} E_x[f(X_T) \mid \widetilde{\mathscr{F}}_T]$$

where

$$\widetilde{\mathscr{F}}_T = \bigvee_{n=1}^{\infty} \mathscr{F}_{T_n}.$$

Infer from question (**1**) that

$$E_x[f(X_T) \mid \widetilde{\mathscr{F}}_T] = f(X_{T-}).$$

3. Show that the conclusion of question (**2**) remains valid if we only assume that $f \in C_0(E)$, and infer that, for every choice of $f, g \in C_0(E)$,

$$E_x[f(X_T)g(X_{T-})] = E_x[f(X_{T-})g(X_{T-})].$$

Conclude that $X_{T-} = X_T$, P_x a.s.

Exercise 6.28 (*Killing operation*) In this exercise, we assume that X has continuous sample paths. Let A be a compact subset of E and

$$T_A = \inf\{t \geq 0 : X_t \in A\}.$$

1. We set, for every $t \geq 0$ and every bounded measurable function φ on E,

$$Q_t^* \varphi(x) = E_x[\varphi(X_t)\,\mathbf{1}_{\{t < T_A\}}]\,, \qquad \forall x \in E.$$

Verify that $Q_{t+s}^* \varphi = Q_t^*(Q_s^* \varphi)$, for every $s, t > 0$.
2. We set $\overline{E} = (E \backslash A) \cup \{\Delta\}$, where Δ is a point added to $E \backslash A$ as an isolated point. For every bounded measurable function φ on $\overline{E}$ and every $t \geq 0$, we set

$$\overline{Q}_t \varphi(x) = E_x[\varphi(X_t)\,\mathbf{1}_{\{t < T_A\}}] + P_x[T_A \leq t]\,\varphi(\Delta)\,, \quad \text{if } x \in E \backslash A$$

and $\overline{Q}_t \varphi(\Delta) = \varphi(\Delta)$. Verify that $(\overline{Q}_t)_{t \geq 0}$ is a transition semigroup on $\overline{E}$. (*The proof of the measurability of the mapping $(t, x) \mapsto \overline{Q}_t \varphi(x)$ will be omitted.*)
3. Show that, under the probability measure P_x, the process $\overline{X}$ defined by

$$\overline{X}_t = \begin{cases} X_t & \text{if } t < T_A \\ \Delta & \text{if } t \geq T_A \end{cases}$$

is a Markov process with semigroup $(\overline{Q}_t)_{t \geq 0}$, with respect to the canonical filtration of X.
4. We take it for granted that the semigroup $(\overline{Q}_t)_{t \geq 0}$ is Feller, and we denote its generator by $\overline{L}$. Let $f \in D(L)$ such that f and Lf vanish on an open set containing A. Write $\overline{f}$ for the restriction of f to $E \backslash A$, and consider $\overline{f}$ as a function on $\overline{E}$ by setting $\overline{f}(\Delta) = 0$. Show that $\overline{f} \in D(\overline{L})$ and $\overline{L}\overline{f}(x) = Lf(x)$ for every $x \in E \backslash A$.

Exercise 6.29 (*Dynkin's formula*)

1. Let $g \in C_0(E)$ and $x \in E$, and let T be a stopping time. Justify the equality

$$E_x\left[\mathbf{1}_{\{T < \infty\}} e^{-\lambda T} \int_0^\infty e^{-\lambda t} g(X_{T+t})\,dt\right] = E_x[\mathbf{1}_{\{T < \infty\}}\,e^{-\lambda T}\,R_\lambda g(X_T)].$$

2. Infer that

$$R_\lambda g(x) = E_x\left[\int_0^T e^{-\lambda t} g(X_t)\,dt\right] + E_x[\mathbf{1}_{\{T<\infty\}}\, e^{-\lambda T} R_\lambda g(X_T)].$$

3. Show that, if $f \in D(L)$,

$$f(x) = E_x\left[\int_0^T e^{-\lambda t}(\lambda f - Lf)(X_t)\,dt\right] + E_x[\mathbf{1}_{\{T<\infty\}}\, e^{-\lambda T} f(X_T)].$$

4. Assuming that $E_x[T] < \infty$, infer from the previous question that

$$E_x\left[\int_0^T Lf(X_t)\,dt\right] = E_x[f(X_T)] - f(x).$$

How could this formula have been established more directly?

5. For every $\varepsilon > 0$, we set $T_{\varepsilon,x} = \inf\{t \geq 0 : d(x, X_t) > \varepsilon\}$. Assume that $E_x[T_{\varepsilon,x}] < \infty$, for every sufficiently small ε. Show that (still under the assumption $f \in D(L)$) one has

$$Lf(x) = \lim_{\varepsilon \downarrow 0} \frac{E_x[f(X_{T_{\varepsilon,x}})] - f(x)}{E_x[T_{\varepsilon,x}]}.$$

6. Show that the assumption $E_x[T_{\varepsilon,x}] < \infty$ for every sufficiently small ε holds if the point x is not absorbing, that is, if there exists a $t > 0$ such that $Q_t(x, \{x\}) < 1$. (*Hint*: Observe that there exists a nonnegative function $h \in C_0(E)$ which vanishes on a ball centered at x and is such that $Q_t h(x) > 0$. Infer that one can choose $\alpha > 0$ and $\eta \in (0, 1)$ such that $P_x(T_{\alpha,x} > nt) \leq (1 - \eta)^n$ for every integer $n \geq 1$.)

Notes and Comments

The theory of Markov processes is a very important area of probability theory. Markov processes have a long history that would be too long to summarize here. Dynkin and Feller played a major role in the development of the theory (see in particular Dynkin's books [20, 21]). We limited our treatment to the minimal material needed for our later applications to stochastic differential equations. Our treatment of Feller processes is inspired by the corresponding chapters in [70] and [71]. We chose to focus on Feller semigroups because this special case allows an easy presentation of key notions such as the generator, and at the same time it includes the main examples we consider in this book. The reader interested in the more general theory of Markov processes may have a look at the classical books of Blumenthal and Getoor [5], Meyer [59] and Sharpe [73]. The idea of characterizing a Markov process by a collection of associated martingales (in the spirit of Theorem 6.14) has led to the theory of martingale problems, for which we refer the reader to the classical book of Stroock and Varadhan [77]. Martingale problems are also discussed in the book [24] of Ethier and Kurtz, which focuses on problems of characterization and convergence of Markov processes, with many examples and applications. Markov processes with a countable state space are treated, along with other topics, in the more recent book [76] of Stroock. We refer to the monograph [3] of Bertoin for a modern presentation of the theory of Lévy processes.

Chapter 7
Brownian Motion and Partial Differential Equations

In this chapter, we use the results of the preceding two chapters to discuss connections between Brownian motion and partial differential equations. After a brief discussion of the heat equation, we focus on the Laplace equation $\Delta u = 0$ and on the relations between Brownian motion and harmonic functions on a domain of $\mathbb{R}^d$. In particular, we give the probabilistic solution of the classical Dirichlet problem in a bounded domain whose boundary satisfies the exterior cone condition. In the case where the domain is a ball, the solution is made explicit by the Poisson kernel, which corresponds to the density of the exit distribution of the ball for Brownian motion. We then discuss recurrence and transience of d-dimensional Brownian motion, and we establish the conformal invariance of planar Brownian motion as a simple corollary of the results of Chap. 5. An important application is the so-called skew-product decomposition of planar Brownian motion, which we use to derive several asymptotic laws, including the celebrated Spitzer theorem on Brownian windings.

7.1 Brownian Motion and the Heat Equation

Throughout this chapter, we let B stand for a d-dimensional Brownian motion that starts from x under the probability measure P_x, for every $x \in \mathbb{R}^d$ (one may use the canonical construction of Sect. 2.2, defining P_x as the image of Wiener measure $W(\mathrm{d}w)$ under the translation $w \mapsto x + w$). Then $(B_t)_{t \geq 0}$ is a Feller process with semigroup

$$Q_t\varphi(x) = \int_{\mathbb{R}^d} (2\pi t)^{-d/2} \exp(-\frac{|y - x|^2}{2t}) \, \varphi(y) \, \mathrm{d}y,$$

© Springer International Publishing Switzerland 2016
J.-F. Le Gall, *Brownian Motion, Martingales, and Stochastic Calculus*,
Graduate Texts in Mathematics 274, DOI 10.1007/978-3-319-31089-3_7

for $\varphi \in B(\mathbb{R}^d)$. We write L for the generator of this Feller process. If ψ is a twice continuously differentiable function on $\mathbb{R}^d$ such that both ψ and $\Delta\psi$ belong to $C_0(\mathbb{R}^d)$, then $\psi \in D(L)$ and $L\psi = \frac{1}{2}\Delta\psi$ (see the end of Sect. 6.2).

If $\varphi \in B(\mathbb{R}^d)$, then, for every fixed $t > 0$, $Q_t\varphi$ can be viewed as the convolution of φ with the C^∞ function

$$p_t(x) = (2\pi t)^{-d/2} \exp(-\frac{|x|^2}{2t}).$$

It follows that $Q_t\varphi$ is also a C^∞ function. Furthermore, if $\varphi \in C_0(\mathbb{R}^d)$, differentiation under the integral sign shows that all derivatives of $Q_t\varphi$ also belong to $C_0(\mathbb{R}^d)$. It follows that we have $Q_t\varphi \in D(L)$ and $L(Q_t\varphi) = \frac{1}{2}\Delta(Q_t\varphi)$, for every $t > 0$.

Theorem 7.1 *Let $\varphi \in C_0(\mathbb{R}^d)$. For every $t > 0$ and $x \in \mathbb{R}^d$, set*

$$u_t(x) = Q_t\varphi(x) = E_x[\varphi(B_t)].$$

Then, the function $(u_t(x))_{t>0,x\in\mathbb{R}^d}$ solves the partial differential equation

$$\frac{\partial u_t}{\partial t} = \frac{1}{2}\Delta u_t,$$

on $(0, \infty) \times \mathbb{R}^d$. Furthermore, for every $x \in \mathbb{R}^d$,

$$\lim_{\substack{s\downarrow 0 \\ y\to x}} u_s(y) = \varphi(x).$$

Proof By the remarks preceding the theorem, we already know that, for every $t > 0$, u_t is a C^∞ function, $u_t \in D(L)$, and $Lu_t = \frac{1}{2}\Delta u_t$. Let $\varepsilon > 0$. By applying Proposition 6.11 to $f = u_\varepsilon$, we get for every $t \geq \varepsilon$,

$$u_t = u_\varepsilon + \int_0^{t-\varepsilon} L(Q_s u_\varepsilon)\, ds = u_\varepsilon + \int_\varepsilon^t Lu_s\, ds.$$

Since $Lu_s = Q_{s-\varepsilon}(Lu_\varepsilon)$ depends continuously on $s \in [\varepsilon, \infty)$, it follows that, for $t \geq \varepsilon$,

$$\frac{\partial u_t}{\partial t} = Lu_t = \frac{1}{2}\Delta u_t.$$

The last assertion is just the fact that $Q_s\varphi \longrightarrow \varphi$ as $s \to 0$. □

Remark We could have proved Theorem 7.1 by direct calculations from the explicit form of $Q_t\varphi$, and these calculations imply that the same statement holds if we only assume that φ is bounded and continuous. The above proof however has the advantage of showing the relation between this result and our general study of

Markov processes. It also indicates that similar results will hold for more general equations of the form $\frac{\partial u}{\partial t} = Au$ provided A can be interpreted as the generator of an appropriate Markov process.

Brownian motion can be used to provide probabilistic representations for solutions of many other parabolic partial differential equations. In particular, solutions of equations of the form

$$\frac{\partial u}{\partial t} = \frac{1}{2}\Delta u - v\,u,$$

where v is a nonnegative function on $\mathbb{R}^d$, are expressed via the so-called Feynman–Kac formula: See Exercise 7.28 below for a precise statement.

7.2 Brownian Motion and Harmonic Functions

Let us now turn to connections between Brownian motion and the Laplace equation $\Delta u = 0$. We start with a classical definition.

Definition 7.2 Let D be a domain of $\mathbb{R}^d$. A function $u : D \longrightarrow \mathbb{R}$ is said to be *harmonic* on D if it is twice continuously differentiable and $\Delta u = 0$ on D.

Let D' be a subdomain of D whose closure is contained in D. Consider the stopping time $T := \inf\{t \geq 0 : B_t \notin D'\}$. An application of Itô's formula (justified by the remark preceding Proposition 5.11) shows that, if u is harmonic on D, then for every $x \in D'$, the process

$$u(B_{t\wedge T}) = u(B_0) + \int_0^{t\wedge T} \nabla u(B_s) \cdot dB_s \tag{7.1}$$

is a local martingale under P_x [here and later, to apply the stochastic calculus results of Chap. 5, we let $(\mathscr{F}_t)$ be the canonical filtration of B completed under P_x].

So, roughly speaking, harmonic functions are functions which when composed with Brownian motion give (local) martingales.

Proposition 7.3 *Let u be harmonic on the domain D. Let D' be a bounded subdomain of D whose closure is contained in D, and consider the stopping time $T := \inf\{t \geq 0 : B_t \notin D'\}$. Then, for every $x \in D'$,*

$$u(x) = E_x[u(B_T)].$$

Proof Since D' is bounded, both u and ∇u are bounded on D', and we also know that $P_x(T < \infty) = 1$ for every $x \in D'$. It follows from (7.1) that $u(B_{t\wedge T})$ is a (true) martingale, and in particular, we have

$$u(x) = E_x[u(B_{t\wedge T})]$$

for every $x \in D'$. By letting $t \to \infty$ and using dominated convergence we get that

$$u(x) = E_x[u(B_T)]$$

for every $x \in D'$.   □

The preceding proposition easily leads to the mean value property for harmonic functions. In order to state this property, first recall that the uniform probability measure on the unit sphere, denoted by $\sigma_1(dy)$, is the unique probability measure on $\{y \in \mathbb{R}^d : |y| = 1\}$ that is invariant under all vector isometries. For every $x \in \mathbb{R}^d$ and $r > 0$, we then let $\sigma_{x,r}(dy)$ be the image of $\sigma_1(dy)$ under the mapping $y \mapsto x + ry$.

Proposition 7.4 (Mean value property) *Suppose that u is harmonic on the domain D. Then, for every $x \in D$ and for every $r > 0$ such that the closed ball of radius r centered at x is contained in D, we have*

$$u(x) = \int \sigma_{x,r}(dy)\, u(y).$$

Proof First observe that, if $T_1 = \inf\{t \geq 0 : |B_t| = 1\}$, the distribution of B_{T_1} under P_0 is invariant under all vector isometries of $\mathbb{R}^d$ (by the invariance properties of Brownian motion stated at the end of Chap. 2) and therefore this distribution is $\sigma_1(dy)$. By scaling and translation invariance, it follows that for every $x \in \mathbb{R}^d$ and $r > 0$, if $T_{x,r} = \inf\{t \geq 0 : |B_t - x| = r\}$, the distribution of $B_{T_{x,r}}$ under P_x is $\sigma_{x,r}$. However, Proposition 7.3 implies that, under the conditions in the proposition, we must have $u(x) = E_x[u(B_{T_{x,r}})]$. The desired result follows.   □

We say that a (locally bounded and measurable) function u on D satisfies the mean value property if the conclusion of Proposition 7.4 holds. It turns out that this property characterizes harmonic functions.

Lemma 7.5 *Let u be a locally bounded and measurable function on D that satisfies the mean value property. Then u is harmonic on D.*

Proof Fix $r_0 > 0$ and let D' be the open subset of D consisting of all points whose distance to D^c is greater than r_0. It is enough to prove that u is twice continuously differentiable and $\Delta u = 0$ on D'. Let $h : \mathbb{R} \longrightarrow \mathbb{R}_+$ be a C^∞ function with compact support contained in $(0, r_0)$ and not identically zero. Then, for every $x \in D'$ and every $r \in (0, r_0)$, we have

$$u(x) = \int \sigma_{x,r}(dy)\, u(y).$$

We multiply both sides of this equality by $r^{d-1}h(r)$ and integrate with respect to Lebesgue measure dr on $(0, r_0)$. Using the formula for integration in polar coordinates, and agreeing for definiteness that $u = 0$ on D^c, we get, for every $x \in D'$,

$$c\, u(x) = \int_{\{|y|<r_0\}} dy\, h(|y|)\, u(x+y) = \int_{\mathbb{R}^d} dz\, h(|z - x|)\, u(z),$$

where $c > 0$ is a constant and we use the fact that $h(|z - x|) = 0$ if $|z - x| \geq r_0$. Since $z \mapsto h(|z|)$ is a C^∞ function, the convolution in the right-hand side of the last display also defines a C^∞ function on D'.

It remains to check that $\Delta u = 0$ on D'. To this end, we use a probabilistic argument (an analytic argument is also easy). By applying Itô's formula to $u(B_t)$ under P_x, we get, for $x \in D'$ and $r \in (0, r_0)$,

$$E_x[u(B_{t \wedge T_{x,r}})] = u(x) + E_x\left[\int_0^{t \wedge T_{x,r}} ds\, \Delta u(B_s)\right].$$

If we let $t \to \infty$, noting that $E_x[T_{x,r}] < \infty$ (cf. example (b) after Corollary 3.24), we get

$$E_x[u(B_{T_{x,r}})] = u(x) + E_x\left[\int_0^{T_{x,r}} ds\, \Delta u(B_s)\right].$$

The mean value property just says that $E_x[u(B_{T_{x,r}})] = u(x)$ and so we have

$$E_x\left[\int_0^{T_{x,r}} ds\, \Delta u(B_s)\right] = 0.$$

Since this holds for any $r \in (0, r_0)$ it follows that $\Delta u(x) = 0$. $\square$

From now on, we assume that the domain D is **bounded**.

Definition 7.6 (Classical Dirichlet problem) Let g be a continuous function on ∂D. A function $u : D \longrightarrow \mathbb{R}$ solves the *Dirichlet problem* in D with boundary condition g, if u is harmonic on D and has boundary condition g, in the sense that, for every $y \in \partial D$, $u(x) \longrightarrow g(y)$ as $x \to y$, $x \in D$.

Recall that $\bar{D}$ stands for the closure of D. If u solves the Dirichlet problem with boundary condition g, the function $\tilde{u}$ defined on $\bar{D}$ by $\tilde{u}(x) = u(x)$ if $x \in D$, and $\tilde{u}(x) = g(x)$ if $x \in \partial D$, is then continuous, hence bounded on $\bar{D}$.

Proposition 7.7 *Let D be a bounded domain, and write $T = \inf\{t \geq 0 : B_t \notin D\}$ for the exit time of Brownian motion from D.*

(i) *Let g be a continuous function on ∂D, and let u be a solution of the Dirichlet problem in D with boundary condition g. Then, for every $x \in D$,*

$$u(x) = E_x[g(B_T)].$$

(ii) *Let g be a bounded measurable function on ∂D. Then the function*

$$u(x) = E_x[g(B_T)], \quad x \in D,$$

is harmonic on D.

Remark Assertion (i) implies that, if a solution to the Dirichlet problem with boundary condition g exists, then it is unique, which is also easy to prove using the mean value property.

Proof

(i) Fix $x \in D$ and $\varepsilon_0 > 0$ such that the ball of radius ε_0 centered at x is contained in D. For every $\varepsilon \in (0, \varepsilon_0)$, let D_ε be the connected component containing x of the open set consisting of all points of D whose distance to D^c is greater than ε. If $T_\varepsilon = \inf\{t \geq 0 : B_t \notin D_\varepsilon\}$, Proposition 7.3 shows that

$$u(x) = E_x[u(B_{T_\varepsilon})].$$

Now observe that $T_\varepsilon \uparrow T$ as $\varepsilon \downarrow 0$ (if T' is the increasing limit of T_ε as $\varepsilon \downarrow 0$, we have $T' \leq T$ and on the other hand $B_{T'} \in \partial D$ by the continuity of sample paths). Using dominated convergence, it follows that $E_x[u(B_{T_\varepsilon})]$ converges to $E_x[g(B_T)]$ as $\varepsilon \to 0$.

(ii) By Lemma 7.5, it is enough to verify that the function $u(x) = E_x[g(B_T)]$ satisfies the mean value property. Recall the notation $T_{x,r} = \inf\{t \geq 0 : |B_t - x| = r\}$ for $x \in \mathbb{R}^d$ and $r > 0$. Fix $x \in D$ and $r > 0$ such that the closed ball of radius r centered at x is contained in D. We apply the strong Markov property in the form given in Theorem 6.17: With the notation of this theorem, we let $\Phi(w)$, for $w \in C(\mathbb{R}_+, \mathbb{R}^d)$ such that $w(0) \in D$, be the value of g at the first exit point of w from D (we take $\Phi(w) = 0$ if w never exits D) and we observe that we have

$$g(B_T) = \Phi(B_t, t \geq 0) = \Phi(B_{T_{x,r}+t}, t \geq 0), \qquad P_x \text{ a.s.}$$

because the paths $(B_t, t \geq 0)$ and $(B_{T_{x,r}+t}, t \geq 0)$ have the same exit point from D. It follows that

$$u(x) = E_x[g(B_T)] = E_x[\Phi(B_{T_{x,r}+t}, t \geq 0)] = E_x[E_{B_{T_{\varepsilon,r}}}[\Phi(B_t, t \geq 0)]] = E_x[u(B_{T_{\varepsilon,r}})].$$

Since we know that the law of $B_{T_{\varepsilon,r}}$ under P_x is $\sigma_{x,r}$, this gives the mean value property.   □

Part (i) of Proposition 7.7 tells us that the solution of the Dirichlet problem with boundary condition g, if it exists, is given by the probabilistic formula $u(x) = E_x[g(B_T)]$. On the other hand, for any choice of the (bounded measurable) function g on ∂D, part (ii) tells us that the probabilistic formula yields a function u that is harmonic on D. Even if g is assumed to be continuous, it is however not clear that the function u has boundary condition g, and this need not be true in general (see Exercises 7.24 and 7.25 for examples where the Dirichlet problem has no solution). We state a theorem that gives a partial answer to this question.

If $y \in \partial D$, we say that D satisfies the *exterior cone condition* at y if there exist a (nonempty) open cone $\mathscr{C}$ with apex y and a real $r > 0$ such that the intersection of $\mathscr{C}$ with the open ball of radius r centered at y is contained in D^c. For instance, a convex domain satisfies the exterior cone condition at every point of its boundary.

Theorem 7.8 (Solution of the Dirichlet problem) *Let D be a bounded domain in $\mathbb{R}^d$. Assume that D satisfies the exterior cone condition at every $y \in \partial D$. Then, for every continuous function g on ∂D, the formula*

$$u(x) = E_x[g(B_T)] , \quad \text{where } T = \inf\{t \geq 0 : B_t \notin D\},$$

gives the unique solution of the Dirichlet problem with boundary condition g.

Proof Thanks to Proposition 7.7 (ii), we only need to verify that, for every fixed $y \in \partial D$,

$$\lim_{x \to y, x \in D} u(x) = g(y). \tag{7.2}$$

Let $\varepsilon > 0$. Since g is continuous, we can find $\delta > 0$ such that we have $|g(z) - g(y)| \leq \varepsilon/3$ whenever $z \in \partial D$ and $|z - y| < \delta$. Let $M > 0$ be such that $|g(z)| \leq M$ for every $z \in \partial D$. Then, for every $\eta > 0$,

$$|u(x) - g(y)| \leq E_x[|g(B_T) - g(y)|\mathbf{1}_{\{T \leq \eta\}}] + E_x[|g(B_T) - g(y)|\mathbf{1}_{\{T > \eta\}}]$$

$$\leq E_x[|g(B_T) - g(y)|\mathbf{1}_{\{T \leq \eta\}}\mathbf{1}_{\{\sup\{|B_t - x| : t \leq \eta\} \leq \delta/2\}}]$$

$$+ 2M P_x\left(\sup_{t \leq \eta} |B_t - x| > \frac{\delta}{2}\right) + 2M P_x(T > \eta).$$

Write A_1, A_2, A_3 for the three terms in the sum in the right-hand side of the last display. We assume that $|y - x| < \delta/2$, and we bound successively these three terms.

First note that we have $|B_T - y| \leq |B_T - x| + |y - x| < \delta$ on the event

$$\{T \leq \eta\} \cap \sup\{|B_t - x| : t \leq \eta\} \leq \delta/2\},$$

and our choice of δ ensures that $A_1 \leq \varepsilon/3$.

Then, translation invariance gives

$$A_2 = 2M P_0\left(\sup_{t \leq \eta} |B_t| > \frac{\delta}{2}\right),$$

which tends to 0 when $\eta > 0$ by the continuity of sample paths. So we can fix $\eta > 0$ so that $A_2 < \varepsilon/3$.

Finally, we claim that we can choose $\alpha \in (0, \delta/2]$ small enough so that we also have $A_3 = 2M P_x(T > \eta) < \varepsilon/3$ whenever $|x - y| < \alpha$. It follows that $|u(x) - g(y)| < \varepsilon$, whenever $|x - y| < \alpha$, thus completing the proof of (7.2). Thus it only remains to prove our claim, which is the goal of the next lemma. □

Lemma 7.9 *Under the exterior cone condition, we have for every $y \in \partial D$ and every $\eta > 0$,*

$$\lim_{x \to y, x \in D} P_x(T > \eta) = 0.$$

Proof For every $u \in \mathbb{R}^d$ with $|u| = 1$ and every $\gamma \in (0, 1)$, consider the circular cone

$$\mathscr{C}(u, \gamma) := \{z \in \mathbb{R}^d : z \cdot u > (1 - \gamma)|z|\},$$

where $z \cdot u$ stands for the usual scalar product. If $y \in \partial D$ is given, the exterior cone condition means that we can fix $r > 0$, u and γ such that

$$y + (\mathscr{C}(u, \gamma) \cap \mathscr{B}_r) \subset D^c,$$

where $\mathscr{B}_r$ denotes the open ball of radius r centered at 0. To simplify notation, we set $\mathscr{C} = \mathscr{C}(u, \gamma) \cap \mathscr{B}_r$, and also

$$\mathscr{C}' = \mathscr{C}(u, \frac{\gamma}{2}) \cap \mathscr{B}_{r/2}$$

which is the intersection of a smaller cone with $\mathscr{B}_{r/2}$.

For every open subset F of $\mathbb{R}^d$, write $T_F = \inf\{t \geq 0 : B_t \in F\}$. An application of Blumenthal's zero-one law (Theorem 2.13, or rather its easy extension to d-dimensional Brownian motion) along the lines of the proof of Proposition 2.14 (i) shows that $P_0(T_{\mathscr{C}(u,\gamma/2)} = 0) = 1$ and hence $P_0(T_{\mathscr{C}'} = 0) = 1$ by the continuity of sample paths. On the other hand, set $\mathscr{C}'_a = \{z \in \mathscr{C}' : |z| > a\}$, for every $a \in (0, r/2)$. The sets $\mathscr{C}'_a$ increase to $\mathscr{C}'$ as $a \downarrow 0$, and thus we have $T_{\mathscr{C}'_a} \downarrow T_{\mathscr{C}'} = 0$ as $a \downarrow 0$, P_0 a.s. Hence, given any $\beta > 0$ we can fix a small enough so that

$$P_0(T_{\mathscr{C}'_a} \leq \eta) \geq 1 - \beta.$$

Recalling that $y + \mathscr{C} \subset D^c$, we have

$$P_x(T \leq \eta) \geq P_x(T_{y+\mathscr{C}} \leq \eta) = P_0(T_{y-x+\mathscr{C}} \leq \eta).$$

However, a simple geometric argument shows that, as soon as $|y-x|$ is small enough, the shifted cone $y - x + \mathscr{C}$ contains $\mathscr{C}'_a$, and therefore

$$P_x(T \leq \eta) \geq P_0(T_{\mathscr{C}'_a} \leq \eta) \geq 1 - \beta.$$

Since β was arbitrary, this completes the proof. □

Remark The exterior cone condition is only a sufficient condition for the existence (and uniqueness) of a solution to the Dirichlet problem. See e.g. [69] for necessary and sufficient conditions that ensure the existence of a solution for any continuous boundary value.

7.3 Harmonic Functions in a Ball and the Poisson Kernel

Consider again a bounded domain D and a continuous function g on ∂D. Let $T = \inf\{t \geq 0 : B_t \notin D\}$ be the exit time of D by Brownian motion. Proposition 7.7 (i) shows that the solution of the Dirichlet problem in D with boundary condition g, if it exists (which is the case under the assumption of Theorem 7.8), is given by

$$u(x) = E_x[g(B_T)] = \int_{\partial D} \omega(x, dy)\, g(y),$$

where, for every $x \in D$, $\omega(x, dy)$ denotes the distribution of B_T under P_x. The measure $\omega(x, dy)$ is a probability measure on ∂D called the *harmonic measure* of D relative to x. In general, it is hopeless to try to find an explicit expression for the measures $\omega(x, dy)$. It turns out that, in the case of balls, such an explicit expression is available and makes the representation of solutions of the Dirichet problem more concrete.

From now on, we suppose that $D = \mathscr{B}_1$ is the open unit ball in $\mathbb{R}^d$. We also assume that $d \geq 2$ to avoid trivialities. The boundary $\partial \mathscr{B}_1$ is the unit sphere in $\mathbb{R}^d$.

Definition 7.10 The *Poisson kernel* (of the unit ball) is the function K defined on $\mathscr{B}_1 \times \partial \mathscr{B}_1$ by

$$K(x, y) = \frac{1 - |x|^2}{|y - x|^d},$$

for every $x \in \mathscr{B}_1$ and $y \in \partial \mathscr{B}_1$.

Lemma 7.11 *For every fixed* $y \in \partial \mathscr{B}_1$, *the function* $x \mapsto K(x, y)$ *is harmonic on* $\mathscr{B}_1$.

Proof Set $K_y(x) = K(x, y)$ for $x \in \mathscr{B}_1$. Then K_y is a C^∞ function on $\mathscr{B}_1$. Moreover a (somewhat tedious) direct calculation left to the reader shows that $\Delta K_y = 0$ on $\mathscr{B}_1$. $\qquad\square$

In view of deriving further properties of the Poisson kernel, the following lemma about radial harmonic functions will be useful.

Lemma 7.12 *Let* $0 \leq r_1 < r_2$ *be two real numbers and let* $h : (r_1, r_2) \to \mathbb{R}$ *be a measurable function. The function* $u(x) = h(|x|)$ *is harmonic on the domain* $\{x \in \mathbb{R}^d : r_1 < |x| < r_2\}$ *if and only if there exist two constants* a *and* b *such that*

$$h(r) = \begin{cases} a + b \log r & \text{if } d = 2, \\ a + b\, r^{2-d} & \text{if } d \geq 3. \end{cases}$$

Proof Suppose that $u(x) = h(|x|)$ is harmonic on $\{x \in \mathbb{R}^d : r_1 < |x| < r_2\}$. Then u is twice continuously differentiable and so is h. From the expression of the

Laplacian of a radial function, we get that $\Delta u = 0$ if and only if

$$h''(r) + \frac{d-1}{r} h'(r) = 0 , \quad r \in (r_1, r_2).$$

The solutions of this second order linear differential equations are the functions of the form given in the statement. The lemma follows. $\qquad\square$

Recall our notation $\sigma_1(dy)$ for the uniform probability measure on the unit sphere $\partial \mathcal{B}_1$.

Lemma 7.13 *For every* $x \in \mathcal{B}_1$,

$$\int_{\partial \mathcal{B}_1} K(x, y)\, \sigma_1(dy) = 1.$$

Proof For every $x \in \mathcal{B}_1$, set

$$F(x) = \int_{\partial \mathcal{B}_1} K(x, y)\, \sigma_1(dy).$$

Then the preceding lemma implies that F is harmonic on $\mathcal{B}_1$. Indeed, if $x \in \mathcal{B}_1$ and $r < 1 - |x|$, Lemma 7.11 and the mean value property imply that, for every $y \in \partial \mathcal{B}_1$,

$$K(x, y) = \int K(z, y)\, \sigma_{x,r}(dz).$$

Hence, using Fubini's theorem,

$$\int F(z)\, \sigma_{x,r}(dz) = \int \left(\int K(z, y)\, \sigma_1(dy) \right) \sigma_{x,r}(dz)$$

$$= \int \left(\int K(z, y)\, \sigma_{x,r}(dz) \right) \sigma_1(dy) = \int K(x, y)\, \sigma_1(dy) = F(x),$$

showing that the mean value property holds for F.

If ψ is a vector isometry of $\mathbb{R}^d$, we have $K(\psi(x), \psi(y)) = K(x, y)$ for every $x \in \mathcal{B}_1$ and $y \in \partial B_1$, and the fact that $\sigma_1(dy)$ is invariant under ψ implies that $F(\psi(x)) = F(x)$ for every $x \in \mathcal{B}_1$. Hence F is a radial harmonic function and Lemma 7.12 (together with the fact that F is bounded in the neighborhood of 0) implies that F is constant. Since $F(0) = 1$, the proof is complete. $\qquad\square$

Theorem 7.14 *Let* g *be a continuous function on* $\partial \mathcal{B}_1$. *The unique solution of the Dirichlet problem in* $\mathcal{B}_1$ *with boundary condition* g *is given by*

$$u(x) = \int_{\partial \mathcal{B}_1} g(y)\, K(x, y)\, \sigma_1(dy) , \quad x \in \mathcal{B}_1.$$

Proof The very same arguments as in the beginning of the proof of Lemma 7.13 show that u is harmonic on $\mathscr{B}_1$. To verify the boundary condition, fix $y_0 \in \partial \mathscr{B}_1$. For every $\delta > 0$, the explicit form of the Poisson kernel shows that, if $x \in \mathscr{B}_1$ and $y \in \partial \mathscr{B}_1$ are such that $|x - y_0| < \delta/2$ and $|y - y_0| > \delta$, then

$$K(x, y) \leq (\frac{2}{\delta})^d (1 - |x|^2).$$

It follows from this bound that, for every $\delta > 0$,

$$\lim_{x \to y_0, x \in \mathscr{B}_1} \int_{\{|y - y_0| > \delta\}} K(x, y)\, \sigma_1(dy) = 0. \tag{7.3}$$

Then, given $\varepsilon > 0$, we can choose $\delta > 0$ sufficiently small so that the conditions $y \in \partial \mathscr{B}_1$ and $|y - y_0| \leq \delta$ imply $|g(y) - g(y_0)| \leq \varepsilon$. If $M = \sup\{|g(y)| : y \in \partial \mathscr{B}_1\}$, it follows that

$$|u(x) - g(y_0)| = \left| \int K(x, y)\, (g(y) - g(y_0))\, \sigma_1(dy) \right|$$

$$\leq 2M \int_{\{|y - y_0| > \delta\}} K(x, y)\, \sigma_1(dy) + \varepsilon,$$

using Lemma 7.13 in the first equality, and then our choice of δ. Thanks to (7.3), we now get

$$\limsup_{x \to y_0, x \in \mathscr{B}_1} |u(x) - g(y_0)| \leq \varepsilon.$$

Since ε was arbitrary, this yields the desired boundary condition. □

The preceding theorem allows us to identify the harmonic measures of the unit ball.

Corollary 7.15 Let $T = \inf\{t \geq 0 : B_t \notin \mathscr{B}_1\}$. For every $x \in \mathscr{B}_1$, the distribution of B_T under P_x has density $K(x, y)$ with respect to $\sigma_1(dy)$.

This is immediate since, by combining Proposition 7.7 (i) with Theorem 7.14, we get that, for any continuous function g on $\partial \mathscr{B}_1$,

$$E_x[g(B_T)] = \int_{\partial \mathscr{B}_1} g(y)\, K(x, y)\, \sigma_1(dy), \quad \forall x \in \mathscr{B}_1.$$

7.4 Transience and Recurrence of Brownian Motion

We consider again a d-dimensional Brownian motion $(B_t)_{t \geq 0}$ that starts from x under the probability measure P_x. We again suppose that $d \geq 2$, since the corresponding results for $d = 1$ have already been derived in the previous chapters.

For every $a \geq 0$, we introduce the stopping time

$$U_a = \inf\{t \geq 0 : |B_t| = a\},$$

with the usual convention $\inf \varnothing = \infty$.

Proposition 7.16 *Suppose that $x \neq 0$, and let ε and R be such that $0 < \varepsilon < |x| < R$. Then,*

$$P_x(U_\varepsilon < U_R) = \begin{cases} \dfrac{\log R - \log |x|}{\log R - \log \varepsilon} & \text{if } d = 2, \\[3mm] \dfrac{R^{2-d} - |x|^{2-d}}{R^{2-d} - \varepsilon^{2-d}} & \text{if } d \geq 3. \end{cases} \tag{7.4}$$

Consequently, we have $P_x(U_0 < \infty) = 0$ and for every $\varepsilon \in (0, |x|)$,

$$P_x(U_\varepsilon < \infty) = \begin{cases} 1 & \text{if } d = 2, \\[3mm] \left(\dfrac{\varepsilon}{|x|}\right)^{d-2} & \text{if } d \geq 3. \end{cases}$$

Proof Write $D_{\varepsilon,R}$ for the annulus $\{y \in \mathbb{R}^d : \varepsilon < |y| < R\}$. Let $u(x)$ be the function defined for $x \in D_{\varepsilon,R}$ that appears in the right-hand side of (7.4). By Lemma 7.12, u is harmonic on $D_{\varepsilon,R}$, and it is also clear that u solves the Dirichlet problem in $D_{\varepsilon,R}$ with boundary condition $g(y) = 0$ if $|y| = R$ and $g(y) = 1$ if $|y| = \varepsilon$. If $T_{\varepsilon,R}$ denotes the first exit time from $D_{\varepsilon,R}$, Proposition 7.7 shows that we must have $u(x) = E_x[g(B_{T_{\varepsilon,R}})]$ for every $x \in D_{\varepsilon,R}$. Formula (7.4) follows since $E_x[g(B_{T_{\varepsilon,R}})] = P_x(U_\varepsilon < U_R)$.

If $R > |x|$ is fixed, the event $\{U_0 < U_R\}$ is (P_x a.s.) contained in $\{U_\varepsilon < U_R\}$, for every $0 < \varepsilon < |x|$. By passing to the limit $\varepsilon \to 0$ in the right-hand side of (7.4), we thus get that $P_x(U_0 < U_R) = 0$. Since $U_R \uparrow \infty$ as $R \uparrow \infty$, it follows that $P_x(U_0 < \infty) = 0$.

Finally, we have also $P_x(U_\varepsilon < \infty) = \lim P_x(U_\varepsilon < U_R)$ as $R \to \infty$, and by letting $R \to \infty$ in the right-hand side of (7.4) we get the stated formula for $P_x(U_\varepsilon < \infty)$. $\square$

Remark The reader will compare formula (7.4) with the exit distribution from an interval for real Brownian motion that was derived in Chap. 4 (example (a) after Corollary 3.24). We could have proved (7.4) in a way similar to what we did for its one-dimensional analog, by applying the optional stopping theorem to the local martingale $\log |B_t|$ (if $d = 2$) or $|B_t|^{2-d}$ (if $d = 3$). See Exercise 5.33.

For every $y \in \mathbb{R}^d$, set $\tau_y = \inf\{t \geq 0 : B_t = y\}$, so that in particular $\tau_0 = U_0$. The property $P_x(\tau_0 < \infty) = 0$ for $x \neq 0$ implies that $P_x(\tau_y < \infty) = 0$ whenever $y \neq x$, by translation invariance. This means that the probability for Brownian motion to visit a fixed point other than its starting point is zero: one says that points are *polar* for d-dimensional Brownian motion with $d \geq 2$ (see Exercise 7.25 for more about polar sets).

If $\mathbf{m}$ denotes Lebesgue measure on $\mathbb{R}^d$, it follows from Fubini's theorem that

$$E_x[\mathbf{m}(\{B_t, t \geq 0\})] = E_x\left[\int_{\mathbb{R}^d} dy\, \mathbf{1}_{\{\tau_y < \infty\}}\right] = \int_{\mathbb{R}^d} dy\, P_x(\tau_y < \infty) = 0,$$

and therefore $\mathbf{m}(\{B_t, t \geq 0\}) = 0$, P_x a.s. One can nonetheless prove that the Hausdorff dimension of the curve $\{B_t, t \geq 0\}$ is equal to 2 in any dimension $d \geq 2$ (see e.g. [62]). In some sense, this shows that the planar Brownian curve is "not so far" from having positive Lebesgue measure.

Theorem 7.17

(i) *In dimension $d = 2$, Brownian motion is recurrent, meaning that almost surely, for every nonempty open subset O of $\mathbb{R}^d$, the set $\{t \geq 0 : B_t \in O\}$ is unbounded.*

(ii) *In dimension $d \geq 3$, Brownian motion is transient, meaning that*

$$\lim_{t \to \infty} |B_t| = \infty, \quad a.s.$$

Proof

(i) It is enough to prove that the statement holds when O is an open ball of rational radius centered at a point with rational coordinates. So it suffices to consider a fixed open ball $\mathcal{B}$ and we may assume that $\mathcal{B}$ is centered at 0 and that the starting point of B is $x \neq 0$. By Proposition 7.16 we know that Brownian motion will never hit 0 (so that $\inf\{|B_r| : 0 \leq r \leq t\} > 0$ for every $t \geq 0$, a.s.) but still will hit any open ball centered at 0. It follows that B must visit $\mathcal{B}$ at arbitrarily large times, a.s.

(ii) Again we can assume that the starting point of B is $x \neq 0$. Since the function $y \mapsto |y|^{2-d}$ is harmonic on $\mathbb{R}^d \backslash \{0\}$, and since we saw that B does not hit 0, we get that $|B_t|^{2-d}$ is a local martingale and hence a supermartingale by Proposition 4.7. By Theorem 3.19 (and the fact that a positive supermartingale is automatically bounded in L^1), we know that $|B_t|^{2-d}$ converges a.s. as $t \to \infty$. The a.s. limit must be zero (otherwise the curve $\{B_t : t \geq 0\}$ would be bounded!) and this says exactly that $|B_t|$ converges to ∞ as $t \to \infty$. $\square$

Remark In dimension $d = 2$, one can (slightly) reinforce the recurrence property by saying that a.s. for every nonempty open subset O of $\mathbb{R}^2$, the Lebesgue measure of $\{t \geq 0 : B_t \in O\}$ is infinite. This follows by a straightforward application of the strong Markov property, and we omit the details.

7.5   Planar Brownian Motion and Holomorphic Functions

In this section, we concentrate on the planar case $d = 2$, and we write $B_t = (X_t, Y_t)$ for a two-dimensional Brownian motion. It will be convenient to identify $\mathbb{R}^2$ with the complex plane $\mathbb{C}$, so that $B_t = X_t + iY_t$, and we sometimes say that B is a *complex Brownian motion*. As previously B starts from z under the probability P_z, for every $z \in \mathbb{C}$.

If $\Phi : \mathbb{C} \longrightarrow \mathbb{C}$ is a holomorphic function, the real and imaginary parts of Φ are harmonic functions, and thus we know that the real and imaginary parts of $\Phi(B_t)$ are continuous local martingales. In fact, much more is true.

Theorem 7.18 *Let $\Phi : \mathbb{C} \longrightarrow \mathbb{C}$ be a nonconstant holomorphic function. For every $t \geq 0$, set*

$$C_t = \int_0^t |\Phi'(B_s)|^2 \, ds.$$

Let $z \in \mathbb{C}$. There exists a complex Brownian motion Γ that starts from $\Phi(z)$ under P_z, such that

$$\Phi(B_t) = \Gamma_{C_t}, \quad \text{for every } t \geq 0, \, P_z \text{ a.s.}$$

In other words, the image of complex Brownian motion under a holomorphic function is a time-changed complex Brownian motion. This is the **conformal invariance property** of planar Brownian motion. It is possible (and useful for many applications) to extend Theorem 7.18 to the case where Φ is defined and holomorphic in a domain D of $\mathbb{C}$ (such that $z \in D$). A similar representation then holds for $\Phi(B_t)$ up to the first exit time of D (see e.g. [18]).

Proof Let g and h stand respectively for the real and imaginary parts of Φ. Since g and h are harmonic, an application of Itô's formula gives under P_z,

$$g(B_t) = g(z) + \int_0^t \frac{\partial g}{\partial x}(B_s) \, dX_s + \int_0^t \frac{\partial g}{\partial y}(B_s) \, dY_s$$

and similarly

$$h(B_t) = h(z) + \int_0^t \frac{\partial h}{\partial x}(B_s) \, dX_s + \int_0^t \frac{\partial h}{\partial y}(B_s) \, dY_s.$$

So $M_t = g(B_t)$ and $N_t = h(B_t)$ are local martingales. Moreover, the Cauchy–Riemann equations

$$\frac{\partial g}{\partial x} = \frac{\partial h}{\partial y}, \; \frac{\partial g}{\partial y} = -\frac{\partial h}{\partial x}$$

give

$$\langle M, N \rangle_t = 0$$

and

$$\langle M, M \rangle_t = \langle N, N \rangle_t = \int_0^t |\Phi'(B_s)|^2 \, ds = C_t.$$

The recurrence of planar Brownian motion implies that $C_\infty = \infty$ a.s. (take a ball $\mathscr{B}$ where $|\Phi'|$ is bounded below by a positive constant, and note that the total time spent by B in the ball $\mathscr{B}$ is a.s. infinite). We can then apply Proposition 5.15 to $M_t - g(z)$ and $N_t - h(z)$ under P_z, and we find two independent real Brownian motions β and γ started from 0 such that $M_t = g(z) + \beta_{C_t}$ and $N_t = h(z) + \gamma_{C_t}$, for every $t \geq 0$, a.s. The desired result follows by setting $\Gamma_t = \Phi(z) + \beta_t + i \gamma_t$. $\square$

We will apply the conformal invariance property of planar Brownian motion to its decomposition in polar coordinates, which is known as the **skew-product representation**.

Theorem 7.19 *Let $z \in \mathbb{C} \backslash \{0\}$ and write $z = \exp(r + i\theta)$ where $r \in \mathbb{R}$ and $\theta \in (-\pi, \pi]$. There exist two independent linear Brownian motions β and γ that start respectively from r and from θ under P_z, such that we have P_z a.s. for every $t \geq 0$,*

$$B_t = \exp(\beta_{H_t} + i \gamma_{H_t}),$$

where

$$H_t = \int_0^t \frac{ds}{|B_s|^2}.$$

Proof The "natural" method for proving Theorem 7.19 would be to apply a generalized version of Theorem 7.18 to a suitable determination of the complex logarithm. This, however, leads to some technical difficulties, and for this reason we will argue differently.

We may assume that $z = 1$ (and thus $r = \theta = 0$). The general case can be reduced to that one using scaling and rotational invariance of Brownian motion. Let $\Gamma_t = \Gamma_t^1 + i \Gamma_t^2$ be a complex Brownian motion started from 0. By Theorem 7.18, we have a.s. for every $t \geq 0$,

$$\exp(\Gamma_t) = Z_{C_t}, \tag{7.5}$$

where Z is a complex Brownian motion started from 1, and for every $t \geq 0$,

$$C_t = \int_0^t \exp(2 \Gamma_s^1) \, ds.$$

Let $(H_s, s \geq 0)$ be the inverse function of $(C_t, t \geq 0)$, so that, by the formula for the derivative of an inverse function,

$$H_s = \int_0^s \exp(-2\,\Gamma^1_{H_u})\,du = \int_0^s \frac{du}{|Z_u|^2},$$

using the fact that $\exp(\Gamma^1_{H_u}) = |Z_u|$ in the last equality. By (7.5) with $t = H_s$, we now get

$$Z_s = \exp(\Gamma^1_{H_s} + i\,\Gamma^2_{H_s}).$$

This is the desired result (since Γ^1 and Γ^2 are independent linear Brownian motions started from 0) except we did not get it for B but for the complex Brownian motion Z introduced in the course of the argument.

To complete the proof, we argue as follows. Write $\arg B_t$ for the continuous determination of the argument of B_t such that $\arg B_0 = 0$ (this makes sense since we know that B does not visit 0, a.s.). The statement of Theorem 7.19 (with $z = 1$) is equivalent to saying that, if we set

$$\beta_t = \log\big|B_{\inf\{s \geq 0:\, \int_0^s |B_u|^{-2}du > t\}}\big|,$$

$$\gamma_t = \arg B_{\inf\{s \geq 0:\, \int_0^s |B_u|^{-2}du > t\}},$$

then β and γ are two independent real Brownian motions started from 0. Note that β and γ are deterministic functions of B, and so their law must be the same if we replace B by the complex Brownian motion Z. This gives the desired result.   □

Let us briefly comment on the skew-product representation. By writing H_t as the inverse of its inverse, we get

$$H_t = \inf\{s \geq 0 : \int_0^s \exp(2\beta_u)\,du > t\}, \qquad (7.6)$$

and it follows that

$$\log|B_t| = \beta_{\inf\{s \geq 0:\, \int_0^s \exp(2\beta_u)\,du > t\}},$$

showing that $|B|$ is completely determined by the linear Brownian motion β. This is related to the fact that $|B_t|$ is a Markov process, namely a two-dimensional Bessel process (cf. Exercise 6.24, and Sect. 8.4.3 for a brief discussion of Bessel processes).

On the other hand, write $\theta_t = \arg B_t = \gamma_{H_t}$. Then θ_t is not a Markov process: At least intuitively, this can be understood by the fact that the past of θ up to time t gives information on the current value of $|B_t|$ (for instance if θ_t oscillates very rapidly just before t this indicates that $|B_t|$ should be small) and therefore on the future evolution of the process θ.

7.6 Asymptotic Laws of Planar Brownian Motion

In this section, we apply the skew-product decomposition to certain asymptotic results for planar Brownian motion. We fix the starting point $z \in \mathbb{C}\backslash\{0\}$ (we will often take $z = 1$) and for simplicity we write P instead of P_z. We keep the notation $\theta_t = \arg B_t$ for a continuous determination of the argument of B_t. Although the process θ_t is not a Markov process, the fact that it can be written as a linear Brownian motion time-changed by an independent increasing process allows one to derive a lot of information about its path properties. For instance, since $H_t \longrightarrow \infty$ as $t \to \infty$, we immediately get from Proposition 2.14 that, a.s.,

$$\limsup_{t \to \infty} \theta_t = +\infty,$$

$$\liminf_{t \to \infty} \theta_t = -\infty.$$

One may then ask about the typical size of θ_t when t is large. This is the celebrated Spitzer theorem on the winding number of planar Brownian motion.

Theorem 7.20 *Let $(\theta_t, t \geq 0)$ be a continuous determination of the argument of the complex Brownian motion B started from $z \in \mathbb{C}\backslash\{0\}$. Then*

$$\frac{2}{\log t}\theta_t$$

converges in distribution as $t \to \infty$ to a standard symmetric Cauchy distribution. In other words, for every $x \in \mathbb{R}$,

$$\lim_{t \to \infty} P\Big(\frac{2}{\log t}\theta_t \leq x\Big) = \int_{-\infty}^{x} \frac{dy}{\pi(1+y^2)}.$$

Before proving the theorem, we will establish a key lemma. Without loss of generality, we may assume that $z = 1$ and $\theta_0 = 0$. We use the notation of Theorem 7.19, so that β and γ are two independent linear Brownian motions started from 0.

Lemma 7.21 *For every $\lambda > 0$, consider the scaled Brownian motion $\beta_t^{(\lambda)} = \frac{1}{\lambda}\beta_{\lambda^2 t}$, for every $t \geq 0$, and set $T_1^{(\lambda)} = \inf\{t \geq 0 : \beta_t^{(\lambda)} = 1\}$. Then*

$$\frac{4}{(\log t)^2} H_t - T_1^{((\log t)/2)} \xrightarrow[t \to \infty]{} 0$$

in probability.

Remark This shows in particular that $4(\log t)^{-2}H_t$ converges in distribution to the law of the hitting time of 1 by a linear Brownian motion started from 0.

Proof For every $a > 0$, set $T_a = \inf\{t \geq 0 : \beta_t = a\}$ and, for every $\lambda > 0$, $T_a^{(\lambda)} = \inf\{t \geq 0 : \beta_t^{(\lambda)} = a\}$. For the sake of simplicity, we write $\lambda_t = (\log t)/2$ throughout the proof and always assume that $t > 1$. We first verify that, for every $\varepsilon > 0$,

$$P\left((\lambda_t)^{-2} H_t > T_{1+\varepsilon}^{(\lambda_t)}\right) \xrightarrow[t \to \infty]{} 0. \tag{7.7}$$

To this end, recall formula (7.6), which shows that

$$\{(\lambda_t)^{-2} H_t > T_{1+\varepsilon}^{(\lambda_t)}\} = \left\{ \int_0^{(\lambda_t)^2 T_{1+\varepsilon}^{(\lambda_t)}} \exp(2\beta_u)\, du < t \right\}$$

$$= \left\{ \frac{1}{2\lambda_t} \log \int_0^{(\lambda_t)^2 T_{1+\varepsilon}^{(\lambda_t)}} \exp(2\beta_u)\, du < 1 \right\}, \tag{7.8}$$

since $2\lambda_t = \log t$. From the change of variables $u = (\lambda_t)^2 v$ in the integral, we get

$$\frac{1}{2\lambda_t} \log \int_0^{(\lambda_t)^2 T_{1+\varepsilon}^{(\lambda_t)}} \exp(2\beta_u)\, du = \frac{\log \lambda_t}{\lambda_t} + \frac{1}{2\lambda_t} \log \int_0^{T_{1+\varepsilon}^{(\lambda_t)}} \exp(2\lambda_t \beta_v^{(\lambda_t)})\, dv.$$

We then note that, for every fixed $t > 1$, the quantity in the right-hand side has the same distribution as

$$\frac{\log \lambda_t}{\lambda_t} + \frac{1}{2\lambda_t} \log \int_0^{T_{1+\varepsilon}} \exp(2\lambda_t \beta_v)\, dv \tag{7.9}$$

since for any $\lambda > 0$ the scaled Brownian motion $(\beta_t^{(\lambda)})_{t \geq 0}$ has the same distribution as $(\beta_t)_{t \geq 0}$. We then use the simple analytic fact stating that, for any continuous function $f : \mathbb{R}_+ \longrightarrow \mathbb{R}$, for any $s > 0$,

$$\frac{1}{2\lambda} \log \int_0^s \exp(2\lambda f(v))\, dv \xrightarrow[\lambda \to \infty]{} \sup_{0 \leq r \leq s} f(r).$$

We leave the proof as an exercise for the reader. It follows that

$$\frac{1}{2\lambda} \log \int_0^{T_{1+\varepsilon}} \exp(2\lambda \beta_v)\, dv \xrightarrow[\lambda \to \infty]{} \sup_{0 \leq r \leq T_{1+\varepsilon}} \beta_r = 1 + \varepsilon,$$

a.s., and so the quantity in (7.9) converges to $1 + \varepsilon$, a.s. as $t \to \infty$. Thus,

$$\frac{1}{2\lambda_t} \log \int_0^{(\lambda_t)^2 T_{1+\varepsilon}^{(\lambda_t)}} \exp(2\beta_u)\, du \xrightarrow[t \to \infty]{} 1 + \varepsilon$$

in probability. Hence the probability of the event in the right-hand side of (7.8) tends to 0, proving that (7.7) holds. The very same arguments show that, for every $\varepsilon \in (0, 1)$,

$$P\left((\lambda_t)^{-2} H_t < T_{1-\varepsilon}^{(\lambda_t)}\right) \underset{t \to \infty}{\longrightarrow} 0.$$

The desired result now follows, noting that $T_{1-\varepsilon}^{(\lambda_t)} < T_1^{(\lambda_t)} < T_{1+\varepsilon}^{(\lambda_t)}$ and that $T_{1+\varepsilon}^{(\lambda_t)} - T_{1-\varepsilon}^{(\lambda_t)}$ has the same distribution as $T_{1+\varepsilon} - T_{1-\varepsilon}$, which tends to 0 in probability when $\varepsilon \to 0$ (clearly, $T_{1-\varepsilon} \uparrow T_1$ as $\varepsilon \to 0$, and on the other hand $T_{1+\varepsilon} \downarrow T_1$ a.s. as $\varepsilon \to 0$, as a consequence of the strong Markov property at time T_1 and Proposition 2.14 (i)). □

Proof of Theorem 7.20 We keep the notation introduced in the preceding proof and also consider, for every $\lambda > 0$, the scaled Brownian motion $\gamma_t^{(\lambda)} = \frac{1}{\lambda} \gamma_{\lambda^2 t}$. Recalling our notation $\lambda_t = (\log t)/2$ for $t > 1$, we have

$$\frac{2}{\log t} \theta_t = \frac{1}{\lambda_t} \gamma_{H_t} = \gamma_{(\lambda_t)^{-2} H_t}^{(\lambda_t)}.$$

It then follows from Lemma 7.21 (using also the fact that the linear Brownian motions $\gamma^{(\lambda)}$ all have the same distribution) that

$$\frac{2}{\log t} \theta_t - \gamma_{T_1^{(\lambda_t)}}^{(\lambda_t)} \underset{t \to \infty}{\longrightarrow} 0,$$

in probability.

To complete the proof, we just have to notice that, for every fixed $\lambda > 0$, $\gamma_{T_1^{(\lambda)}}^{(\lambda)}$ has the standard symmetric Cauchy distribution. Indeed, since $(\beta^{(\lambda)}, \gamma^{(\lambda)})$ is a pair of independent linear Brownian motions started from 0, this variable has the same distribution as γ_{T_1}, and its characteristic distribution is computed by conditioning first with respect to T_1, and then using the Laplace transform of T_1 found in Example (c) after Corollary 3.24,

$$E[\exp(i\xi \gamma_{T_1})] = E[\exp(-\frac{1}{2} \xi^2 T_1)] = \exp(-|\xi|),$$

which we recognize as the characteristic function of the Cauchy distribution. □

The skew-product decomposition and Lemma 7.21 can be used to derive other asymptotic laws. We know that the planar Brownian motion B started from $z \neq 0$ does not hit 0 a.s., but on the other hand the recurrence property ensures that $\min\{|B_s| : 0 \leq s \leq t\}$ tends to 0 as $t \to \infty$. One may then ask about the typical size of $\min\{|B_s| : 0 \leq s \leq t\}$ when t is large: In other words, at which speed does planar Brownian motion approach a point different from its starting point?

Proposition 7.22 *Consider the planar Brownian motion B started from* $z \neq 0$. *Then, for every* $a > 0$,

$$\lim_{t \to \infty} P\left(\min_{0 \leq s \leq t} |B_s| \leq t^{-a/2} \right) = \frac{1}{1 + a}.$$

For instance, the probability that Brownian motion started from a nonzero initial value comes within distance $1/t$ from the origin before time t converges to $1/3$ as $t \to \infty$, a result which was not so easy to guess!

Proof Without loss of generality, we take $z = 1$. We keep the notation introduced in the proofs of Lemma 7.21 and Theorem 7.20. We observe that

$$\log \left(\min_{0 \leq s \leq t} |B_s| \right) = \min_{0 \leq s \leq t} \beta_{H_s} = \min_{0 \leq s \leq H_t} \beta_s.$$

It follows that

$$\frac{2}{\log t} \log \left(\min_{0 \leq s \leq t} |B_s| \right) = \frac{1}{\lambda_t} \min_{0 \leq s \leq H_t} \beta_s = \min_{0 \leq s \leq (\lambda_t)^{-2}} \beta_s^{(\lambda_t)}.$$

By Lemma 7.21,

$$\min_{0 \leq s \leq (\lambda_t)^{-2}} \beta_s^{(\lambda_t)} - \min_{0 \leq s \leq T_1^{(\lambda_t)}} \beta_s^{(\lambda_t)} \xrightarrow[t \to \infty]{} 0$$

in probability. We conclude that we have the following convergence in distribution,

$$\frac{2}{\log t} \log \left(\min_{0 \leq s \leq t} |B_s| \right) \xrightarrow[t \to \infty]{} \min_{0 \leq s \leq T_1} \beta_s,$$

where β is a linear Brownian motion started from 0 and $T_1 = \inf\{s \geq 0 : \beta_s = 1\}$. To complete the argument, note that

$$P\left(\min_{0 \leq s \leq T_1} \beta_s \leq -a \right) = P(T_{-a} < T_1),$$

if $T_{-a} = \inf\{s \geq 0 : \beta_s = -a\}$, and that $P(T_{-a} < T_1) = (1 + a)^{-1}$ (cf. Sect. 3.4). □

As a last application of the skew-product decomposition, we state the Kallianpur–Robbins asymptotic law for the time spent by Brownian motion in a ball. Here the initial value can be arbitrary.

Theorem 7.23 *Let $z \in \mathbb{C}$ and $R > 0$. Then, under P_z,*

$$\frac{2}{\log t} \int_0^t \mathbf{1}_{\{|B_s| < R\}} \, ds$$

converges in distribution as $t \to \infty$ to an exponential distribution with mean R^2.

We postpone our proof of Theorem 7.23 to the end of Chap. 9 since it relies in part on the theory of local times developed in that chapter.

Exercises

In all exercises, $(B_t)_{t \geq 0}$ is a d-dimensional Brownian motion starting from x under the probability measure P_x. Except in Exercise 7.28, we always assume that $d \geq 2$.

Exercise 7.24 Let $\mathscr{B}_1$ be the open unit ball of $\mathbb{R}^d$ ($d \geq 2$), and $\mathscr{B}_1^* = \mathscr{B}_1 \backslash \{0\}$. Let g be the continuous function defined on $\partial \mathscr{B}_1^*$ by $g(x) = 0$ if $|x| = 1$ and $g(0) = 1$. Prove that the Dirichlet problem in $\mathscr{B}_1^*$ with boundary condition g has no solution.

Exercise 7.25 (*Polar sets*) Throughout this exercise, we consider a nonempty compact subset K of $\mathbb{R}^d$ ($d \geq 2$). We set $T_K = \inf\{t \geq 0 : B_t \in K\}$. We say that K is *polar* if there exists an $x \in K^c$ such that $P_x(T_K < \infty) = 0$.

1. Using the strong Markov property as in the proof of Proposition 7.7 (ii), prove that the function $x \mapsto P_x(T_K < \infty)$ is harmonic on every connected component of K^c.
2. From now on until question 4., we assume that K is polar. Prove that K^c is connected, and that the property $P_x(T_K < \infty) = 0$ holds for every $x \in K^c$. (*Hint*: Observe that $\{x \in K^c : P_x(T_K < \infty) = 0\}$ is both open and closed).
3. Let D be a bounded domain containing K, and $D' = D \backslash K$. Prove that any bounded harmonic function h on D' can be extended to a harmonic function on D. Does this remain true if the word "bounded" is replaced by "positive"?
4. Set $g(x) = 0$ if $x \in \partial D$ and $g(x) = 1$ if $x \in \partial D' \backslash \partial D$. Prove that the Dirichlet problem in D' with boundary condition g has no solution. (Note that this generalizes the result of Exercise 7.24.)
5. If $\alpha \in (0, d]$, we say that the compact set K has zero α-dimensional Hausdorff measure if, for every $\varepsilon > 0$, we can find an integer $N_\varepsilon \geq 1$ and N_ε open balls $\mathscr{B}_{(1)}, \ldots, \mathscr{B}_{(N_\varepsilon)}$ with respective radii $r_{(1)}, \ldots r_{(N_\varepsilon)}$, such that K is contained in the union $\mathscr{B}_{(1)} \cup \cdots \cup \mathscr{B}_{(N_\varepsilon)}$, and

$$\sum_{j=1}^{N_\varepsilon} (r_j)^\alpha \leq \varepsilon.$$

Prove that if $d \geq 3$ and K has zero $d - 2$-dimensional Hausdorff measure then K is polar.

Exercise 7.26 In this exercise, $d \geq 3$. Let K be a compact subset of the open unit ball of $\mathbb{R}^d$, and $T_K := \inf\{t \geq 0 : B_t \in K\}$. We assume that $D := \mathbb{R}^d \backslash K$ is connected. We also consider a function g defined and continuous on K. The goal of the exercise is to determine all functions $u : \bar{D} \to \mathbb{R}$ that satisfy:

(P) u is bounded and continuous on $\bar{D}$, harmonic on D, and $u(y) = g(y)$ if $y \in \partial D$.

(This is the Dirichlet problem in D, but in contrast with Sect. 7.3 above, D is unbounded here.) We fix an increasing sequence $(R_n)_{n \geq 1}$ of reals, with $R_1 \geq 1$ and $R_n \uparrow \infty$ as $n \to \infty$. For every $n \geq 1$, we set $T_{(n)} := \inf\{t \geq 0 : |B_t| \geq R_n\}$.

1. Suppose that u satisfies (P). Prove that, for every $n \geq 1$ and every $x \in D$ such that $|x| < R_n$,

$$u(x) = E_x[g(B_{T_K}) \mathbf{1}_{\{T_K \leq T_{(n)}\}}] + E_x[u(B_{T_{(n)}}) \mathbf{1}_{\{T_{(n)} \leq T_K\}}].$$

2. Show that, by replacing the sequence $(R_n)_{n \geq 1}$ with a subsequence if necessary, we may assume that there exists a constant $\alpha \in \mathbb{R}$ such that, for every $x \in D$,

$$\lim_{n \to \infty} E_x[u(B_{T_{(n)}})] = \alpha,$$

and that we then have

$$\lim_{|x| \to \infty} u(x) = \alpha.$$

3. Show that, for every $x \in D$,

$$u(x) = E_x[g(B_{T_K}) \mathbf{1}_{\{T_K < \infty\}}] + \alpha\, P_x(T_K = \infty).$$

4. Assume that D satisfies the exterior cone condition at every $y \in \partial D$ (this is defined in the same way as when D is bounded). Show that, for any choice of $\alpha \in \mathbb{R}$, the formula of question 3. gives a solution of the problem (P).

Exercise 7.27 Let $f : \mathbb{C} \to \mathbb{C}$ be a nonconstant holomorphic function. Use planar Brownian motion to prove that the set $\{f(z) : z \in \mathbb{C}\}$ is dense in $\mathbb{C}$. (Much more is true, since Picard's little theorem asserts that the complement of $\{f(z) : z \in \mathbb{C}\}$ in $\mathbb{C}$ contains at most one point: This can also be proved using Brownian motion, but the argument is more involved, see [12].)

Exercise 7.28 (*Feynman–Kac formula for Brownian motion*) This is a continuation of Exercise 6.26 in Chap. 6. With the notation of this exercise, we assume that $E = \mathbb{R}^d$ and $X_t = B_t$. Let v be a nonnegative function in $C_0(\mathbb{R}^d)$, and assume that v is continuously differentiable with bounded first derivatives. As in Exercise 6.26, set, for every $\varphi \in B(\mathbb{R}^d)$,

$$Q_t^* \varphi(x) = E_x\left[\varphi(X_t) \exp\left(-\int_0^t v(X_s)\, ds\right)\right].$$

1. Using the formula derived in question **2.** of Exercise 6.26, prove that, for every $t > 0$, and every $\varphi \in C_0(\mathbb{R}^d)$, the function $Q_t^* \varphi$ is twice continuously differentiable on $\mathbb{R}^d$, and that $Q_t^* \varphi$ and its partial derivatives up to order 2 belong to $C_0(\mathbb{R}^d)$. Conclude that $Q_t^* \varphi \in D(L)$.

2. Let $\varphi \in C_0(\mathbb{R}^d)$ and set $u_t(x) = Q_t^* \varphi(x)$ for every $t > 0$ and $x \in \mathbb{R}^d$. Using question **3.** of Exercise 6.26, prove that, for every $x \in \mathbb{R}^d$, the function $t \mapsto u_t(x)$ is continuously differentiable on $(0, \infty)$, and

$$\frac{\partial u_t}{\partial t} = \frac{1}{2} \Delta u_t - v \, u_t.$$

Exercise 7.29 In this exercise $d = 2$ and $\mathbb{R}^2$ is identified with the complex plane $\mathbb{C}$. Let $\alpha \in (0, \pi)$, and consider the open cone

$$\mathscr{C}_\alpha = \{ r e^{i\theta} : r > 0, \theta \in (-\alpha, \alpha) \}.$$

Set $T := \inf\{t \geq 0 : B_t \notin \mathscr{C}_\alpha\}$.

1. Show that the law of $\log |B_T|$ under P_1 is the law of $\beta_{\inf\{t \geq 0 : |\gamma_t| = \alpha\}}$, where β and γ are two independent linear Brownian motions started from 0.

2. Verify that, for every $\lambda \in \mathbb{R}$,

$$E_1[e^{i\lambda \log |B_T|}] = \frac{1}{\cosh(\alpha\lambda)}.$$

Notes and Comments

Connections between Brownian motion and partial differential equations have been known for a long time and motivated the study of this random process. A survey of the partial differential equations that can be solved in terms of Brownian motion can be found in the book of Durrett [18, Chapter 8]. The representation of Theorem 7.1 (written in terms of the Gaussian density) goes back to the ninetieth century and the work of Fourier and Laplace – see the references in [49]. The beautiful relations between Brownian motion and harmonic functions were discovered and studied by Kakutani [45, 46], and Hunt [33, 34] later studied the connections between potential theory and transient Markov processes (see the Blumenthal–Getoor book [5] for more on this topic). Nice accounts of the links between Brownian motion and classical potential theory can be found in the books by Port and Stone [69] and Doob [16] (see also Itô and McKean [42], Chung [9], and Chapters 3 and 8 of [62]). The conformal invariance of planar Brownian motion was stated by Lévy [54] with a very sketchy proof. Davis' paper [12] is a nice survey of relations between planar Brownian motion and analytic functions, see also Durrett's book [18], and the paper [28] by Getoor and Sharpe for a notion of conformal martingale that plays in martingale theory a role similar to that of analytic functions. Spitzer's theorem was obtained in the classical paper [75], and the Kallianpur–Robbins law was derived in [48]. A number of remarkable properties of planar Brownian motion had already been observed by Lévy [53] in 1940. We refer to Pitman and Yor [68] for a systematic study of asymptotic laws of planar Brownian motion. Our presentation closely follows [52], where other applications of the skew-product decomposition can be found.

Chapter 8
Stochastic Differential Equations

This chapter is devoted to stochastic differential equations, which motivated Itô's construction of stochastic integrals. After giving the general definitions, we provide a detailed treatment of the Lipschitz case, where strong existence and uniqueness statements hold. Still in the Lipschitz case, we show that the solution of a stochastic differential equation is a Markov process with a Feller semigroup, whose generator is a second-order differential operator. By results of Chap. 6, the Feller property immediately gives the strong Markov property of solutions of stochastic differential equations. The last section presents a few important examples. This chapter can be read independently of Chap. 7.

8.1 Motivation and General Definitions

The goal of stochastic differential equations is to provide a model for a differential equation perturbed by a random noise. Consider an ordinary differential equation of the form

$$y'(t) = b(y(t)),$$

or, in differential form,

$$dy_t = b(y_t)\, dt.$$

Such an equation is used to model the evolution of a physical system. If we take random perturbations of the system into account, we add a noise term, which is typically of the form $\sigma\, dB_t$, where B denotes a Brownian motion, and σ is a constant corresponding to the intensity of the noise. Note that the use of Brownian motion here is justified by its property of independence of increments, corresponding to the

© Springer International Publishing Switzerland 2016

J.-F. Le Gall, *Brownian Motion, Martingales, and Stochastic Calculus*,
Graduate Texts in Mathematics 274, DOI 10.1007/978-3-319-31089-3_8

fact that the random perturbations affecting disjoint time intervals are assumed to be independent.

In this way, we arrive at a stochastic differential equation of the form

$$dy_t = b(y_t)\, dt + \sigma\, dB_t,$$

or in integral form, the only one with a rigorous mathematical meaning,

$$y_t = y_0 + \int_0^t b(y_s)\, ds + \sigma\, B_t.$$

We generalize the preceding equation by allowing σ to depend on the state of the system at time t :

$$dy_t = b(y_t)\, dt + \sigma(y_t)\, dB_t,$$

or, in integral form,

$$y_t = y_0 + \int_0^t b(y_s)\, ds + \int_0^t \sigma(y_s)\, dB_s.$$

Because of the integral in dB_s, the preceding equation only makes sense thanks to the theory of stochastic integrals developed in Chap. 5. We can still generalize the preceding equation by allowing σ and b to depend on the time parameter t. This leads to the following definition.

Definition 8.1 Let d and m be positive integers, and let σ and b be locally bounded measurable functions defined on $\mathbb{R}_+ \times \mathbb{R}^d$ and taking values in $M_{d\times m}(\mathbb{R})$ and in $\mathbb{R}^d$ respectively, where $M_{d\times m}(\mathbb{R})$ is the set of all $d \times m$ matrices with real coefficients. We write $\sigma = (\sigma_{ij})_{1\le i\le d, 1\le j\le m}$ and $b = (b_i)_{1\le i\le d}$.

A *solution of the stochastic differential equation*

$$E(\sigma, b) \qquad\qquad dX_t = \sigma(t, X_t)\, dB_t + b(t, X_t)\, dt$$

consists of:

- a filtered probability space $(\Omega, \mathcal{F}, (\mathcal{F}_t)_{t\in[0,\infty]}, P)$ (where the filtration is always assumed to be complete);
- an m-dimensional $(\mathcal{F}_t)$-Brownian motion $B = (B^1, \dots, B^m)$ started from 0;
- an $(\mathcal{F}_t)$-adapted process $X = (X^1, \dots, X^d)$ with values in $\mathbb{R}^d$, with continuous sample paths, such that

$$X_t = X_0 + \int_0^t \sigma(s, X_s)\, dB_s + \int_0^t b(s, X_s)\, ds,$$

meaning that, for every $i \in \{1, \ldots, d\}$,

$$X_t^i = X_0^i + \sum_{j=1}^{m} \int_0^t \sigma_{ij}(s, X_s) \, dB_s^j + \int_0^t b_i(s, X_s) \, ds.$$

If additionally $X_0 = x \in \mathbb{R}^d$, we say that X is a solution of $E_x(\sigma, b)$.

Note that, when we speak about a solution of $E(\sigma, b)$, we do not fix a priori the filtered probability space and the Brownian motion B. When we fix these objects, we will say so explicitly.

There are several notions of existence and uniqueness for stochastic differential equations.

Definition 8.2 For the equation $E(\sigma, b)$ we say that there is

- *weak existence* if, for every $x \in \mathbb{R}^d$, there exists a solution of $E_x(\sigma, b)$;
- *weak existence and weak uniqueness* if in addition, for every $x \in \mathbb{R}^d$, all solutions of $E_x(\sigma, b)$ have the same law;
- *pathwise uniqueness* if, whenever the filtered probability space $(\Omega, \mathscr{F}, (\mathscr{F}_t), P)$ and the $(\mathscr{F}_t)$-Brownian motion B are fixed, two solutions X and X' such that $X_0 = X_0'$ a.s. are indistinguishable.

Furthermore, we say that a solution X of $E_x(\sigma, b)$ is a *strong solution* if X is adapted with respect to the completed canonical filtration of B.

Remark It may happen that weak existence and weak uniqueness hold but pathwise uniqueness fails. For a simple example, consider a real Brownian motion β started from $\beta_0 = y$, and set

$$B_t = \int_0^t \operatorname{sgn}(\beta_s) \, d\beta_s,$$

where $\operatorname{sgn}(x) = 1$ if $x > 0$ and $\operatorname{sgn}(x) = -1$ if $x \le 0$. Then, one immediately gets from the "associativity" of stochastic integrals that

$$\beta_t = y + \int_0^t \operatorname{sgn}(\beta_s) \, dB_s.$$

Moreover, B is a continuous martingale with quadratic variation $\langle B, B \rangle_t = t$, and Theorem 5.12 shows that B is a Brownian motion started from 0. We thus see that β solves the stochastic differential equation

$$dX_t = \operatorname{sgn}(X_t) \, dB_t, \qquad X_0 = y,$$

and it follows that weak existence holds for this equation. Theorem 5.12 again shows that any other solution of this equation must be a Brownian motion, which gives weak uniqueness. On the other hand, pathwise uniqueness fails. In fact, taking $y = 0$ in the construction, one easily sees that both β and $-\beta$ solve the preceding stochastic differential equation with the same Brownian motion B and initial value 0 (note that $\int_0^t \mathbf{1}_{\{\beta_s=0\}} \, ds = 0$, which implies $\int_0^t \mathbf{1}_{\{\beta_s=0\}} \, dB_s = 0$). One can also show that β is not a strong solution: One verifies that the canonical filtration of B coincides with the canonical filtration of $|\beta|$, which is strictly smaller than that of β (we omit the proof, which is a simple application of formula (9.18) in Chap. 9).

The next theorem links the different notions of existence and uniqueness.

Theorem (Yamada–Watanabe) *If both weak existence and pathwise uniqueness hold, then weak uniqueness also holds. Moreover, for any choice of the filtered probability space $(\Omega, \mathscr{F}, (\mathscr{F}_t), P)$ and of the $(\mathscr{F}_t)$-Brownian motion B, there exists for every $x \in \mathbb{R}^d$ a (unique) strong solution of $E_x(\sigma, b)$.*

We omit the proof (see Yamada and Watanabe [83]) because we will not need this result. In the Lipschitz case that we will consider, we will establish directly the properties given by the Yamada–Watanabe theorem.

8.2 The Lipschitz Case

In this section, we work under the following assumptions.

Assumptions The functions σ are b are continuous on $\mathbb{R}_+ \times \mathbb{R}^d$ and Lipschitz in the variable x: There exists a constant K such that, for every $t \geq 0$, $x, y \in \mathbb{R}^d$,

$$|\sigma(t, x) - \sigma(t, y)| \leq K |x - y|,$$

$$|b(t, x) - b(t, y)| \leq K |x - y|.$$

Theorem 8.3 *Under the preceding assumptions, pathwise uniqueness holds for $E(\sigma, b)$, and, for every choice of the filtered probability space $(\Omega, \mathscr{F}, (\mathscr{F}_t), P)$ and of the $(\mathscr{F}_t)$-Brownian motion B, for every $x \in \mathbb{R}^d$, there exists a (unique) strong solution of $E_x(\sigma, b)$.*

The theorem implies in particular that weak existence holds for $E(\sigma, b)$. Weak uniqueness will follow from the next theorem (it can also be deduced from pathwise uniqueness using the Yamada–Watanabe theorem).

Remark One can "localize" the Lipschitz assumption on σ and b, meaning that the constant K may depend on the compact set on which the parameters t and x, y are considered. In that case, it is, however, necessary to keep a condition of linear growth of the form

$$|\sigma(t, x)| \leq K(1 + |x|), \quad |b(t, x)| \leq K(1 + |x|).$$

This kind of condition, which avoids the blow-up of solutions, already appears in ordinary differential equations.

Proof For the sake of simplicity, we consider only the case $d = m = 1$. The reader will be able to check that the general case follows from exactly the same arguments, at the cost of a heavier notation. Let us start by proving pathwise uniqueness. We consider (on the same filtered probability space, with the same Brownian motion B) two solutions X and X' such that $X_0 = X_0'$. Fix $M > 0$ and set

$$\tau = \inf\{t \geq 0 : |X_t| \geq M \text{ or } |X_t'| \geq M\}.$$

Then, for every $t \geq 0$,

$$X_{t \wedge \tau} = X_0 + \int_0^{t \wedge \tau} \sigma(s, X_s) \, dB_s + \int_0^{t \wedge \tau} b(s, X_s) \, ds$$

and an analogous equation holds for $X_{t \wedge \tau}'$. Fix a constant $T > 0$. By considering the difference between the two equations and using the bound (5.14), we get, for $t \in [0, T]$,

$$E[(X_{t \wedge \tau} - X_{t \wedge \tau}')^2]$$

$$\leq 2E\Big[\Big(\int_0^{t \wedge \tau} (\sigma(s, X_s) - \sigma(s, X_s')) \, dB_s\Big)^2\Big] + 2E\Big[\Big(\int_0^{t \wedge \tau} (b(s, X_s) - b(s, X_s')) \, ds\Big)^2\Big]$$

$$\leq 2\Big(E\Big[\int_0^{t \wedge \tau} (\sigma(s, X_s) - \sigma(s, X_s'))^2 ds\Big] + T E\Big[\int_0^{t \wedge \tau} (b(s, X_s) - b(s, X_s'))^2 ds\Big]\Big)$$

$$\leq 2K^2(1 + T) E\Big[\int_0^{t \wedge \tau} (X_s - X_s')^2 ds\Big]$$

$$\leq 2K^2(1 + T) E\Big[\int_0^t (X_{s \wedge \tau} - X_{s \wedge \tau}')^2 ds\Big].$$

Hence the function $h(t) = E[(X_{t \wedge \tau} - X_{t \wedge \tau}')^2]$ satisfies

$$h(t) \leq C \int_0^t h(s) \, ds$$

for every $t \in [0, T]$, with $C = 2K^2(1 + T)$.

Lemma 8.4 (Gronwall's lemma) *Let $T > 0$ and let g be a nonnegative bounded measurable function on $[0, T]$. Assume that there exist two constants $a \geq 0$ and $b \geq 0$ such that, for every $t \in [0, T]$,*

$$g(t) \leq a + b \int_0^t g(s) \, ds.$$

Then, we also have, for every $t \in [0, T]$,

$$g(t) \leq a \exp(bt).$$

Proof of the lemma By iterating the condition on g, we get,

$$g(t) \leq a + a(bt) + b^2 \int_0^t \mathrm{d}s \int_0^s \mathrm{d}r \, g(r)$$

$$\leq a + a(bt) + a\frac{(bt)^2}{2} + \cdots + a\frac{(bt)^n}{n!} + b^{n+1} \int_0^t \mathrm{d}s_1 \int_0^{s_1} \mathrm{d}s_2 \cdots \int_0^{s_n} \mathrm{d}s_{n+1} g(s_{n+1}),$$

for every $n \geq 1$. If A is a constant such that $0 \leq g \leq A$, the last term in the right-hand side is bounded above by $A(bt)^{n+1}/(n+1)!$, hence tends to 0 as $n \to \infty$. The desired result now follows. $\square$

Let us return to the proof of the theorem. The function h is bounded above by $4M^2$ and the assumption of the lemma holds with $a = 0, b = C$. We thus get $h = 0$, so that $X_{t \wedge \tau} = X'_{t \wedge \tau}$. By letting M tend to ∞, we get $X_t = X'_t$, which completes the proof of pathwise uniqueness.

For the second assertion, we construct a solution using Picard's approximation method. We define by induction

$$X_t^0 = x,$$

$$X_t^1 = x + \int_0^t \sigma(s, x) \, \mathrm{d}B_s + \int_0^t b(s, x) \, \mathrm{d}s,$$

$$X_t^n = x + \int_0^t \sigma(s, X_s^{n-1}) \, \mathrm{d}B_s + \int_0^t b(s, X_s^{n-1}) \, \mathrm{d}s.$$

The stochastic integrals are well defined since one verifies by induction that, for every n, the process X^n is adapted and has continuous sample paths.

It is enough to show that, for every $T > 0$, there is a strong solution of $E_x(\sigma, b)$ on the time interval $[0, T]$. Indeed, the uniqueness part of the argument will then allow us to get a (unique) strong solution on $\mathbb{R}_+$ that will coincide with the solution on $[0, T]$ up to time T.

We fix $T > 0$ and, for every $n \geq 1$ and every $t \in [0, T]$, we set

$$g_n(t) = E\left[\sup_{0 \leq s \leq t} |X_s^n - X_s^{n-1}|^2 \right].$$

We will bound the functions g_n by induction on n (at present, it is not yet clear that these functions are finite). The fact that the functions $\sigma(\cdot, x)$ and $b(\cdot, x)$ are continuous, hence bounded, over $[0, T]$ implies that there exists a constant C'_T such that $g_1(t) \leq C'_T$ for every $t \in [0, T]$ (use Doob's inequality in L^2 for the stochastic integral term).

Then we observe that

$$X_t^{n+1} - X_t^n = \int_0^t (\sigma(s, X_s^n) - \sigma(s, X_s^{n-1}))\, dB_s + \int_0^t (b(s, X_s^n) - b(s, X_s^{n-1}))\, ds.$$

Hence, using the case $p = 2$ of the Burkholder–Davis–Gundy inequalities in the second bound (and writing $C_{(2)}$ for the constant in this inequality),

$$
\begin{aligned}
E\Big[\sup_{0 \le s \le t} |X_s^{n+1} - X_s^n|^2\Big] &\le 2\, E\Big[\sup_{0 \le s \le t}\Big|\int_0^s (\sigma(u, X_u^n) - \sigma(u, X_u^{n-1}))\, dB_u\Big|^2 \\
&\qquad + \sup_{0 \le s \le t}\Big|\int_0^s (b(u, X_u^n) - b(u, X_u^{n-1}))\, du\Big|^2\Big] \\
&\le 2\Big(C_{(2)}\, E\Big[\int_0^t (\sigma(u, X_u^n) - \sigma(u, X_u^{n-1}))^2\, du\Big] \\
&\qquad + T\, E\Big[\int_0^t (b(u, X_u^n) - b(u, X_u^{n-1}))^2\, du\Big]\Big) \\
&\le 2(C_{(2)} + T)K^2\, E\Big[\int_0^t |X_u^n - X_u^{n-1}|^2\, du\Big] \\
&\le C_T\, E\Big[\int_0^t \sup_{0 \le r \le u} |X_r^n - X_r^{n-1}|^2\, du\Big]
\end{aligned}
$$

where $C_T = 2(C_{(2)} + T)K^2$. We have thus obtained that, for every $n \ge 1$,

$$g_{n+1}(t) \le C_T \int_0^t g_n(u)\, du. \tag{8.1}$$

Recalling that $g_1(t) \le C_T'$, an induction argument using (8.1) shows that, for every $n \ge 1$ and $t \in [0, T]$,

$$g_n(t) \le C_T'(C_T)^{n-1} \frac{t^{n-1}}{(n-1)!}.$$

In particular, $\sum_{n=1}^\infty g_n(T)^{1/2} < \infty$, which implies that

$$\sum_{n=0}^\infty \sup_{0 \le t \le T} |X_t^{n+1} - X_t^n| < \infty, \quad \text{a.s.}$$

Hence the sequence $(X_t^n, 0 \le t \le T)$ converges uniformly on $[0, T]$, a.s., to a limiting process $(X_t, 0 \le t \le T)$, which has continuous sample paths. By induction, one also verifies that, for every n, X^n is adapted with respect to the (completed) canonical filtration of B, and the same holds for X.

Finally, from the fact that σ and b are Lipschitz in the variable x, we also get that, for every $t \in [0, T]$,

$$\lim_{n \to \infty} \left(\int_0^t \sigma(s, X_s) \, dB_s - \int_0^t \sigma(s, X_s^n) \, dB_s \right) = 0,$$

$$\lim_{n \to \infty} \left(\int_0^t b(s, X_s) \, ds - \int_0^t b(s, X_s^n) \, ds \right) = 0,$$

in probability (to deal with the stochastic integrals, we may use Proposition 5.8, noting that $|X_s^n - X_s|$ is dominated by $\sum_{k=0}^{\infty} \sup_{0 \le r \le s} |X_r^{k+1} - X_r^k|$). By passing to the limit in the induction equation defining X^n, we get that X solves $E_x(\sigma, b)$ on $[0, T]$. This completes the proof of the theorem. $\square$

In the following statement, $W(dw)$ stands for the Wiener measure on the canonical space $C(\mathbb{R}_+, \mathbb{R}^m)$ of all continuous functions from $\mathbb{R}_+$ into $\mathbb{R}^m$ ($W(dw)$ is the law of $(B_t, t \ge 0)$ if B is an m-dimensional Brownian motion started from 0).

Theorem 8.5 *Under the assumptions of the preceding theorem, there exists, for every $x \in \mathbb{R}$, a mapping $F_x : C(\mathbb{R}_+, \mathbb{R}^m) \longrightarrow C(\mathbb{R}_+, \mathbb{R}^d)$, which is measurable when $C(\mathbb{R}_+, \mathbb{R}^m)$ is equipped with the Borel σ-field completed by the W-negligible sets, and $C(\mathbb{R}_+, \mathbb{R}^d)$ is equipped with the Borel σ-field, such that the following properties hold:*

(i) *for every $t \ge 0$, $F_x(w)_t$ coincides $W(dw)$ a.s. with a measurable function of $(w(r), 0 \le r \le t)$;*

(ii) *for every $w \in C(\mathbb{R}_+, \mathbb{R}^m)$, the mapping $x \mapsto F_x(w)$ is continuous;*

(iii) *for every $x \in \mathbb{R}^d$, for every choice of the (complete) filtered probability space $(\Omega, \mathscr{F}, (\mathscr{F}_t), P)$ and of the m-dimensional $(\mathscr{F}_t)$-Brownian motion B with $B_0 = 0$, the process X_t defined $X_t = F_x(B)_t$ is the unique solution of $E_x(\sigma, b)$; furthermore, if U is an $\mathscr{F}_0$-measurable real random variable, the process $F_U(B)_t$ is the unique solution with $X_0 = U$.*

Remark Assertion (iii) implies in particular that weak uniqueness holds for $E(\sigma, b)$: any solution of $E_x(\sigma, b)$ must be of the form $F_x(B)$ and its law is thus uniquely determined as the image of $W(dw)$ under F_x.

Proof Again we consider only the case $d = m = 1$. Let $\mathscr{N}$ be the class of all W-negligible sets in $C(\mathbb{R}_+, \mathbb{R})$, and, for every $t \in [0, \infty]$, set

$$\mathscr{G}_t = \sigma(w(s), 0 \le s \le t) \vee \mathscr{N}.$$

For every $x \in \mathbb{R}$, we write X^x for the solution of $E_x(\sigma, b)$ corresponding to the filtered probability space $(C(\mathbb{R}_+, \mathbb{R}), \mathscr{G}_\infty, (\mathscr{G}_t), W)$ and the (canonical) Brownian motion $B_t(w) = w(t)$. This solution exists and is unique (up to indistinguishability) by Theorem 8.3, noting that the filtration $(\mathscr{G}_t)$ is complete by construction.

Let $x, y \in \mathbb{R}$ and let T_n be the stopping time defined by

$$T_n = \inf\{t \geq 0 : |X_t^x| \geq n \text{ or } |X_t^y| \geq n\}.$$

Let $p \geq 2$ and $T \geq 1$. Using the Burkholder–Davis–Gundy inequalities (Theorem 5.16) and then the Hölder inequality, we get, for $t \in [0, T]$,

$$E\Big[\sup_{s \leq t} |X_{s \wedge T_n}^x - X_{s \wedge T_n}^y|^p\Big]$$

$$\leq C_p\Big(|x - y|^p + E\Big[\sup_{s \leq t}\Big|\int_0^{s \wedge T_n} (\sigma(r, X_r^x) - \sigma(r, X_r^y))\, dB_r\Big|^p\Big]$$

$$+ E\Big[\sup_{s \leq t}\Big|\int_0^{s \wedge T_n} (b(r, X_r^x) - b(r, X_r^y))\, dr\Big|^p\Big]\Big)$$

$$\leq C_p\Big(|x - y|^p + C_p' E\Big[\Big(\int_0^{t \wedge T_n} (\sigma(r, X_r^x) - \sigma(r, X_r^y))^2 dr\Big)^{p/2}\Big]$$

$$+ E\Big[\Big(\int_0^{t \wedge T_n} |b(r, X_r^x) - b(r, X_r^y)|\, dr\Big)^p\Big]\Big)$$

$$\leq C_p\Big(|x - y|^p + C_p' t^{\frac{p}{2}-1} E\Big[\int_0^t |\sigma(r \wedge T_n, X_{r \wedge T_n}^x) - \sigma(r \wedge T_n, X_{r \wedge T_n}^y)|^p dr\Big]$$

$$+ t^{p-1} E\Big[\int_0^t |b(r \wedge T_n, X_{r \wedge T_n}^x) - b(r \wedge T_n, X_{r \wedge T_n}^y)|^p dr\Big]\Big)$$

$$\leq C_p''\Big(|x - y|^p + T^p \int_0^t E[|X_{r \wedge T_n}^x - X_{r \wedge T_n}^y|^p]\, dr\Big),$$

where the constants $C_p, C_p', C_p'' < \infty$ depend on p (and on the constant K appearing in our assumption on σ and b) but not on n or on x, y and T.

As the function $t \mapsto E\Big[\sup_{s \leq t} |X_{s \wedge T_n}^x - X_{s \wedge T_n}^y|^p\Big]$ is bounded, Lemma 8.4 implies that, for $t \in [0, T]$,

$$E\Big[\sup_{s \leq t} |X_{s \wedge T_n}^x - X_{s \wedge T_n}^y|^p\Big] \leq C_p''|x - y|^p \exp(C_p'' T^p t),$$

hence, letting n tend to ∞,

$$E\Big[\sup_{s \leq t} |X_s^x - X_s^y|^p\Big] \leq C_p''|x - y|^p \exp(C_p'' T^p t).$$

The topology on the space $C(\mathbb{R}_+, \mathbb{R})$ is defined by the distance

$$\mathbf{d}(w, w') = \sum_{k=1}^{\infty} \alpha_k\Big(\sup_{s \leq k} |w(s) - w'(s)| \wedge 1\Big),$$

where the sequence of positive reals α_k can be chosen in an arbitrary way, provided that the series $\sum \alpha_k$ converges. We may choose the coefficients α_k so that

$$\sum_{k=1}^{\infty} \alpha_k \exp(C_p'' k^{p+1}) < \infty.$$

For every $x \in \mathbb{R}$, we consider X^x as a random variable with values in $C(\mathbb{R}_+, \mathbb{R})$. The preceding estimates and Jensen's inequality then show that

$$E[\mathbf{d}(X^x, X^y)^p] \leq \left(\sum_{k=1}^{\infty} \alpha_k \right)^{p-1} \sum_{k=1}^{\infty} \alpha_k E\left[\sup_{s \leq k} |X_s^x - X_s^y|^p \right] \leq \bar{C}_p |x - y|^p,$$

with a constant $\bar{C}_p$ independent of x and y. By Kolmogorov's lemma (Theorem 2.9), applied to the process $(X^x, x \in \mathbb{R})$ with values in the space $E = C(\mathbb{R}_+, \mathbb{R})$ equipped with the distance $\mathbf{d}$, we get that $(X^x, x \in \mathbb{R})$ has a modification with continuous sample paths, which we denote by $(\tilde{X}^x, x \in \mathbb{R})$. We set $F_x(\mathrm{w}) = \tilde{X}^x(\mathrm{w}) = (\tilde{X}_t^x(\mathrm{w}))_{t \geq 0}$. Property (ii) is then obvious.

The mapping $\mathrm{w} \mapsto F_x(\mathrm{w})$ is measurable from $C(\mathbb{R}_+, \mathbb{R})$ equipped with the σ-field $\mathscr{G}_\infty$ into $C(\mathbb{R}_+, \mathbb{R})$ equipped with the Borel σ-field $\mathscr{C} = \sigma(\mathrm{w}(s), s \geq 0)$. Moreover, for every $t \geq 0$, $F_x(\mathrm{w})_t = \tilde{X}_t^x(\mathrm{w}) \overset{\mathrm{a.s.}}{=} X_t^x(\mathrm{w})$ is $\mathscr{G}_t$-measurable hence coincides $W(d\mathrm{w})$ a.s. with a measurable function of $(\mathrm{w}(s), 0 \leq s \leq t)$. Thus property (i) holds.

Let us now prove the first part of assertion (iii). To this end, we fix the filtered probability space $(\Omega, \mathscr{F}, (\mathscr{F}_t), P)$ and the $(\mathscr{F}_t)$-Brownian motion B. We need to verify that the process $(F_x(B)_t)_{t \geq 0}$ then solves $E_x(\sigma, b)$. This process (trivially) has continuous sample paths, and is also adapted since $F_x(B)_t$ coincides a.s. with a measurable function of $(B_r, 0 \leq r \leq t)$, by (i), and since the filtration $(\mathscr{F}_t)$ is complete. On the other hand, by the construction of F_x (and because $\tilde{X}^x = X^x$ a.s.), we have, for every $t \geq 0$, $W(d\mathrm{w})$ a.s.

$$F_x(\mathrm{w})_t = x + \int_0^t \sigma(s, F_x(\mathrm{w})_s) \, d\mathrm{w}(s) + \int_0^t b(s, F_x(\mathrm{w})_s) \, ds,$$

where the stochastic integral $\int_0^t \sigma(s, F_x(\mathrm{w})_s) d\mathrm{w}(s)$ can be defined by

$$\int_0^t \sigma(s, F_x(\mathrm{w})_s) \, d\mathrm{w}(s) = \lim_{k \to \infty} \sum_{i=0}^{2^{n_k}-1} \sigma\left(\frac{it}{2^{n_k}}, F_x(\mathrm{w})_{it/2^{n_k}}\right) \left(\mathrm{w}\left(\frac{(i+1)t}{2^{n_k}}\right) - \mathrm{w}\left(\frac{it}{2^{n_k}}\right) \right),$$

$W(\mathrm{dw})$ a.s. Here $(n_k)_{k\geq 1}$ is a suitable subsequence, and we used Proposition 5.9. We can now replace w by B (whose distribution is $W(\mathrm{dw})$!) and get a.s.

$$F_x(B)_t = x + \lim_{k\to\infty} \sum_{i=0}^{2^{n_k}-1} \sigma\left(\frac{it}{2^{n_k}}, F_x(B)_{it/2^{n_k}}\right)(B_{(i+1)t/2^{n_k}} - B_{it/2^{n_k}}) + \int_0^t b(s, F_x(B)_s)\mathrm{d}s$$

$$= x + \int_0^t \sigma(s, F_x(B)_s)\mathrm{d}B_s + \int_0^t b(s, F_x(B)_s)\mathrm{d}s,$$

again thanks to Proposition 5.9. We thus obtain that $F_x(B)$ is the desired solution.

We still have to prove the second part of assertion (iii). We again fix the filtered probability space $(\Omega, \mathscr{F}, (\mathscr{F}_t), P)$ and the $(\mathscr{F}_t)$-Brownian motion B. Let U be an $\mathscr{F}_0$-measurable random variable. If in the stochastic integral equation satisfied by $F_x(B)$ we formally substitute U for x, we obtain that $F_U(B)$ solves $E(\sigma, b)$ with initial value U. However, this formal substitution is not so easy to justify, and we will argue with some care.

We first observe that the mapping $(x, \omega) \mapsto F_x(B)_t$ is continuous with respect to the variable x (if ω is fixed) and $\mathscr{F}_t$-measurable with respect to ω (if x is fixed). It easily follows that this mapping is measurable for the σ-field $\mathscr{B}(\mathbb{R}) \otimes \mathscr{F}_t$. Since U is $\mathscr{F}_0$-measurable, we get that $F_U(B)_t$ is $\mathscr{F}_t$-measurable. Hence the process $F_U(B)$ is adapted. For $x \in \mathbb{R}$ and $\mathrm{w} \in C(\mathbb{R}_+, \mathbb{R})$, we define $G(x, \mathrm{w}) \in C(\mathbb{R}_+, \mathbb{R})$ by the formula

$$G(x, \mathrm{w})_t = \int_0^t b(s, F_x(\mathrm{w})_s)\, \mathrm{d}s.$$

We also set $H(x, \mathrm{w}) = F_x(\mathrm{w}) - x - G(x, \mathrm{w})$. We have already seen that, for every $x \in \mathbb{R}$, we have $W(\mathrm{dw})$ a.s.,

$$H(x, \mathrm{w})_t = \int_0^t \sigma(s, F_x(\mathrm{w})_s)\, \mathrm{dw}(s).$$

Hence, if

$$H_n(x, \mathrm{w})_t = \sum_{i=0}^{2^n-1} \sigma\left(\frac{it}{2^n}, F_x(\mathrm{w})_{it/2^n}\right)\left(\mathrm{w}\left(\frac{(i+1)t}{2^n}\right) - \mathrm{w}\left(\frac{it}{2^n}\right)\right),$$

Proposition 5.9 shows that

$$H(x, \mathrm{w})_t = \lim_{n\to\infty} H_n(x, \mathrm{w})_t,$$

in probability under $W(\mathrm{dw})$, for every $x \in \mathbb{R}$. Using the fact that U and B are independent (because U is $\mathscr{F}_0$-measurable), we infer from the latter convergence that

$$H(U, B)_t = \lim_{n \to \infty} H_n(U, B)_t$$

in probability. Thanks again to Proposition 5.9, the limit must be the stochastic integral

$$\int_0^t \sigma(s, F_U(B)_s) \, dB_s.$$

We have thus proved that

$$\int_0^t \sigma(s, F_U(B)_s) \, dB_s = H(U, B)_t = F_U(B)_t - U - \int_0^t b(s, F_U(B)_s) \, ds,$$

which shows that $F_U(B)$ solves $E(\sigma, b)$ with initial value U.   □

A consequence of Theorem 8.5, especially of property (ii) in this theorem, is the continuity of solutions with respect to the initial value. Given the filtered probability space $(\Omega, \mathscr{F}, (\mathscr{F}_t), P)$ and the $(\mathscr{F}_t)$-Brownian motion B, one can construct, for every $x \in \mathbb{R}^d$, the solution X^x of $E_x(\sigma, b)$ in such a way that, for every $\omega \in \Omega$, the mapping $x \mapsto X^x(\omega)$ is continuous. More precisely, the arguments of the previous proof give, for every $\varepsilon \in (0, 1)$ and for every choice of the constants $A > 0$ and $T > 0$, a (random) constant $C_{\varepsilon,A,T}(\omega)$ such that, if $|x|, |y| \leq A$,

$$\sup_{t \leq T} |X_t^x(\omega) - X_t^y(\omega)| \leq C_{\varepsilon,A,T}(\omega) \, |x - y|^{1-\varepsilon}$$

(in fact the version of Kolmogorov's lemma in Theorem 2.9 gives this only for $d = 1$, but there is an analogous version of Kolmogorov's lemma for processes indexed by a multidimensional parameter, see [70, Theorem I.2.1]).

8.3 Solutions of Stochastic Differential Equations as Markov Processes

In this section, we consider the homogeneous case where $\sigma(t, y) = \sigma(y)$ and $b(t, y) = b(y)$. As in the previous section, we assume that σ and b are Lipschitz: There exists a constant K such that, for every $x, y \in \mathbb{R}^d$,

$$|\sigma(x) - \sigma(y)| \leq K|x - y|, \quad |b(x) - b(y)| \leq K|x - y|.$$

Let $x \in \mathbb{R}^d$, and let X^x be a solution of $E_x(\sigma, b)$. Since weak uniqueness holds, for every $t \geq 0$, the law of X_t^x does not depend on the choice of the solution. In fact, this law is the image of Wiener measure on $C(\mathbb{R}_+, \mathbb{R}^d)$ under the mapping $w \mapsto F_x(w)_t$, where the mappings F_x were introduced in Theorem 8.5. The latter theorem also shows that the law of X_t^x depends continuously on the pair (x, t).

Theorem 8.6 *Assume that $(X_t)_{t \geq 0}$ is a solution of $E(\sigma, b)$ on a (complete) filtered probability space $(\Omega, \mathcal{F}, (\mathcal{F}_t), P)$. Then $(X_t)_{t \geq 0}$ is a Markov process with respect to the filtration $(\mathcal{F}_t)$, with semigroup $(Q_t)_{t \geq 0}$ defined by*

$$Q_t f(x) = E[f(X_t^x)],$$

where X^x is an arbitrary solution of $E_x(\sigma, b)$.

Remark With the notation of Theorem 8.5, we have also

$$Q_t f(x) = \int f(F_x(w)_t) \, W(dw). \tag{8.2}$$

Proof We first verify that, for any bounded measurable function f on $\mathbb{R}^d$, and for every $s, t \geq 0$, we have

$$E[f(X_{s+t}) \mid \mathcal{F}_s] = Q_t f(X_s),$$

where $Q_t f$ is defined by (8.2). To this end, we fix $s \geq 0$ and we write, for every $t \geq 0$,

$$X_{s+t} = X_s + \int_s^{s+t} \sigma(X_r) \, dB_r + \int_s^{s+t} b(X_r) \, dr \tag{8.3}$$

where B is an $(\mathcal{F}_t)$-Brownian motion starting from 0. We then set, for every $t \geq 0$,

$$X_t' = X_{s+t}, \quad \mathcal{F}_t' = \mathcal{F}_{s+t}, \quad B_t' = B_{s+t} - B_s.$$

We observe that the filtration $(\mathcal{F}_t')$ is complete (of course $\mathcal{F}_\infty' = \mathcal{F}_\infty$), that the process X' is adapted to $(\mathcal{F}_t')$, and that B' is an m-dimensional $(\mathcal{F}_t')$-Brownian motion. Furthermore, using the approximation results for the stochastic integral of adapted processes with continuous sample paths (Proposition 5.9), one easily verifies that, a.s. for every $t \geq 0$,

$$\int_s^{s+t} \sigma(X_r) \, dB_r = \int_0^t \sigma(X_u') \, dB_u'$$

where the stochastic integral in the right-hand side is computed in the filtration $(\mathcal{F}_t')$. It follows from (8.3) that

$$X_t' = X_s + \int_0^t \sigma(X_u') \, dB_u' + \int_0^t b(X_u') \, du.$$

Hence X' solves $E(\sigma, b)$, on the space $(\Omega, \mathscr{F}, (\mathscr{F}_t'), P)$ and with the Brownian motion B', with initial value $X_0' = X_s$ (note that X_s is $\mathscr{F}_0'$-measurable). By the last assertion of Theorem 8.5, we must have $X' = F_{X_s}(B')$, a.s.

Consequently, for every $t \geq 0$,

$$E[f(X_{s+t})|\mathscr{F}_s] = E[f(X_t')|\mathscr{F}_s] = E[f(F_{X_s}(B')_t)|\mathscr{F}_s] = \int f(F_{X_s}(w)_t)\, W(dw)$$
$$= Q_t f(X_s),$$

by the definition of $Q_t f$. In the third equality, we used the fact that B' is independent of $\mathscr{F}_s$, and distributed according to $W(dw)$, whereas X_s is $\mathscr{F}_s$-measurable.

We still have to verify that $(Q_t)_{t \geq 0}$ is a transition semigroup. Properties (i) and (iii) of the definition are immediate (for (iii), we use the fact that the law of X_t^x depends continuously on the pair (x, t)). For the Chapman–Kolmogorov relation, we observe that, by applying the preceding considerations to X^x, we have, for every $s, t \geq 0$,

$$Q_{t+s} f(x) = E[f(X_{s+t}^x)] = E[E[f(X_{s+t}^x)|\mathscr{F}_s]] = E[Q_t f(X_s^x)] = \int Q_s(x, dy) Q_t f(y).$$

This completes the proof.   □

We write $C_c^2(\mathbb{R}^d)$ for the space of all twice continuously differentiable functions with compact support on $\mathbb{R}^d$.

Theorem 8.7 *The semigroup $(Q_t)_{t \geq 0}$ is Feller. Furthermore, its generator L is such that*

$$C_c^2(\mathbb{R}^d) \subset D(L)$$

and, for every $f \in C_c^2(\mathbb{R}^d)$,

$$Lf(x) = \frac{1}{2} \sum_{i,j=1}^d (\sigma \sigma^*)_{ij}(x) \frac{\partial^2 f}{\partial x_i \partial x_j}(x) + \sum_{i=1}^d b_i(x) \frac{\partial f}{\partial x_i}(x)$$

where σ^* denotes the transpose of the matrix σ.

Proof For the sake of simplicity, we give the proof only in the case when σ and b are bounded. We fix $f \in C_0(\mathbb{R}^d)$ and we first verify that $Q_t f \in C_0(\mathbb{R}^d)$. Since the mappings $x \mapsto F_x(w)$ are continuous, formula (8.2) and dominated convergence show that $Q_t f$ is continuous. Then, since

$$X_t^x = x + \int_0^t \sigma(X_s^x)\, dB_s + \int_0^t b(X_s^x)\, ds,$$

and σ and b are assumed to be bounded, we get the existence of a constant C, which does not depend on t, x, such that

$$E[(X_t^x - x)^2] \le C(t + t^2). \tag{8.4}$$

Using Markov's inequality, we have thus, for every $t \ge 0$,

$$\sup_{x \in \mathbb{R}^d} P(|X_t^x - x| > A) \xrightarrow[A \to \infty]{} 0.$$

Writing

$$|Q_t f(x)| = |E[f(X_t^x)]| \le |E[f(X_t^x)\, \mathbf{1}_{\{|X_t^x - x| \le A\}}]| + \|f\|\, P(|X_t^x - x| > A),$$

we get, using our assumption $f \in C_0(\mathbb{R}^d)$,

$$\limsup_{x \to \infty} |Q_t f(x)| \le \|f\|\, \sup_{x \in \mathbb{R}^d} P(|X_t^x - x| > A),$$

and thus, since A was arbitrary,

$$\lim_{x \to \infty} Q_t f(x) = 0,$$

which completes the proof of the property $Q_t f \in C_0(\mathbb{R}^d)$.

Let us show similarly that $Q_t f \longrightarrow f$ when $t \to 0$. For every $\varepsilon > 0$,

$$\sup_{x \in \mathbb{R}^d} |E[f(X_t^x)] - f(x)| \le \sup_{x, y \in \mathbb{R}^d, |x-y| \le \varepsilon} |f(x) - f(y)| + 2\|f\|\, \sup_{x \in \mathbb{R}^d} P(|X_t^x - x| > \varepsilon).$$

However, using (8.4) and Markov's inequality again, we get

$$\sup_{x \in \mathbb{R}^d} P(|X_t^x - x| > \varepsilon) \xrightarrow[t \to 0]{} 0,$$

hence

$$\limsup_{t \to 0} \|Q_t f - f\| = \limsup_{t \to 0} \left(\sup_{x \in \mathbb{R}^d} |E[f(X_t^x)] - f(x)| \right) \le \sup_{x, y \in \mathbb{R}^d, |x-y| \le \varepsilon} |f(x) - f(y)|$$

which can be made arbitrarily close to 0 by taking ε small.

Let us prove the second assertion of the theorem. Let $f \in C_c^2(\mathbb{R}^d)$. We apply Itô's formula to $f(X_t^x)$, recalling that, if $X_t^x = (X_t^{x,1}, \ldots, X_t^{x,d})$, we have, for every $i \in \{1, \ldots, d\}$,

$$X_t^{x,i} = x_i + \sum_{j=1}^{m} \int_0^t \sigma_{ij}(X_s^x)\, dB_s^j + \int_0^t b_i(X_s^x)\, ds.$$

We get

$$f(X_t^x) = f(x) + M_t + \sum_{i=1}^{d} \int_0^t b_i(X_s^x) \frac{\partial f}{\partial x_i}(X_s^x) ds + \frac{1}{2} \sum_{i,i'=1}^{d} \int_0^t \frac{\partial^2 f}{\partial x_i \partial x_{i'}}(X_s^x) d\langle X^{x,i}, X^{x,i'} \rangle_s$$

where M is a continuous local martingale. Moreover, if $i, i' \in \{1, \ldots, d\}$,

$$d\langle X^{x,i}, X^{x,i'} \rangle_s = \sum_{j=1}^{m} \sigma_{ij}(X_s^x) \sigma_{i'j}(X_s^x) ds = (\sigma \sigma^*)_{ii'}(X_s^x) ds.$$

We thus see that, if g is the function defined by

$$g(x) = \frac{1}{2} \sum_{i,i'=1}^{d} (\sigma \sigma^*)_{ii'}(x) \frac{\partial^2 f}{\partial x_i \partial x_{i'}}(x) + \sum_{i=1}^{d} b_i(x) \frac{\partial f}{\partial x_i}(x),$$

the process

$$M_t = f(X_t^x) - f(x) - \int_0^t g(X_s^x) ds$$

is a continuous local martingale. Since f and g are bounded, Proposition 4.7 (ii) shows that M is a martingale. It now follows from Theorem 6.14 that $f \in D(L)$ and $Lf = g$.   □

Corollary 8.8 *Suppose that $(X_t)_{t \geq 0}$ solves $E(\sigma, b)$ on a filtered probability space $(\Omega, \mathscr{F}, (\mathscr{F}_t), P)$. Then $(X_t)_{t \geq 0}$ satisfies the strong Markov property: If T is a stopping time and if Φ is a Borel measurable function from $C(\mathbb{R}_+, \mathbb{R}^d)$ into $\mathbb{R}_+$,*

$$E[\mathbf{1}_{\{T < \infty\}} \Phi(X_{T+t}, t \geq 0) \mid \mathscr{F}_T] = \mathbf{1}_{\{T < \infty\}} \mathbb{E}_{X_T}[\Phi],$$

where, for every $x \in \mathbb{R}^d$, $\mathbb{P}_x$ denotes the law on $C(\mathbb{R}_+, \mathbb{R}^d)$ of an arbitrary solution of $E_x(\sigma, b)$.

Proof It suffices to apply Theorem 6.17. Alternatively, we could also argue in a similar manner as in the proof of Theorem 8.6, letting the stopping time T play the same role as the deterministic time s in the latter proof, and using the strong Markov property of Brownian motion.   □

Markov processes with continuous sample paths that are obtained as solutions of stochastic differential equations are sometimes called *diffusion processes* (certain authors call a diffusion process any strong Markov process with continuous sample paths in $\mathbb{R}^d$ or on a manifold). Note that, even in the Lipschitz setting considered here, Theorem 8.7 does not completely identify the generator L, but only its action on a subset of the domain $D(L)$: As we already mentioned in Chap. 6, it is often very difficult to give a complete description of the domain. However, in many instances, one can show that a partial knowledge of the generator, such as the one given by Theorem 8.7, suffices to characterize the law of the process. This observation is at the core of the powerful theory of martingale problems, which is developed in the classical book [77] by Stroock and Varadhan.

At least when restricted to $C_c^2(\mathbb{R}^d)$, the generator L is a second order differential operator. The stochastic differential equation $E(\sigma, b)$ allows one to give a probabilistic approach (as well as an interpretation) to many analytic results concerning this differential operator, in the spirit of the connections between Brownian motion and the Laplace operator described in the previous chapter. We refer to Durrett [18, Chapter 9] and Friedman [26, 27] for more about links between stochastic differential equations and partial differential equations. These connections between probability and analysis were an important motivation for the definition and study of stochastic differential equations.

8.4   A Few Examples of Stochastic Differential Equations

In this section, we briefly discuss three important examples, all in dimension one. In the first two examples, one can obtain an explicit formula for the solution, which is of course not the case in general.

8.4.1   The Ornstein–Uhlenbeck Process

Let $\lambda > 0$. The (one-dimensional) Ornstein–Uhlenbeck process is the solution of the stochastic differential equation

$$dX_t = dB_t - \lambda X_t \, dt.$$

This equation is solved by applying Itô's formula to $e^{\lambda t} X_t$, and we get

$$X_t = X_0 e^{-\lambda t} + \int_0^t e^{-\lambda(t-s)} \, dB_s.$$

Note that the stochastic integral is a Wiener integral (the integrand is deterministic), which thus belongs to the Gaussian space generated by B.

First consider the case where $X_0 = x \in \mathbb{R}$. By the previous remark, X is a (non-centered) Gaussian process, whose mean function is $m(t) = E[X_t] = x e^{-\lambda t}$, and whose covariance function is also easy to compute:

$$K(s, t) = \operatorname{cov}(X_s, X_t) = \frac{e^{-\lambda|t-s|} - e^{-\lambda(t+s)}}{2\lambda}.$$

It is also interesting to consider the case when X_0 is distributed according to $\mathcal{N}(0, \frac{1}{2\lambda})$. In that case, X is a centered Gaussian process with covariance function

$$\frac{1}{2\lambda} e^{-\lambda|t-s|}.$$

Notice that this is a stationary covariance function. In that case, the Ornstein–Uhlenbeck process X is both a stationary Gaussian process (indexed by $\mathbb{R}_+$) and a Markov process.

8.4.2   Geometric Brownian Motion

Let $\sigma > 0$ and $r \in \mathbb{R}$. The geometric Brownian motion with parameters σ and r is the solution of the stochastic differential equation

$$dX_t = \sigma X_t \, dB_t + r X_t \, dt.$$

One solves this equation by applying Itô's formula to $\log X_t$ (say in the case where $X_0 > 0$), and it follows that:

$$X_t = X_0 \exp\left(\sigma B_t + (r - \frac{\sigma^2}{2})t\right).$$

Note in particular that, if the initial value X_0 is (strictly) positive, the solution remains so at every time $t \geq 0$. Geometric Brownian motion is used in the celebrated Black–Scholes model of financial mathematics. The reason for the use of this process comes from an economic assumption of independence of the successive ratios

$$\frac{X_{t_2} - X_{t_1}}{X_{t_1}}, \frac{X_{t_3} - X_{t_2}}{X_{t_2}}, \dots, \frac{X_{t_n} - X_{t_{n-1}}}{X_{t_{n-1}}}$$

corresponding to disjoint time intervals: From the explicit formula for X_t, we see that this is nothing but the property of independence of increments of Brownian motion.

8.4.3 Bessel Processes

Let $m \geq 0$ be a real number. The m-dimensional squared Bessel process is the real process taking nonnegative values that solves the stochastic differential equation

$$dX_t = 2\sqrt{X_t}\, dB_t + m\, dt \,. \tag{8.5}$$

Notice that this equation does not fit into the Lipschitz setting studied in this chapter, because the function $\sigma(x) = 2\sqrt{x}$ is not Lipschitz over $\mathbb{R}_+$ (one might also observe that this function is only defined on $\mathbb{R}_+$ and not on $\mathbb{R}$, but this is a minor point because one can replace $2\sqrt{x}$ by $2\sqrt{|x|}$ and check a posteriori that a solution starting from a nonnegative value stays nonnegative). However, there exist (especially in dimension one) criteria weaker than our Lipschitz continuity assumptions, which apply to (8.5) and give the existence and pathwise uniqueness of solutions of (8.5). See in particular Exercise 8.14 for a criterion of pathwise uniqueness that applies to (8.5).

One of the main reasons for studying Bessel processes comes from the following observation. If $d \geq 1$ is an integer and $\beta = (\beta^1, \ldots, \beta^d)$ is a d-dimensional Brownian motion, an application of Itô's formula shows that the process

$$|\beta_t|^2 = (\beta_t^1)^2 + \cdots + (\beta_t^d)^2$$

is a d-dimensional squared Bessel process: See Exercise 5.33. Furthermore, one can also check that, when $m = 0$, the process $(\frac{1}{2}X_t)_{t\geq 0}$ has the same distribution as Feller's branching diffusion discussed at the end of Chap. 6 (see Exercise 8.11).

Suppose from now on that $m > 0$ and $X_0 = x > 0$. For every $r \geq 0$, set $T_r := \inf\{t \geq 0 : X_t = r\}$. If $r > x$, we have $P(T_r < \infty) = 1$. To get this, use (8.5) to see that $X_{t \wedge T_r} = x + m(t \wedge T_r) + Y_{t \wedge T_r}$, where $E[(Y_{t \wedge T_r})^2] \leq 4\,r t$. By Markov's inequality, $P(Y_{t \wedge T_r} > t^{3/4}) \longrightarrow 0$ as $t \to \infty$, and if we assume that $P(T_r = \infty) > 0$ the preceding expression for $X_{t \wedge T_r}$ gives a contradiction.

Set, for every $t \in [0, T_0)$,

$$M_t = \begin{cases} (X_t)^{1-\frac{m}{2}} & \text{if } m \neq 2, \\ \log(X_t) & \text{if } m = 2. \end{cases}$$

It follows from Itô's formula that, for every $\varepsilon \in (0, x)$, $M_{t \wedge T_\varepsilon}$ is a continuous local martingale. This continuous local martingale is bounded over the time interval $[0, T_\varepsilon \wedge T_A]$, for every $A > x$, and an application of the optional stopping theorem (using the fact that $T_A < \infty$ a.s.) gives, if $m \neq 2$,

$$P(T_\varepsilon < T_A) = \frac{A^{1-\frac{m}{2}} - x^{1-\frac{m}{2}}}{A^{1-\frac{m}{2}} - \varepsilon^{1-\frac{m}{2}}},$$

and if $m = 2$,

$$P(T_\varepsilon < T_A) = \frac{\log A - \log x}{\log A - \log \varepsilon}.$$

When $m = d$ is an integer, we recover the formulas of Proposition 7.16.

Let us finally concentrate on the case $m \geq 2$. Letting ε go to 0 in the preceding formulas, we obtain that $P(T_0 < \infty) = 0$. If we let A tend to ∞, we also get that $P(T_\varepsilon < \infty) = 1$ if $m = 2$ (as we already noticed in Chap. 7) and $P(T_\varepsilon < \infty) = (\varepsilon/x)^{(m/2)-1}$ if $m > 2$.

It then follows from the property $P(T_0 < \infty) = 0$ that the process M_t is well-defined for every $t \geq 0$ and is a continuous local martingale. When $m > 2$, M_t takes nonnegative values and is thus a supermartingale (Proposition 4.7 (i)), which converges a.s. as $t \to \infty$ (Proposition 3.19). The limit must be 0, since we already noticed that $P(T_A < \infty) = 1$ for every $A > x$, and we conclude that X_t converges a.s. to ∞ as $t \to \infty$ when $m > 2$. One can show that the continuous local martingale M_t is not a (true) martingale (cf. Question **8.** in Exercise 5.33 in the case $m = 3$).

Exercise 5.31 in Chap. 5 gives a number of important calculations related to squared Bessel processes. We refer to Chapter XI in [70] for a thorough study of this class of processes.

Remark The m-dimensional Bessel process is (of course) obtained by taking $Y_t = \sqrt{X_t}$, and, when $m = d$ is a positive integer, it corresponds to the norm of d-dimensional Brownian motion. When $m > 1$, the process Y also satisfies a stochastic differential equation, which is however less tractable than (8.5): See Exercise 8.13 below.

Exercises

Exercise 8.9 (*Time change method*) We consider the stochastic differential equation

$E(\sigma, 0)$ $\qquad\qquad\qquad\qquad dX_t = \sigma(X_t)\, dB_t$

where the function $\sigma : \mathbb{R} \longrightarrow \mathbb{R}$ is continuous and there exist constants $\varepsilon > 0$ and M such that $\varepsilon \leq \sigma \leq M$.

1. In this question and the next one, we assume that X solves $E(\sigma, 0)$ with $X_0 = x$. We set, for every $t \geq 0$,

$$A_t = \int_0^t \sigma(X_s)^2\, ds \quad , \quad \tau_t = \inf\{s \geq 0 : A_s > t\}.$$

Justify the equalities

$$\tau_t = \int_0^t \frac{dr}{\sigma(X_{\tau_r})^2}, \quad A_t = \inf\{s \geq 0 : \int_0^s \frac{dr}{\sigma(X_{\tau_r})^2} > t\}.$$

2. Show that there exists a real Brownian motion $\beta = (\beta_t)_{t\geq 0}$ started from x such that, a.s. for every $t \geq 0$,

$$X_t = \beta_{\inf\{s\geq 0: \int_0^s \sigma(\beta_r)^{-2}dr>t\}}.$$

3. Show that weak existence and weak uniqueness hold for $E(\sigma, 0)$. (*Hint:* For the existence part, observe that, if X is defined from a Brownian motion β by the formula of question 2., X is (in an appropriate filtration) a continuous local martingale with quadratic variation $\langle X, X \rangle_t = \int_0^t \sigma(X_s)^2 ds$.)

Exercise 8.10 We consider the stochastic differential equation
$E(\sigma, b)$ $\qquad\qquad\qquad dX_t = \sigma(X_t)\,dB_t + b(X_t)\,dt$
where the functions $\sigma, b : \mathbb{R} \longrightarrow \mathbb{R}$ are bounded and continuous, and such that $\int_{\mathbb{R}} |b(x)|dx < \infty$ and $\sigma \geq \varepsilon$ for some constant $\varepsilon > 0$.

1. Let X be a solution of $E(\sigma, b)$. Show that there exists a monotone increasing function $F : \mathbb{R} \longrightarrow \mathbb{R}$, which is also twice continuously differentiable, such that $F(X_t)$ is a martingale. Give an explicit formula for F in terms of σ and b.
2. Show that the process $Y_t = F(X_t)$ solves a stochastic differential equation of the form $dY_t = \sigma'(Y_t)\,dB_t$, with a function σ' to be determined.
3. Using the result of the preceding exercise, show that weak existence and weak uniqueness hold for $E(\sigma, b)$. Show that pathwise uniqueness also holds if σ is Lipschitz.

Exercise 8.11 We suppose that, for every $x \in \mathbb{R}_+$, one can construct on the same filtered probability space $(\Omega, \mathscr{F}, (\mathscr{F}_t), P)$ a process X^x taking nonnegative values, which solves the stochastic differential equation

$$\begin{cases} dX_t = \sqrt{2X_t}\,dB_t \\ X_0 = x \end{cases}$$

and that the processes X^x are Markov processes with values in $\mathbb{R}_+$, with the same semigroup $(Q_t)_{t\geq 0}$, with respect to the filtration $(\mathscr{F}_t)$. (This is, of course, close to Theorem 8.6, which however cannot be applied directly because the function $\sqrt{2x}$ is not Lipschitz.)

1. We fix $x \in \mathbb{R}_+$, and a real $T > 0$. We set, for every $t \in [0, T]$

$$M_t = \exp\left(-\frac{\lambda X_t^x}{1 + \lambda(T - t)}\right).$$

Show that the process $(M_{t\wedge T})_{t\geq 0}$ is a martingale.

2. Show that $(Q_t)_{t\geq 0}$ is the semigroup of Feller's branching diffusion (see the end of Chap. 6).

Exercise 8.12 We consider two sequences $(\sigma_n)_{n\geq 1}$ and $(b_n)_{n\geq 1}$ of real functions defined on $\mathbb{R}$. We assume that:

(i) There exists a constant $C > 0$ such that $|\sigma_n(x)| \leq C$ and $|b_n(x)| \leq C$ for every $n \geq 1$ and $x \in \mathbb{R}$.
(ii) There exists a constant $K > 0$ such that, for every $n \geq 1$ and $x, y \in \mathbb{R}$,

$$|\sigma_n(x) - \sigma_n(y)| \leq K|x - y| \quad , \quad |b_n(x) - b_n(y)| \leq K|x - y|.$$

Let B be an $(\mathscr{F}_t)$-Brownian motion and, for every $n \geq 1$, let X^n be the unique adapted process satisfying

$$X_t^n = \int_0^t \sigma_n(X_s^n)\,dB_s + \int_0^t b_n(X_s^n)\,ds.$$

1. Let $T > 0$. Show that there exists a constant $A > 0$ such that, for every real $M > 0$ and for every $n \geq 1$,

$$P\left(\sup_{t\leq T} |X_t^n| \geq M\right) \leq \frac{A}{M^2}.$$

2. We assume that the sequences (σ_n) and (b_n) converge uniformly on every compact subset of $\mathbb{R}$ to limiting functions denoted by σ and b respectively. Justify the existence of an adapted process $X = (X_t)_{t\geq 0}$ with continuous sample paths, such that

$$X_t = \int_0^t \sigma(X_s)\,dB_s + \int_0^t b(X_s)\,ds,$$

then show that there exists a constant A' such that, for every real $M > 0$, for every $t \in [0, T]$ and $n \geq 1$,

$$E\left[\sup_{s\leq t}(X_s^n - X_s)^2\right] \leq 4(4 + T)K^2 \int_0^t E[(X_s^n - X_s)^2]\,ds + \frac{A'}{M^2}$$

$$+ 4T\left(4 \sup_{|x|\leq M} (\sigma_n(x) - \sigma(x))^2 + T \sup_{|x|\leq M} (b_n(x) - b(x))^2\right).$$

3. Infer from the preceding question that

$$\lim_{n\to\infty} E\left[\sup_{s\leq T}(X_s^n - X_s)^2\right] = 0.$$

Exercise 8.13 Let $\beta = (\beta_t)_{t \geq 0}$ be an $(\mathscr{F}_t)$-Brownian motion started from 0. We fix two real parameters α and r, with $\alpha > 1/2$ and $r > 0$. For every integer $n \geq 1$ and every $x \in \mathbb{R}$, we set

$$f_n(x) = \frac{1}{|x|} \wedge n .$$

1. Let $n \geq 1$. Justify the existence of the unique semimartingale Z^n that solves the equation

$$Z^n_t = r + \beta_t + \alpha \int_0^t f_n(Z^n_s) \, ds.$$

2. We set $S_n = \inf\{t \geq 0 : Z^n_t \leq 1/n\}$. After observing that, for $t \leq S_n \wedge S_{n+1}$,

$$Z^{n+1}_t - Z^n_t = \alpha \int_0^t \left(\frac{1}{Z^{n+1}_s} - \frac{1}{Z^n_s} \right) ds ,$$

   show that $Z^{n+1}_t = Z^n_t$ for every $t \in [0, S_n \wedge S_{n+1}]$, a.s. Infer that $S_{n+1} \geq S_n$.

3. Let g be a twice continuously differentiable function on $\mathbb{R}$. Show that the process

$$g(Z^n_t) - g(r) - \int_0^t \left(\alpha g'(Z^n_s) f_n(Z^n_s) + \frac{1}{2} g''(Z^n_s) \right) ds$$

   is a continuous local martingale.

4. We set $h(x) = x^{1-2\alpha}$ for every $x > 0$. Show that, for every integer $n \geq 1$, $h(Z^n_{t \wedge S_n})$ is a bounded martingale. Infer that, for every $t \geq 0$, $P(S_n \leq t)$ tends to 0 as $n \to \infty$, and consequently $S_n \to \infty$ a.s. as $n \to \infty$.

5. Infer from questions **2.** and **4.** that there exists a unique positive semimartingale Z such that, for every $t \geq 0$,

$$Z_t = r + \beta_t + \alpha \int_0^t \frac{ds}{Z_s}.$$

6. Let $d \geq 3$ and let B be a d-dimensional Brownian motion started from $y \in \mathbb{R}^d \backslash \{0\}$. Show that $Y_t = |B_t|$ satisfies the stochastic equation in question **5.** (with an appropriate choice of β) with $r = |y|$ and $\alpha = (d-1)/2$. One may use the results of Exercise 5.33.

Exercise 8.14 (*Yamada–Watanabe uniqueness criterion*)   The goal of the exercise is to get pathwise uniqueness for the one-dimensional stochastic differential equation

$$dX_t = \sigma(X_t) dB_t + b(X_t) dt$$

when the functions σ and b satisfy the conditions

$$|\sigma(x) - \sigma(y)| \le K \sqrt{|x - y|}, \quad |b(x) - b(y)| \le K |x - y|,$$

for every $x, y \in \mathbb{R}$, with a constant $K < \infty$.

1. **Preliminary question.** Let Z be a semimartingale such that $\langle Z, Z \rangle_t = \int_0^t h_s \, ds$, where $0 \le h_s \le C |Z_s|$, with a constant $C < \infty$. Show that, for every $t \ge 0$,

$$\lim_{n \to \infty} n E\left[\int_0^t \mathbf{1}_{\{0 < |Z_s| \le 1/n\}} \, d\langle Z, Z \rangle_s \right] = 0.$$

(*Hint*: Observe that, for every $n \ge 1$,

$$E\left[\int_0^t |Z_s|^{-1} \mathbf{1}_{\{0 < |Z_s| \le 1\}} \, d\langle Z, Z \rangle_s \right] \le C t < \infty.)$$

2. For every integer $n \ge 1$, let φ_n be the function defined on $\mathbb{R}$ by

$$\varphi_n(x) = \begin{cases} 0 & \text{if } |x| \ge 1/n, \\ 2n(1 - nx) & \text{if } 0 \le x \le 1/n, \\ 2n(1 + nx) & \text{if } -1/n \le x \le 0. \end{cases}$$

Also write F_n for the unique twice continuously differentiable function on $\mathbb{R}$ such that $F_n(0) = F_n'(0) = 0$ and $F_n'' = \varphi_n$. Note that, for every $x \in \mathbb{R}$, one has $F_n(x) \longrightarrow |x|$ and $F_n'(x) \longrightarrow \text{sgn}(x) := \mathbf{1}_{\{x > 0\}} - \mathbf{1}_{\{x < 0\}}$ when $n \to \infty$.

Let X and X' be two solutions of $E(\sigma, b)$ on the same filtered probability space and with the same Brownian motion B. Infer from question **1.** that

$$\lim_{n \to \infty} E\left[\int_0^t \varphi_n(X_s - X_s') \, d\langle X - X', X - X' \rangle_s \right] = 0.$$

3. Let T be a stopping time such that the semimartingale $X_{t \wedge T} - X_{t \wedge T}'$ is bounded. By applying Itô's formula to $F_n(X_{t \wedge T} - X_{t \wedge T}')$, show that

$$E[|X_{t \wedge T} - X_{t \wedge T}'|] = E[|X_0 - X_0'|] + E\left[\int_0^{t \wedge T} (b(X_s) - b(X_s')) \, \text{sgn}(X_s - X_s') \, ds \right].$$

4. Using Gronwall's lemma, show that, if $X_0 = X_0'$, one has $X_t = X_t'$ for every $t \ge 0$, a.s.

Notes and Comments

As already mentioned, the treatment of stochastic differential equations motivated Itô's invention of stochastic differential equations. For further reading on this topic, the reader may consult the classical books of Ikeda and Watanabe [43] and Stroock and Varadhan [77], the latter studying stochastic differential equations in connection with martingale problems. Øksendal's book [66] emphasizes the applications of stochastic differential equations in other fields. The books [26, 27] of Friedman focus on connections with partial differential equations. We have chosen to concentrate on the Lipschitz case, where the main results of existence and uniqueness were already obtained by Itô [37, 38]. In dimension one, the criteria ensuring pathwise uniqueness can be weakened significantly (see in particular Yamada and Watanabe [83], which inspired Exercise 8.14) but this is no longer the case in higher dimensions. Chapter XI of [70] contains a lot of information about Bessel processes.

Chapter 9
Local Times

In this chapter, we apply stochastic calculus to the theory of local times of continuous semimartingales. Roughly speaking, the local time at level a of a semimartingale X is an increasing process that measures the "number of visits" of X at level a. We use the classical Tanaka formulas to construct local times and then to study their regularity properties with respect to the space variable. We show how local times can be used to obtain a generalized version of Itô's formula, and we establish the so-called density of occupation time formula. We also give several approximations of local times. We then focus on the case of Brownian motion, where we state the classical Trotter theorem as a corollary of our results for general semimartingales, and we derive the famous Lévy theorem identifying the law of the Brownian local time process at level 0. In the last section, we use Brownian local times to prove the Kallianpur–Robbins law that was stated at the end of Chap. 7. This chapter can be read independently of Chaps. 6, 7 and 8, except for the last section that relies on Chap. 7.

9.1 Tanaka's Formula and the Definition of Local Times

Throughout this chapter, we argue on a filtered probability space $(\Omega, \mathscr{F}, (\mathscr{F}_t)_{t\geq0}, P)$, and the filtration $(\mathscr{F}_t)_{t\geq0}$ is assumed to be complete. Let X be a continuous semimartingale. If f is a twice continuously differentiable function defined on $\mathbb{R}$, Itô's formula asserts that $f(X_t)$ is still a continuous semimartingale, and

$$f(X_t) = f(X_0) + \int_0^t f'(X_s)\,dX_s + \frac{1}{2}\int_0^t f''(X_s)\,d\langle X, X\rangle_s.$$

The next proposition shows that this formula can be extended to the case when f is a convex function.

© Springer International Publishing Switzerland 2016
J.-F. Le Gall, *Brownian Motion, Martingales, and Stochastic Calculus*,
Graduate Texts in Mathematics 274, DOI 10.1007/978-3-319-31089-3_9

Proposition 9.1 *Let f be a convex function on $\mathbb{R}$. Then $f(X_t)$ is a semimartingale, and, more precisely, there exists an increasing process A^f such that, for every $t \geq 0$,*

$$f(X_t) = f(X_0) + \int_0^t f'_-(X_s)\, dX_s + A^f_t,$$

where $f'_-(x)$ denotes the left-derivative of f at x.

More generally, $f(X_t)$ is a semimartingale if f is a difference of convex functions.

Proof Let h be a nonnegative continuous function on $\mathbb{R}$ such that $h(x) = 0$ if $x \notin [0, 1]$ and $\int_0^1 h(x)\, dx = 1$. For every integer $n \geq 1$, set $h_n(x) = n\, h(nx)$. Define a function $\varphi_n : \mathbb{R} \longrightarrow \mathbb{R}$ by

$$\varphi_n(x) = h_n * f(x) = \int_{\mathbb{R}} h_n(y) f(x - y)\, dy.$$

Then it is elementary to verify that φ_n is twice continuously differentiable on $\mathbb{R}$, $\varphi'_n = h_n * f'_-$, and $\varphi_n(x) \longrightarrow f(x)$, $\varphi'_n(x) \longrightarrow f'_-(x)$ as $n \to \infty$, for every $x \in \mathbb{R}$. Furthermore, the functions φ_n are also convex, so that $\varphi''_n \geq 0$.

Let $X = M + V$ be the canonical decomposition of the semimartingale X, and consider an integer $K \geq 1$. Introduce the stopping time

$$T_K := \inf\{t \geq 0 : |X_t| + \langle M, M \rangle_t + \int_0^t |dV_s| \geq K\}.$$

From Itô's formula, we have

$$\varphi_n(X_{t \wedge T_K}) = \varphi_n(X_0) + \int_0^{t \wedge T_K} \varphi'_n(X_s)\, dX_s + \frac{1}{2} \int_0^{t \wedge T_K} \varphi''_n(X_s)\, d\langle M, M \rangle_s. \tag{9.1}$$

From the definition of T_K, we have $\langle M, M \rangle_{T_K} \leq K$. Noting that the functions φ'_n are uniformly bounded over any compact interval, we get, by a simple application of Proposition 5.8,

$$\int_0^{t \wedge T_K} \varphi'_n(X_s)\, dX_s \xrightarrow[n \to \infty]{} \int_0^{t \wedge T_K} f'_-(X_s)\, dX_s, \tag{9.2}$$

in probability. For every $t \geq 0$, set

$$A^{f,K}_t := f(X_{t \wedge T_K}) - f(X_0) - \int_0^{t \wedge T_K} f'_-(X_s)\, dX_s. \tag{9.3}$$

Since $\varphi_n(X_0) \to f(X_0)$ and $\varphi_n(X_{t \wedge T_K}) \to f(X_{t \wedge T_K})$ as $n \to \infty$, we deduce from (9.2) and (9.1) that

$$\frac{1}{2} \int_0^{t \wedge T_K} \varphi_n''(X_s) \, d\langle M, M \rangle_s \xrightarrow[n \to \infty]{} A_t^{f,K}, \tag{9.4}$$

in probability. By (9.3), the process $(A_t^{f,K})_{t \geq 0}$ has continuous sample paths, and $A_0^{f,K} = 0$. Since $\varphi_n'' \geq 0$, it follows from the convergence (9.4) that the sample paths of $(A_t^{f,K})_{t \geq 0}$ are also nondecreasing. Hence $A^{f,K}$ is an increasing process. Finally, one gets from (9.4) that $A_t^{f,K} = A_{t \wedge T_K}^{f,K'}$ if $K \leq K'$. It follows that there exists an increasing process A^f such that $A_t^{f,K} = A_{t \wedge T_K}^f$ for every $t \geq 0$ and $K \geq 1$. We then get the formula of the proposition by letting $K \to \infty$ in (9.3).   □

Remark Write f_+' for the right-derivative of f. An argument similar to the preceding proof shows that there exists an increasing process $\tilde{A}^f$ such that

$$f(X_t) = f(X_0) + \int_0^t f_+'(X_s) \, dX_s + \tilde{A}_t^f.$$

If f is twice continuously differentiable, $A_t^f = \tilde{A}_t^f = \frac{1}{2} \int_0^t f''(X_s) \, d\langle X, X \rangle_s$. In general, however, we may have $\tilde{A}_t^f \neq A_t^f$.

The previous proposition leads to an easy definition of the local times of a semimartingale. For every $x \in \mathbb{R}$, we set $\mathrm{sgn}(x) := \mathbf{1}_{\{x > 0\}} - \mathbf{1}_{\{x \leq 0\}}$ (the fact that we define $\mathrm{sgn}(0) = -1$ here plays a significant role).

Proposition 9.2 *Let X be a continuous semimartingale and $a \in \mathbb{R}$. There exists an increasing process $(L_t^a(X))_{t \geq 0}$ such that the following three identities hold:*

$$|X_t - a| = |X_0 - a| + \int_0^t \mathrm{sgn}(X_s - a) \, dX_s + L_t^a(X), \tag{9.5}$$

$$(X_t - a)^+ = (X_0 - a)^+ + \int_0^t \mathbf{1}_{\{X_s > a\}} \, dX_s + \frac{1}{2} L_t^a(X), \tag{9.6}$$

$$(X_t - a)^- = (X_0 - a)^- - \int_0^t \mathbf{1}_{\{X_s \leq a\}} \, dX_s + \frac{1}{2} L_t^a(X). \tag{9.7}$$

The increasing process $(L_t^a(X))_{t \geq 0}$ is called the local time of X at level a. Furthermore, for every stopping time T, we have $L_t^a(X^T) = L_{t \wedge T}^a(X)$.

We will refer to any of the identities (9.5), (9.6), (9.7) as Tanaka's formula.

Proof We apply Proposition 9.1 to the convex function $f(x) = |x - a|$, noting that $f'_-(x) = \text{sgn}(x - a)$. It follows from Proposition 9.1 that the process $(L^a_t(X))_{t \geq 0}$ defined by

$$L^a_t(X) := |X_t - a| - |X_0 - a| - \int_0^t \text{sgn}(X_s - a)\, dX_s$$

is an increasing process. We then need to verify that (9.6) and (9.7) also hold. To this end, we apply Proposition 9.1 to the convex functions $f(x) = (x - a)^+$ and $f(x) = (x - a)^-$. It follows that there exist two increasing processes $A^{a,(+)}$ and $A^{a,(-)}$ such that

$$(X_t - a)^+ = (X_0 - a)^+ + \int_0^t \mathbf{1}_{\{X_s > a\}}\, dX_s + A^{a,(+)}_t,$$

and

$$(X_t - a)^- = (X_0 - a)^- - \int_0^t \mathbf{1}_{\{X_s \leq a\}}\, dX_s + A^{a,(-)}_t.$$

By considering the difference between the last two displays, we immediately get that $A^{a,(+)} = A^{a,(-)}$. On the other hand, if we add these two displays and compare with (9.5), we get $A^{a,(+)}_t + A^{a,(-)}_t = L^a_t(X)$. Hence $A^{a,(+)}_t = A^{a,(-)}_t = \frac{1}{2}L^a_t(X)$.

The last assertion immediately follows from (9.5) since $\int_0^{t \wedge T} \text{sgn}(X_s - a)\, dX_s = \int_0^t \text{sgn}(X^T_s - a)\, dX^T_s$ by properties of the stochastic integral. $\square$

Let us state the key property of local times. We use the notation $d_s L^a_s(X)$ for the random measure associated with the increasing function $s \mapsto L^a_s(X)$ (i.e. $\int_{[0,t]} d_s L^a_s(X) = L^a_t(X)$).

Proposition 9.3 *Let X be a continuous semimartingale and let $a \in \mathbb{R}$. Then a.s. the random measure $d_s L^a_s(X)$ is supported on $\{s \geq 0 : X_s = a\}$.*

Proof Set $W_t = |X_t - a|$ and note that (9.5) gives $\langle W, W \rangle_t = \langle X, X \rangle_t$ since $|\text{sgn}(x)| = 1$ for every $x \in \mathbb{R}$. By applying Itô's formula to $(W_t)^2$, we get

$$(X_t - a)^2 = W^2_t = (X_0 - a)^2 + 2\int_0^t (X_s - a)\, dX_s + 2\int_0^t |X_s - a|\, d_s L^a_s(X) + \langle X, X \rangle_t.$$

Comparing with the result of a direct application of Itô's formula to $(X_t - a)^2$, we get

$$\int_0^t |X_s - a| \, d_s L_s^a(X) = 0,$$

which gives the desired result. □

Proposition 9.3 shows that the function $t \mapsto L_t^a(X)$ may only increase when $X_t = a$. So in some sense, $L_t^a(X)$ measures the "number of visits" of the process X at level a before time t (the results of Sect. 9.3 give rigorous versions of this intuitive statement). This also justifies the name "local time".

9.2 Continuity of Local Times and the Generalized Itô Formula

We consider a continuous semimartingale X and write $X = M + V$ for its canonical decomposition. Our first goal is to study the continuity of the local times of X with respect to the space variable a.

It is convenient to write $L^a(X)$ for the random continuous function $(L_t^a(X))_{t \geq 0}$, which we view as a random variable with values in the space $C(\mathbb{R}_+, \mathbb{R}_+)$. As usual, the latter space is equipped with the topology of uniform convergence on every compact set.

Theorem 9.4 *The process $(L^a(X), a \in \mathbb{R})$ with values in $C(\mathbb{R}_+, \mathbb{R}_+)$ has a càdlàg modification, which we consider from now on and for which we keep the same notation $(L^a(X), a \in \mathbb{R})$. Furthermore, if $L^{a-}(X) = (L_t^{a-}(X))_{t \geq 0}$ denotes the left limit of $b \longrightarrow L^b(X)$ at a, we have for every $t \geq 0$,*

$$L_t^a(X) - L_t^{a-}(X) = 2 \int_0^t 1_{\{X_s = a\}} \, dV_s. \tag{9.8}$$

In particular, if X is a continuous local martingale, the process $(L_t^a(X))_{a \in \mathbb{R}, t \geq 0}$ has jointly continuous sample paths.

The proof of the theorem relies on Tanaka's formula and the following technical lemma.

Lemma 9.5 *Let $p \geq 1$. There exists a constant C_p, which only depends on p, such that for every $a, b \in \mathbb{R}$ with $a < b$, we have*

$$E\left[\left(\int_0^t 1_{\{a < X_s \leq b\}} d\langle M, M \rangle_s \right)^p\right] \leq C_p(b - a)^p \left(E[(\langle M, M \rangle_t)^{p/2}] + E\left[\left(\int_0^t |dV_s|\right)^p\right]\right).$$

For every $a \in \mathbb{R}$, write $Y^a = (Y^a_t)_{t\geq 0}$ for the random variable with values in $C(\mathbb{R}_+, \mathbb{R})$ defined by

$$Y^a_t = \int_0^t \mathbf{1}_{\{X_s > a\}} dM_s.$$

The process $(Y^a, a \in \mathbb{R})$ has a continuous modification.

Proof Let us start with the first assertion. It is enough to prove that the stated bound holds when $a = -u$ and $b = u$ for some $u > 0$ (then take $u = (b-a)/2$ and replace X by $X - (b+a)/2$). Let f be the unique twice continuously differentiable function such that

$$f''(x) = (2 - \frac{|x|}{u})^+,$$

and $f(0) = f'(0) = 0$. Note that we then have $|f'(x)| \leq 2u$ for every $x \in \mathbb{R}$. Since $f'' \geq 0$ and $f''(x) \geq 1$ if $-u \leq x \leq u$, we have

$$\int_0^t \mathbf{1}_{\{-u < X_s \leq u\}} d\langle M, M \rangle_s \leq \int_0^t f''(X_s) d\langle M, M \rangle_s. \tag{9.9}$$

However, by Itô's formula,

$$\frac{1}{2} \int_0^t f''(X_s) d\langle M, M \rangle_s = f(X_t) - f(X_0) - \int_0^t f'(X_s)\, dX_s. \tag{9.10}$$

Recalling that $|f'| \leq 2u$, we have

$$E[|f(X_t) - f(X_0)|^p] \leq (2u)^p E[|X_t - X_0|^p]$$

$$\leq (2u)^p E\left[\left(|M_t - M_0| + \int_0^t |dV_s|\right)^p\right]$$

$$\leq C_p (2u)^p \left(E[(\langle M, M \rangle_t)^{p/2}] + E\left[\left(\int_0^t |dV_s|\right)^p\right]\right),$$

using the Burkholder–Davis–Gundy inequalities (Theorem 5.16). Here and below, C_p stands for a constant that depends only on p, which may vary from line to line. Then,

$$\int_0^t f'(X_s)\, dX_s = \int_0^t f'(X_s)\, dM_s + \int_0^t f'(X_s)\, dV_s.$$

We have

$$E\Big[\Big|\int_0^t f'(X_s)\,dV_s\Big|^p\Big] \le (2u)^p\,E\Big[\Big(\int_0^t |dV_s|\Big)^p\Big]$$

and, using the Burkholder–Davis–Gundy inequalities once again,

$$E\Big[\Big|\int_0^t f'(X_s)\,dM_s\Big|^p\Big] \le C_p\,E\Big[\Big(\int_0^t f'(X_s)^2\,d\langle M,M\rangle_s\Big)^{p/2}\Big]$$

$$\le C_p(2u)^p\,E[(\langle M,M\rangle_t)^{p/2}].$$

The first assertion of the lemma follows by combining the previous bounds, using (9.9) and (9.10).

Let us turn to the second assertion. We fix $p > 2$. By the Burkholder–Davis–Gundy inequalities, we have for every $a < b$ and every $t \ge 0$,

$$E\Big[\sup_{s\le t} |Y_s^b - Y_s^a|^p\Big] \le C_p\,E\Big[\Big(\int_0^t \mathbf{1}_{\{a<X_s\le b\}}d\langle M,M\rangle_s\Big)^{p/2}\Big], \tag{9.11}$$

and the right-hand side can be estimated from the first assertion of the lemma. More precisely, for every integer $n \ge 1$, introduce the stopping time

$$T_n := \inf\{t \ge 0 : \langle M,M\rangle_t + \int_0^t |dV_s| \ge n\}.$$

From the first assertion of the lemma with X replaced by the stopped process X^{T_n}, we have, for every $t \ge 0$,

$$E\Big[\Big(\int_0^{t\wedge T_n} \mathbf{1}_{\{a<X_s\le b\}}d\langle M,M\rangle_s\Big)^{p/2}\Big] \le C_p(n^{p/4} + n^{p/2})\,(b-a)^{p/2}.$$

Using (9.11), again with X replaced by X^{T_n}, and letting $t \to \infty$, we obtain

$$E\Big[\sup_{s\ge 0} |Y_{s\wedge T_n}^b - Y_{s\wedge T_n}^a|^p\Big] \le C_p(n^{p/4} + n^{p/2})\,(b-a)^{p/2}.$$

Since $p > 2$, we see that we can apply Kolmogorov's lemma (Theorem 2.9) to get the existence of a continuous modification of the process $a \to (Y_{s\wedge T_n}^a)_{s\ge 0}$, with values in $C(\mathbb{R}_+, \mathbb{R})$. Write $(Y_s^{(n),a})_{s\ge 0}$ for this continuous modification.

Then, if $1 \le n < m$, for every fixed a, we have $Y_s^{(n),a} = Y_{s\wedge T_n}^{(m),a}$ for every $s \ge 0$, a.s. By a continuity argument, the latter equality holds simultaneously for every $a \in \mathbb{R}$ and every $s \ge 0$, outside a single set of probability zero. It follows that we can define a process $(\tilde{Y}^a, a \in \mathbb{R})$ with values in $C(\mathbb{R}_+, \mathbb{R})$, with continuous sample paths, such that, for every $n \geq 1$, $Y_s^{(n),a} = \tilde{Y}_{s \wedge T_n}^a$ for every $a \in \mathbb{R}$ and every $s \geq 0$, a.s. The process $(\tilde{Y}^a, a \in \mathbb{R})$ is the desired continuous modification.   □

Remark By applying the bound of Lemma 9.5 to X^{T_n} (with T_n as in the previous proof) and letting a tend to b, we get that, for every $b \in \mathbb{R}$,

$$\int_0^t \mathbf{1}_{\{X_s=b\}} \, d\langle M, M \rangle_s = 0$$

for every $t \geq 0$, a.s. Consequently, using Proposition 4.12, we also have

$$\int_0^t \mathbf{1}_{\{X_s=b\}} dM_s = 0, \tag{9.12}$$

for every $t \geq 0$, a.s.

Proof of Theorem 9.4 With a slight abuse of notation, we still write $(Y^a, a \in \mathbb{R})$ for the continuous modification obtained in the second assertion of Lemma 9.5. We also let $(Z^a, a \in \mathbb{R})$ be the process with values in $C(\mathbb{R}_+, \mathbb{R})$ defined by

$$Z_t^a = \int_0^t \mathbf{1}_{\{X_s>a\}} \, dV_s.$$

By Tanaka's formula, we have for every fixed $a \in \mathbb{R}$,

$$L_t^a = 2\Big((X_t - a)^+ - (X_0 - a)^+ - Y_t^a - Z_t^a\Big), \text{ for every } t \geq 0, \text{ a.s.}$$

The right-hand side of the last display provides the desired càdlàg modification. Indeed, the process

$$a \mapsto \Big((X_t - a)^+ - (X_0 - a)^+ - Y_t^a\Big)_{t \geq 0}$$

has continuous sample paths, and on the other hand the process $a \mapsto Z^a$ has càdlàg sample paths: For every $a_0 \in \mathbb{R}$, the dominated convergence theorem shows that

$$\int_0^t \mathbf{1}_{\{X_s>a\}} \, dV_s \xrightarrow[a \downarrow a_0]{} \int_0^t \mathbf{1}_{\{X_s>a_0\}} \, dV_s,$$

$$\int_0^t \mathbf{1}_{\{X_s>a\}} \, dV_s \xrightarrow[a \uparrow a_0, a<a_0]{} \int_0^t \mathbf{1}_{\{X_s \geq a_0\}} \, dV_s,$$

uniformly on every compact time interval. The previous display also shows that the jump $Z^{a_0} - Z^{a_0-}$ is given by

$$Z_t^{a_0} - Z_t^{a_0-} = -\int_0^t \mathbf{1}_{\{X_s = a_0\}} \, dV_s,$$

and this completes the proof of the theorem.   □

From now on, we only deal with the càdlàg modification of local times obtained in Theorem 9.4.

Remark To illustrate Theorem 9.4, set $W_t = |X_t|$, which is also a semimartingale by Tanaka's formula (9.5). By (9.6) applied to W_t, we have

$$W_t = (W_t)^+ = |X_0| + \int_0^t \mathbf{1}_{\{|X_s|>0\}} (\mathrm{sgn}(X_s)dX_s + dL_s^0(X)) + \frac{1}{2}L_t^0(W)$$

$$= |X_0| + \int_0^t \mathrm{sgn}(X_s)dX_s + \int_0^t \mathbf{1}_{\{X_s=0\}}dX_s + \frac{1}{2}L_t^0(W),$$

noting that $\int_0^t \mathbf{1}_{\{|X_s|>0\}}dL_s^0(X) = 0$ by the support property of local time (Proposition 9.3). Comparing the resulting formula with (9.5) written with $a = 0$, we get

$$L_t^0(W) = 2L_t^0(X) - 2\int_0^t \mathbf{1}_{\{X_s=0\}}dX_s = L_t^0(X) + L_t^{0-}(X),$$

using (9.8). The formula $L_t^0(W) = L_t^0(X) + L_t^{0-}(X)$ is a special case of the more general formula $L_t^a(W) = L_t^a(X) + L_t^{(-a)-}(X)$, for every $a \geq 0$, which is easily deduced from Corollary 9.7 below. We note that the support property of local time implies $L_t^a(W) = 0$ for every $a < 0$, and in particular $L_t^{0-}(W) = 0$. We leave it as an exercise for the reader to verify that formula (9.8) applied to $L_t^0(W) - L_t^{0-}(W)$ gives a result which is consistent with the preceding expression for $L_t^0(W)$.

We will now give an extension of Itô's formula (in the case where it is applied to a function of a single semimartingale). If f is a convex function on $\mathbb{R}$, the left derivative f'_- is a left-continuous monotone nondecreasing function, and there exists a unique Radon measure $f''(dy)$ on $\mathbb{R}_+$ such that $f''([a, b)) = f'_-(b) - f'_-(a)$, for every $a < b$. One can also interpret f'' as the second derivative of f in the sense of distributions. Note that $f''(da) = f''(a)da$ if f is twice continuously differentiable. If f is now a difference of convex functions, that is, $f = f_1 - f_2$ where both f_1 and f_2 are convex, we can still make sense of $\int f''(dy)\, \varphi(y) = \int f_1''(dy)\, \varphi(y) - \int f_2''(dy)\, \varphi(y)$ for any bounded measurable function φ supported on a compact interval of $\mathbb{R}$.

The next theorem identifies the increasing process A_t^f that appeared in Proposition 9.1.

Theorem 9.6 (Generalized Itô formula) *Let f be a difference of convex functions on $\mathbb{R}$. Then, for every $t \geq 0$,*

$$f(X_t) = f(X_0) + \int_0^t f'_-(X_s)\,dX_s + \frac{1}{2}\int_{\mathbb{R}} L_t^a(X) f''(da).$$

Remark By Proposition 9.2 and a continuity argument, we have

$$L_t^a(X) = 0 \text{ for every } a \notin \left[\min_{0 \leq s \leq t} X_s, \max_{0 \leq s \leq t} X_s\right], \quad \text{a.s.}$$

and furthermore the function $a \mapsto L_t^a(X)$ is bounded. Together with the observations preceding the statement of the theorem, this shows that the integral $\int_{\mathbb{R}} L_t^a(X) f''(da)$ makes sense.

Proof By linearity, it suffices to treat the case when f is convex. Furthermore, by simple "localization" arguments, we can assume that f'' is a finite measure supported on the interval $[-K, K]$ for some $K > 0$. By adding an affine function to f, we can also assume that $f = 0$ on $(-\infty, -K]$. Then, it is elementary to verify that, for every $x \in \mathbb{R}$,

$$f(x) = \int_{\mathbb{R}} (x - a)^+ f''(da),$$

and

$$f'_-(x) = \int \mathbf{1}_{\{a < x\}} f''(da). \tag{9.13}$$

Tanaka's formula gives, for every $a \in \mathbb{R}$,

$$(X_t - a)^+ = (X_0 - a)^+ + Y_t^a + Z_t^a + \frac{1}{2} L_t^a(X),$$

where we use the notation of the proof of Theorem 9.4 (and we recall that $(Y^a, a \in \mathbb{R})$ stands for the continuous modification obtained in Lemma 9.5). We can integrate the latter equality with respect to the finite measure $f''(da)$ and we get

$$f(X_t) = f(X_0) + \int Y_t^a f''(da) + \int Z_t^a f''(da) + \frac{1}{2}\int L_t^a(X) f''(da).$$

By Fubini's theorem,

$$\int Z_t^a f''(da) = \int \left(\int_0^t \mathbf{1}_{\{X_s > a\}} dV_s\right) f''(da) = \int_0^t \left(\int \mathbf{1}_{\{X_s > a\}} f''(da)\right) dV_s$$

$$= \int_0^t f'_-(X_s)\,dV_s.$$

So the proof will be complete if we can also verify that

$$\int Y_t^a f''(\mathrm{d}a) = \int_0^t f'_-(X_s)\,\mathrm{d}M_s. \tag{9.14}$$

This identity should be viewed as a kind of Fubini theorem involving a stochastic integral. To provide a rigorous justification, it is convenient to introduce the stopping times $T_n := \inf\{s \geq 0 : \langle M, M\rangle_s \geq n\}$, for every $n \geq 1$. Recalling (9.13), we see that our claim (9.14) will follow if we can verify that, for every $n \geq 1$, we have a.s.

$$\int \left(\int_0^{t\wedge T_n} \mathbf{1}_{\{X_s > a\}}\mathrm{d}M_s\right) f''(\mathrm{d}a) = \int_0^{t\wedge T_n} \left(\int \mathbf{1}_{\{X_s > a\}} f''(\mathrm{d}a)\right)\mathrm{d}M_s, \tag{9.15}$$

where in the left-hand side we agree that we consider the continuous modification of $a \mapsto \int_0^{t\wedge T_n} \mathbf{1}_{\{X_s > a\}}\mathrm{d}M_s$ provided by Lemma 9.5. It is straightforward to verify that the left-hand side of (9.15) defines a martingale M_t^f in $\mathbb{H}^2$, and furthermore, for any other martingale N in $\mathbb{H}^2$,

$$E[\langle M^f, N\rangle_\infty] = E[M_\infty^f N_\infty] = E\left[\int \left(\int_0^{T_n} \mathbf{1}_{\{X_s > a\}}\mathrm{d}\langle M, N\rangle_s\right) f''(\mathrm{d}a)\right]$$

$$= E\left[\int_0^{T_n} \left(\int \mathbf{1}_{\{X_s > a\}} f''(\mathrm{d}a)\right)\mathrm{d}\langle M, N\rangle_s\right]$$

$$= E\left[\left(\int_0^{T_n} \left(\int \mathbf{1}_{\{X_s > a\}} f''(\mathrm{d}a)\right)\mathrm{d}M_s\right) N_\infty\right].$$

By a duality argument in $\mathbb{H}^2$, this suffices to verify that M_t^f coincides with the martingale of $\mathbb{H}^2$ in the right-hand side of (9.15). This completes the proof. □

The following corollary is even more important than the preceding theorem.

Corollary 9.7 (Density of occupation time formula) *We have almost surely, for every $t \geq 0$ and every nonnegative measurable function φ on $\mathbb{R}$,*

$$\int_0^t \varphi(X_s)\,\mathrm{d}\langle X, X\rangle_s = \int_\mathbb{R} \varphi(a)\,L_t^a(X)\,\mathrm{d}a.$$

More generally, we have a.s. for any nonnegative measurable function F on $\mathbb{R}_+ \times \mathbb{R}$,

$$\int_0^\infty F(s, X_s)\,\mathrm{d}\langle X, X\rangle_s = \int_\mathbb{R} \mathrm{d}a \int_0^\infty F(s, a)\,\mathrm{d}_s L_s^a(X).$$

Proof Fix $t \geq 0$ and consider a nonnegative continuous function φ on $\mathbb{R}$ with compact support. Let f be a twice continuously differentiable function on $\mathbb{R}$ such that $f'' = \varphi$. Note that f is convex since $\varphi \geq 0$. By comparing Itô's formula applied to $f(X_t)$ and the formula of Theorem 9.6, we immediately get that a.s.

$$\int_0^t \varphi(X_s)\,\mathrm{d}\langle X, X\rangle_s = \int_{\mathbb{R}} \varphi(a)\,L_t^a(X)\,\mathrm{d}a.$$

This formula holds simultaneously (outside a set of probability zero) for every $t \geq 0$ (by a continuity argument) and for every function φ belonging to a countable dense subset of the set of all nonnegative continuous functions on $\mathbb{R}$ with compact support. This suffices to conclude that a.s. for every $t \geq 0$, the random measure

$$A \mapsto \int_0^t \mathbf{1}_A(X_s)\,\mathrm{d}\langle X, X\rangle_s$$

has density $(L_t^a(X))_{a \in \mathbb{R}}$ with respect to Lebesgue measure on $\mathbb{R}$. This gives the first assertion of the corollary. It follows that the formula in the second assertion holds when F is of the type

$$F(s, a) = \mathbf{1}_{[u,v]}(s)\,\mathbf{1}_A(a)$$

where $0 \leq u \leq v$ and A is a Borel subset of $\mathbb{R}$. Hence, a.s. the σ-finite measures

$$B \longrightarrow \int_0^\infty \mathbf{1}_B(s, X_s)\,\mathrm{d}\langle X, X\rangle_s$$

and

$$B \longrightarrow \int_{\mathbb{R}} \mathrm{d}a \int_0^\infty \mathbf{1}_B(s, a)\,\mathrm{d}_s L_s^a(X)$$

take the same value for B of the form $B = [u, v] \times A$, and this implies that the two measures coincide. $\qquad\square$

If $X = M + V$ is a continuous semimartingale, then an immediate application of the density of occupation time formula gives, for every $b \in \mathbb{R}$,

$$\int_0^t \mathbf{1}_{\{X_s=b\}}\mathrm{d}\langle M, M\rangle_s = \int_{\mathbb{R}} \mathbf{1}_{\{b\}}(a)\,L_t^a(X)\,\mathrm{d}a = 0.$$

This property has already been derived after the proof of Lemma 9.5. On the other hand, there may exist values of b such that

$$\int_0^t \mathbf{1}_{\{X_s=b\}}\mathrm{d}V_s \neq 0,$$

and these values of b correspond to discontinuities of the local time with respect to the space variable, as shown by Theorem 9.4.

Corollary 9.8 *If X is of the form $X_t = X_0 + V_t$, where V is a finite variation process, then $L_t^a(X) = 0$ for all $a \in \mathbb{R}$ and $t \geq 0$.*

Proof From the density of occupation time formula and the fact that $\langle X, X \rangle = 0$, we get $\int_{\mathbb{R}} \varphi(a) L_t^a(X) \, da = 0$ for any nonnegative measurable function φ, and the desired result follows.   □

Remark We could have derived the last corollary directly from Tanaka's formula.

9.3 Approximations of Local Times

Our first approximation result is an easy consequence of the density of occupation time formula.

Proposition 9.9 *Let X be a continuous semimartingale. Then a.s. for every $a \in \mathbb{R}$ and $t \geq 0$,*

$$L_t^a(X) = \lim_{\varepsilon \to 0} \frac{1}{\varepsilon} \int_0^t \mathbf{1}_{\{a \leq X_s \leq a+\varepsilon\}} \, d\langle X, X \rangle_s.$$

Proof By the density of occupation time formula,

$$\frac{1}{\varepsilon} \int_0^t \mathbf{1}_{\{a \leq X_s \leq a+\varepsilon\}} \, d\langle X, X \rangle_s = \frac{1}{\varepsilon} \int_a^{a+\varepsilon} L_t^b(X) \, db,$$

and the result follows from the right-continuity of $b \mapsto L_t^b(X)$ at a (Theorem 9.4).   □

Remark The same argument gives

$$\lim_{\varepsilon \to 0} \frac{1}{2\varepsilon} \int_0^t \mathbf{1}_{\{a-\varepsilon \leq X_s \leq a+\varepsilon\}} \, d\langle X, X \rangle_s = \frac{1}{2} \left(L_t^a(X) + L_t^{a-}(X) \right).$$

The quantity $\tilde{L}_t^a(X) := \frac{1}{2} \left(L_t^a(X) + L_t^{a-}(X) \right)$ is sometimes called the symmetric local time of the semimartingale X. Note that the density of occupation time formula remains true if $L_t^a(X)$ is replaced by $\tilde{L}_t^a(X)$ (indeed, $\tilde{L}_t^a(X)$ and $L_t^a(X)$ may differ in at most countably many values of a). The generalized Itô formula (Theorem 9.6) also remains true if $L_t^a(X)$ is replaced by $\tilde{L}_t^a(X)$, provided the left-derivative f_-' is replaced by $\frac{1}{2}(f_+' + f_-')$. Similar observations apply to Tanaka's formulas.

As a consequence of the preceding proposition and Lemma 9.5, we derive a useful bound on moments of local times.

Corollary 9.10 *Let $p \geq 1$. There exists a constant C_p such that, for any continuous semimartingale X with canonical decomposition $X = M + V$, we have for every $a \in \mathbb{R}$ and $t \geq 0$,*

$$E[(L_t^a(X))^p] \leq C_p \Big(E[(\langle M, M \rangle_t)^{p/2}] + E\Big[\Big(\int_0^t |dV_s|\Big)^p\Big]\Big).$$

Proof This readily follows from the bound of Lemma 9.5, using the approximation of $L_t^a(X)$ in Proposition 9.9 and Fatou's lemma. $\qquad\square$

We next turn to the upcrossing approximation of local time. We first need to introduce some notation. We let X be a continuous semimartingale, and $\varepsilon > 0$. We then introduce two sequences $(\sigma_n^\varepsilon)_{n \geq 1}$ and $(\tau_n^\varepsilon)_{n \geq 1}$ of stopping times, which are defined inductively by

$$\sigma_1^\varepsilon := \inf\{t \geq 0 : X_t = 0\}, \quad \tau_1^\varepsilon := \inf\{t \geq \sigma_1^\varepsilon : X_t = \varepsilon\},$$

and, for every $n \geq 1$,

$$\sigma_{n+1}^\varepsilon := \inf\{t \geq \tau_n^\varepsilon : X_t = 0\}, \quad \tau_{n+1}^\varepsilon := \inf\{t > \sigma_n^\varepsilon : X_t = \varepsilon\}.$$

We then define the upcrossing number of X along $[0, \varepsilon]$ before time t by

$$N_\varepsilon^X(t) = \mathrm{Card}\{n \geq 1 : \tau_n^\varepsilon \leq t\}.$$

This notion has already been introduced in Sect. 3.3 with a slightly different presentation.

Proposition 9.11 *We have, for every $t \geq 0$,*

$$\varepsilon N_\varepsilon^X(t) \underset{\varepsilon \to 0}{\longrightarrow} \frac{1}{2} L_t^0(X)$$

in probability.

Proof To simplify notation, we write L_s^0 instead of $L_s^0(X)$ in this proof. We first use Tanaka's formula to get, for every $n \geq 1$,

$$(X_{\tau_n^\varepsilon \wedge t})^+ - (X_{\sigma_n^\varepsilon \wedge t})^+ = \int_{\sigma_n^\varepsilon \wedge t}^{\tau_n^\varepsilon \wedge t} \mathbf{1}_{\{X_s > 0\}} dX_s + \frac{1}{2}(L_{\tau_n^\varepsilon \wedge t}^0 - L_{\sigma_n^\varepsilon \wedge t}^0).$$

We sum the last identity over all $n \geq 1$ to get

$$\sum_{n=1}^\infty \big((X_{\tau_n^\varepsilon \wedge t})^+ - (X_{\sigma_n^\varepsilon \wedge t})^+\big) = \int_0^t \Big(\sum_{n=1}^\infty \mathbf{1}_{(\sigma_n^\varepsilon, \tau_n^\varepsilon]}(s)\Big) \mathbf{1}_{\{X_s > 0\}} dX_s + \frac{1}{2}\sum_{n=1}^\infty (L_{\tau_n^\varepsilon \wedge t}^0 - L_{\sigma_n^\varepsilon \wedge t}^0).$$

$$\tag{9.16}$$

Note that there are only finitely many values of n such that $\tau_\varepsilon^n \leq t$, and that the interversion of the series and the stochastic integral is justified by approximating the series with finite sums and using Proposition 5.8 (the required domination is obvious since the integrands are bounded by 1).

Consider the different terms in (9.16). Since the local time L^0 does not increase on intervals of the type $[\tau_n^\varepsilon, \sigma_{n+1}^\varepsilon)$ (nor on $[0, \sigma_1^\varepsilon)$), we have

$$\sum_{n=1}^\infty (L^0_{\tau_n^\varepsilon \wedge t} - L^0_{\sigma_n^\varepsilon \wedge t}) = \sum_{n=1}^\infty (L^0_{\sigma_{n+1}^\varepsilon \wedge t} - L^0_{\sigma_n^\varepsilon \wedge t}) = L^0_t.$$

Then, noting that $(X_{\tau_n^\varepsilon \wedge t})^+ - (X_{\sigma_n^\varepsilon \wedge t})^+ = \varepsilon$ if $\tau_n^\varepsilon \leq t$, we have

$$\sum_{n=1}^\infty \left((X_{\tau_n^\varepsilon \wedge t})^+ - (X_{\sigma_n^\varepsilon \wedge t})^+\right) = \varepsilon\, N_\varepsilon^X(t) + u(\varepsilon),$$

where $0 \leq u(\varepsilon) \leq \varepsilon$.

From (9.16) and the last two displays, the result of the proposition will follow if we can verify that

$$\int_0^t \left(\sum_{n=1}^\infty \mathbf{1}_{(\sigma_n^\varepsilon, \tau_n^\varepsilon]}(s)\right) \mathbf{1}_{\{X_s > 0\}} dX_s \xrightarrow[\varepsilon \to 0]{} 0$$

in probability. This is again a consequence of Proposition 5.8, since

$$0 \leq \left(\sum_{n=1}^\infty \mathbf{1}_{(\sigma_n^\varepsilon, \tau_n^\varepsilon]}(s)\right) \mathbf{1}_{\{X_s > 0\}} \leq \mathbf{1}_{\{0 < X_s \leq \varepsilon\}}$$

and $\mathbf{1}_{\{0 < X_s \leq \varepsilon\}} \longrightarrow 0$ as $\varepsilon \to 0$. $\qquad\square$

9.4   The Local Time of Linear Brownian Motion

Throughout this section, $(B_t)_{t \geq 0}$ is a real Brownian motion started from 0 and $(\mathscr{F}_t)$ is the (completed) canonical filtration of B.

The following theorem, which is known as Trotter's theorem, is essentially a restatement of the results of the previous sections in the special case of real Brownian motion. Still the importance of the result justifies this repetition. We write $\mathrm{supp}(\mu)$ for the topological support of a finite measure μ on $\mathbb{R}_+$.

Theorem 9.12 (Trotter)   *There exists a (unique) process $(L_t^a(B))_{a \in \mathbb{R}, t \geq 0}$, whose sample paths are continuous functions of the pair (a, t), such that, for every fixed $a \in \mathbb{R}$, $(L_t^a(B))_{t \geq 0}$ is an increasing process, and, a.s. for every $t \geq 0$, for every* *nonnegative measurable function φ on $\mathbb{R}$,*

$$\int_0^t \varphi(B_s)\, ds = \int_{\mathbb{R}} \varphi(a)\, L_t^a(B)\, da.$$

Furthermore, a.s. for every $a \in \mathbb{R}$,

$$\mathrm{supp}(d_s L_s^a(B)) \subset \{s \geq 0 : B_s = a\}, \tag{9.17}$$

and this inclusion is an equality with probability one if a is fixed.

Proof The first assertion follows by applying Theorem 9.4 and Corollary 9.7 to $X = B$, noting that $\langle B, B \rangle_t = t$. We have already seen that the inclusion (9.17) holds with probability one if a is fixed, hence simultaneously for all rationals, a.s. A continuity argument allows us to get that (9.17) holds simultaneously for all $a \in \mathbb{R}$ outside a single set of probability zero. Indeed, suppose that for some $a \in \mathbb{R}$ and $0 \leq s < t$ we have $L_t^a(B) > L_s^a(B)$ and $B_r \neq a$ for every $r \in [s, t]$. Then we can find a rational $b \in \mathbb{R}$ sufficiently close to a such that the same properties hold when a is replaced by b, giving a contradiction.

Finally, let us verify that (9.17) is an a.s. equality if $a \in \mathbb{R}$ is fixed. So let us fix $a \in \mathbb{R}$, and for every rational $q \geq 0$, set

$$H_q := \inf\{t \geq q : B_t = a\}.$$

Our claim will follow if we can verify that a.s. for every $\varepsilon > 0$, $L_{H_q+\varepsilon}^a(B) > L_{H_q}^a(B)$. Using the strong Markov property at time H_q, it suffices to prove that, if B' is a real Brownian motion started from a, we have $L_\varepsilon^a(B') > 0$, for every $\varepsilon > 0$, a.s. Clearly we can take $a = 0$. We then observe that we have

$$L_\varepsilon^0(B) \overset{(d)}{=} \sqrt{\varepsilon}\, L_1^0(B),$$

by an easy scaling argument (use for instance the approximations of the previous section). Also $P(L_1^0(B) > 0) > 0$ since $E[L_1^0(B)] = E[|B_1|]$ by Tanaka's formula. An application of Blumenthal's zero-one law (Theorem 2.13) to the event

$$A := \bigcap_{n=1}^{\infty} \{L_{2^{-n}}^0(B) > 0\} = \lim_{n\uparrow\infty} \uparrow \{L_{2^{-n}}^0(B) > 0\}$$

completes the proof.   □

Remark Theorem 9.12 remains true with a similar proof for an arbitrary (possibly random) initial value B_0.

We now turn to distributional properties of local times of Brownian motion.

Proposition 9.13 (i) *Let $a \in \mathbb{R} \setminus \{0\}$ and $T_a := \inf\{t \geq 0 : B_t = a\}$. Then $L^0_{T_a}(B)$ has an exponential distribution with mean $2|a|$.*

(ii) *Let $a > 0$ and $U_a := \inf\{t \geq 0 : |B_t| = a\}$. Then $L^0_{U_a}(B)$ has an exponential distribution with mean a.*

Proof

(i) By simple scaling and symmetry arguments, it is enough to take $a = 1$. We then observe that $L^0_\infty(B) = \infty$ a.s. Indeed, the scaling argument of the preceding proof shows that $L^0_\infty(B)$ has the same distribution as $\lambda L^0_\infty(B)$, for any $\lambda > 0$, and we have also seen that $L^0_\infty(B) > 0$ a.s. Fix $s > 0$ and set

$$\tau := \inf\{t \geq 0 : L^0_t(B) \geq s\},$$

so that τ is a stopping time of the filtration $(\mathscr{F}_t)$. Furthermore, $B_\tau = 0$ by the support property of local time. By the strong Markov property,

$$B'_t := B_{\tau+t}$$

is a Brownian motion started from 0, which is also independent of $\mathscr{F}_\tau$. Proposition 9.9 gives, for every $t \geq 0$,

$$L^0_t(B') = L^0_{\tau+t}(B) - s.$$

On the event $\{L^0_{T_1}(B) \geq s\} = \{\tau \leq T_1\}$, we thus have

$$L^0_{T_1}(B) - s = L^0_{T_1-\tau}(B') = L^0_{T'_1}(B'),$$

where $T'_1 := \inf\{t \geq 0 : B'_t = 1\}$. Since the event $\{\tau \leq T_1\}$ is $\mathscr{F}_\tau$-measurable and B' is independent of $\mathscr{F}_\tau$, we get that the conditional distribution of $L^0_{T_1}(B) - s$ knowing that $L^0_{T_1}(B) \geq s$ is the same as the unconditional distribution of $L^0_{T_1}(B)$. This implies that the distribution of $L^0_{T_1}(B)$ is exponential.

Finally, Tanaka's formula (9.6) shows that $\frac{1}{2}E[L^0_{t \wedge T_1}] = E[(B_{t \wedge T_1})^+]$. As $t \to \infty$, $E[L^0_{t \wedge T_1}]$ converges to $E[L^0_{T_1}]$ by monotone convergence and $E[(B_{t \wedge T_1})^+]$ converges to $E[(B_{T_1})^+]$ by dominated convergence, since $0 \leq (B_{t \wedge T_1})^+ \leq 1$. This shows that $E[L^0_{t \wedge T_1}] = 2$, as desired.

(ii) The argument is exactly similar. We now use Tanaka's formula (9.5) to verify that $E[L^0_{U_a}(B)] = a$. $\qquad\square$

Remark One can give an alternative proof of the proposition using stochastic calculus. To get (ii), for instance, use Itô's formula to verify that, for every $\lambda > 0$,

$$(1 + \lambda|B_t|) \exp(-\lambda L^0_t(B))$$

is a continuous local martingale, which is bounded on $[0, U_a]$. An application of the optional stopping theorem then shows that $E[\exp(-\lambda L^0_{U_a}(B))] = (1 + \lambda a)^{-1}$.

The previous proof has the advantage of explaining the appearance of the exponential distribution.

For every $t \geq 0$, we set

$$S_t := \sup_{0 \leq s \leq t} B_s , \quad I_t := \inf_{0 \leq s \leq t} B_s .$$

Theorem 9.14 (Lévy) *The two processes $(S_t, S_t - B_t)_{t \geq 0}$ and $(L_t^0(B), |B_t|)_{t \geq 0}$ have the same distribution.*

Remark By an obvious symmetry argument, the pair $(-I_t, B_t - I_t)_{t \geq 0}$ also has the same distribution as $(S_t, S_t - B_t)_{t \geq 0}$.

Proof By Tanaka's formula, for every $t \geq 0$,

$$|B_t| = -\beta_t + L_t^0(B), \tag{9.18}$$

where

$$\beta_t = -\int_0^t \operatorname{sgn}(B_s) \, dB_s.$$

Since $\langle \beta, \beta \rangle_t = t$, Theorem 5.12 ensures that β is a real Brownian motion started from 0. We then claim that, for every $t \geq 0$,

$$L_t^0(B) = \sup\{\beta_s : s \leq t\}.$$

The fact that $L_t^0(B) \geq \sup\{\beta_s : s \leq t\}$ is immediate since (9.18) shows that $L_t^0(B) \geq L_s^0(B) \geq \beta_s$ for every $s \in [0, t]$. To get the reverse inequality, write γ_t for the last zero of B before time t. By the support property of local time, $L_t^0(B) = L_{\gamma_t}^0(B)$, and using (9.18), $L_{\gamma_t}^0(B) = \beta_{\gamma_t} \leq \sup\{\beta_s : s \leq t\}$.

We have thus proved a.s.

$$(L_t^0(B), |B_t|)_{t \geq 0} = (\sup\{\beta_s : s \leq t\}, \sup\{\beta_s : s \leq t\} - \beta_t)_{t \geq 0},$$

and since $(\beta_s)_{s \geq 0}$ and $(B_s)_{s \geq 0}$ have the same distribution, the pair in the right-hand side has the same distribution as $(S_t, S_t - B_t)_{t \geq 0}$. $\square$

Theorem 9.14 has several interesting consequences. For every $t \geq 0$, S_t has the same law as $|B_t|$ (Theorem 2.21), and thus the same holds for $L_t^0(B)$. From the explicit formula (2.2) for the density of (S_t, B_t), we also get the density of the pair $(L_t^0(B), B_t)$.

For every $s \geq 0$, set

$$\tau_s := \inf\{t \geq 0 : L_t^0(B) > s\}.$$

The process $(\tau_s)_{s\geq0}$ is called the inverse local time (at 0) of the Brownian motion B. By construction, $(\tau_s)_{s\geq0}$ has càdlàg increasing sample paths. From Lévy's Theorem 9.14, one gets that

$$(\tau_s)_{s\geq0} \overset{(d)}{=} (\tilde{T}_s)_{s\geq0},$$

where, for every $s \geq 0$, $\tilde{T}_s := \inf\{t \geq 0 : B_t > s\}$ (it is easy to verify that, for every $s \geq 0$, $\tilde{T}_s = T_s$ a.s., but $(\tilde{T}_s)_{s\geq0}$ has càdlàg sample paths, which is not the case for $(T_s)_{s\geq0}$).

The same application of the strong Markov property as in the proof of Proposition 9.13 shows that $(\tau_s)_{s\geq0}$ has stationary independent increments – compare with Exercise 2.26. Furthermore, using the invariance of Brownian motion under scaling, we have for every $\lambda > 0$,

$$(\tau_{\lambda s})_{s\geq0} \overset{(d)}{=} (\lambda^2 \tau_s)_{s\geq0}.$$

The preceding properties can be summarized by saying that $(\tau_s)_{s\geq0}$ is a stable subordinator with index $1/2$ (a subordinator is a Lévy process with nondecreasing sample paths).

The interest of considering the process $(\tau_s)_{s\geq0}$ comes in part from the following proposition.

Proposition 9.15 *We have a.s.*

$$\{t \geq 0 : B_t = 0\} = \{\tau_s : s \geq 0\} \cup \{\tau_{s-} : s \in D\}$$

where D is the countable set of jump times of $(\tau_s)_{s\geq0}$.

Proof We know from (9.17) that a.s.

$$\mathrm{supp}(d_t L_t^0(B)) \subset \{t \geq 0 : B_t = 0\}.$$

It follows that any time t of the form $t = \tau_s$ or $t = \tau_{s-}$ must belong to the zero set of B. Conversely, recalling that (9.17) is an a.s. equality for $a = 0$, we also get that, a.s. for every t such that $B_t = 0$, we have either $L_{t+\varepsilon}^0(B) > L_t^0(B)$ for every $\varepsilon > 0$, or, if $t > 0$, $L_t^0(B) > L_{t-\varepsilon}^0(B)$ for every $\varepsilon > 0$ with $\varepsilon \leq t$ (or both simultaneously), which implies that we have $t = \tau_{L_t^0(B)}$ or $t = \tau_{L_t^0(B)-}$. $\square$

As a consequence of Proposition 9.15, the connected components of the complement of the zero set $\{t \geq 0 : B_t = 0\}$ are exactly the intervals (τ_{s-}, τ_s) for $s \in D$. These connected components are called the *excursion intervals* (away from 0). For every $s \in D$, the associated *excursion* is defined by

$$e_s(t) := B_{(\tau_{s-}+t)\wedge\tau_s}, \quad t \geq 0.$$

The goal of excursion theory is to describe the distribution of the excursion process, that is, of the collection $(e_s)_{s \in D}$. This study, however, goes beyond the scope of the present book, see in particular [4, 70, 72].

9.5   The Kallianpur–Robbins Law

In this section, we use local times to give a short proof of the Kallianpur–Robbins law for planar Brownian motion, which was stated at the end of Chap. 7 as Theorem 7.23. Let us recall the notation we need. We let B stand for a complex Brownian motion, and for simplicity we assume that $B_0 = 1$ (the general case will then follow, for instance by applying the strong Markov property at the first hitting time of the unit circle). According to Theorem 7.19, we can write $|B_t| = \exp(\beta_{H_t})$, where β is a real Brownian motion started from 0, and

$$H_t = \int_0^t \frac{ds}{|B_s|^2} = \inf\{s \geq 0 : \int_0^s \exp(2\beta_u) \, du > t\}.$$

For every $\lambda > 0$, we also consider the scaled Brownian motion $\beta_t^{(\lambda)} = \frac{1}{\lambda} \beta_{\lambda^2 t}$, and for $t > 1$ we use the notation $\lambda_t = (\log t)/2$.

We aim at proving that, for every $R > 0$,

$$\frac{2}{\log t} \int_0^t \mathbf{1}_{\{|B_s| < R\}} \, ds$$

converges in distribution as $t \to \infty$ to an exponential distribution with mean R^2. To this end, we write, for every fixed $t > 1$,

$$\frac{2}{\log t} \int_0^t \mathbf{1}_{\{|B_s| < R\}} \, ds = \frac{1}{\lambda_t} \int_0^t \mathbf{1}_{\{\beta_{H_s} < \log R\}} \, ds$$

$$= \frac{1}{\lambda_t} \int_0^{H_t} \mathbf{1}_{\{\beta_u < \log R\}} \exp(2\beta_u) \, du$$

$$= \lambda_t \int_0^{(\lambda_t)^{-2} H_t} \mathbf{1}_{\{\beta_u^{(\lambda_t)} < (\lambda_t)^{-1} \log R\}} \exp(2\lambda_t \beta_u^{(\lambda_t)}) \, du$$

$$= \lambda_t \int_{-\infty}^{(\lambda_t)^{-1} \log R} \exp(2\lambda_t a) L_{(\lambda_t)^{-2} H_t}^a (\beta^{(\lambda_t)}) \, da$$

$$= \int_0^R L_{(\lambda_t)^{-2} H_t}^{(\lambda_t)^{-1} \log r} (\beta^{(\lambda_t)}) \, r \, dr.$$

In the second last equality, we applied the density of occupation time formula (Corollary 9.7) to the Brownian motion $\beta^{(\lambda_t)}$, and in the last one we used the change of variables $r = e^{\lambda_t a}$. As $t \to \infty$, $(\lambda_t)^{-1} \log r \longrightarrow 0$, for every $r > 0$, and Lemma 7.21 also tells us that $(\lambda_t)^{-2} H_t - T_1^{(\lambda_t)}$ converges in probability to 0, with the notation $T_1^{(\lambda)} = \inf\{s \geq 0 : \beta_s^{(\lambda)} = 1\}$. From the joint continuity of Brownian local times (Theorem 9.12), we then get that, for every $\varepsilon \in (0, R)$,

$$\sup_{\varepsilon \leq r \leq R} \left| L_{(\lambda_t)^{-2} H_t}^{(\lambda_t)^{-1} \log r}(\beta^{(\lambda_t)}) - L_{T_1^{(\lambda_t)}}^0(\beta^{(\lambda_t)}) \right| \underset{t \to \infty}{\longrightarrow} 0,$$

in probability. By combining this with the previous display, we obtain that

$$\left| \frac{2}{\log t} \int_0^t \mathbf{1}_{\{|B_s| < R\}} \, ds - \frac{R^2}{2} L_{T_1^{(\lambda_t)}}^0(\beta^{(\lambda_t)}) \right| \underset{t \to \infty}{\longrightarrow} 0,$$

in probability. To complete the proof, we just note that the law of $L_{T_1^{(\lambda)}}^0(\beta^{(\lambda)})$ does not depend on $\lambda > 0$, and is exponential with mean 2, by Proposition 9.13.

Remark The preceding proof shows that the limiting exponential variable in Theorem 7.23 does not depend on the choice of R, in the sense that we can obtain a joint convergence by taking several values of R, with the same exponential variable in the limit, up to multiplicative constants. This can also be deduced from the Chacon–Ornstein ergodic theorem, which implies that the same limit in distribution holds more generally for the occupation time of an arbitrary compact subset K of $\mathbb{C}$, the constant R^2 then being replaced by π^{-1} times the Lebesgue measure of K. Our method of proof also shows that the convergence in the Kallianpur–Robbins theorem holds jointly with that of windings in Spitzer's theorem (Theorem 7.20) and the limiting joint distribution is the law of $(\frac{R^2}{2} L_{T_1}^0(\beta), \gamma_{T_1})$, where β and γ are independent real Brownian motions started from 0, and $T_1 = \inf\{t \geq 0 : \beta_t = 1\}$.

Exercises

Exercise 9.16 Let $f : \mathbb{R} \longrightarrow \mathbb{R}$ be a monotone increasing function, and assume that f is a difference of convex functions. Let X be a semimartingale and consider the semimartingale $Y_t = f(X_t)$. Prove that, for every $a \in \mathbb{R}$,

$$L_t^a(Y) = f'_+(a) L_t^a(X), \quad L_t^{a-}(Y) = f'_-(a) L_t^{a-}(X).$$

In particular, if X is a Brownian motion, the local times of $f(X)$ are continuous in the space variable if and only if f is continuously differentiable.

Exercise 9.17 Let M be a continuous local martingale such that $\langle M, M \rangle_\infty = \infty$ a.s., and let B be the Brownian motion associated with M via the Dambis–Dubins–Schwarz theorem (Theorem 5.13). Prove that, a.s. for every $a \geq 0$ and $t \geq 0$,

$$L_t^a(M) = L_{\langle M, M \rangle_t}^a(B).$$

Exercise 9.18 Let X be a continuous semimartingale, and assume that X can be written in the form

$$X_t = X_0 + \int_0^t \sigma(\omega, s)\, dB_s + \int_0^t b(\omega, s)\, ds,$$

where B is a Brownian motion and σ and b are progressive and locally bounded. Assume that $\sigma(\omega, s) \neq 0$ for Lebesgue a.e. $s \geq 0$, a.s. Show that the local times $L_t^a(X)$ are jointly continuous in the pair (a, t).

Exercise 9.19 Let X be a continuous semimartingale. Show that the property

$$\mathrm{supp}(d_s L_s^a(X)) \subset \{s \geq 0 : X_s = a\}$$

holds simultaneously for all $a \in \mathbb{R}$, outside a single set of probability zero.

Exercise 9.20 Let B be a Brownian motion started from 0. Show that a.s. there exists an $a \in \mathbb{R}$ such that the inclusion $\mathrm{supp}(d_s L_s^a(B)) \subset \{s \geq 0 : B_s = a\}$ is not an equality. (*Hint*: Consider the maximal value of B over $[0, 1]$.)

Exercise 9.21 Let B be a Brownian motion started from 0. Note that

$$\int_0^\infty \mathbf{1}_{\{B_s > 0\}} ds = \infty$$

a.s. and set, for every $t \geq 0$,

$$A_t = \int_0^t \mathbf{1}_{\{B_s > 0\}} ds\,, \quad \sigma_t = \inf\{s \geq 0 : A_s > t\}.$$

1. Verify that the process

$$\gamma_t = \int_0^{\sigma_t} \mathbf{1}_{\{B_s > 0\}} dB_s$$

is a Brownian motion in an appropriate filtration.
2. Show that the process $\Lambda_t = L_{\sigma_t}^0(B)$ has nondecreasing and continuous sample paths, and that the support of the measure $d_s \Lambda_s$ is contained in $\{s : B_{\sigma_s} = 0\}$.
3. Show that the process $(B_{\sigma_t})_{t \geq 0}$ has the same distribution as $(|B_t|)_{t \geq 0}$.

Exercise 9.22 (*Skew Brownian motion*) Let $\alpha, \beta > 0$ and consider the function $g(x) = \alpha 1_{\{x \geq 0\}} - \beta 1_{\{x < 0\}}$. Let X be a continuous semimartingale such that

$$X_t = \int_0^t g(X_s)\, dB_s, \qquad (9.19)$$

where B is a Brownian motion.

1. Set $\varphi(x) = \frac{1}{\alpha} x 1_{\{x \geq 0\}} - \frac{1}{\beta} x 1_{\{x < 0\}}$, and $Y_t = \varphi(X_t)$. Prove that $Y_t = \varphi(X_t)$ solves the equation

$$Y_t = B_t + \frac{1}{2}(1 - \frac{\alpha}{\beta}) L_t^0(Y)$$

   (use the result of Exercise 9.16).
2. Compute $L_t^0(Y) - L_t^{0-}(Y)$ in terms of $L_t^0(Y)$, in two different ways.
3. Starting from a Brownian motion β with $\beta_0 = 0$, set

$$A_t = \int_0^t \frac{ds}{g(\beta_s)^2}, \qquad \sigma_t = \inf\{s \geq 0 : A_s > t\}.$$

   Verify that the process $X_t = \beta_{\sigma_t}$ satisfies the equation (9.19) in an appropriate filtration and with an appropriate Brownian motion B.

Exercise 9.23 Let $g : \mathbb{R} \longrightarrow \mathbb{R}$ be a real integrable function ($\int_{\mathbb{R}} |g(x)|\, dx < \infty$). Let B be a Brownian motion started from 0, and set

$$A_t = \int_0^t g(B_s)\, ds.$$

1. Justify the fact that the integral defining A_t makes sense, and verify that, for every $c > 0$ and every $u \geq 0$, $A_{c^2 u}$ has the same distribution as

$$c^2 \int_0^u g(c\, B_s)\, ds.$$

2. Prove that

$$\frac{1}{\sqrt{t}} A_t \xrightarrow[t \to \infty]{(d)} \left(\int g(x)\, dx\right) |N|,$$

   where N is $\mathcal{N}(0, 1)$.

Exercise 9.24 Let σ and b be two locally bounded measurable functions on $\mathbb{R}_+ \times \mathbb{R}$, and consider the stochastic differential equation

$$E(\sigma, b) \qquad\qquad dX_t = \sigma(t, X_t)\, dB_t + b(t, X_t)\, dt.$$

Let X and X' be two solutions of $E(\sigma, b)$, on the same filtered probability space and with the same Brownian motion B.

1. Suppose that $L_t^0(X - X') = 0$ for every $t \geq 0$. Show that both $X_t \vee X_t'$ and $X_t \wedge X_t'$ are also solutions of $E(\sigma, b)$. (*Hint:* Write $X_t \vee X_t' = X_t + (X_t' - X_t)^+$, and use Tanaka's formula.)
2. Suppose that $\sigma(t, x) = 1$ for every t, x. Show that the assumption in question 1. holds automatically. Suppose in addition that weak uniqueness holds for $E(\sigma, b)$. Show that, if $X_0 = X_0' = x \in \mathbb{R}$, the two processes X and X' are indistinguishable.

Exercise 9.25 (*Another look at the Yamada–Watanabe criterion*) Let ρ be a nondecreasing function from $[0, \infty)$ into $[0, \infty)$ such that, for every $\varepsilon > 0$,

$$\int_0^\varepsilon \frac{du}{\rho(u)} = \infty.$$

Consider then the one-dimensional stochastic differential equation

$$E(\sigma, b) \qquad\qquad\qquad dX_t = \sigma(X_t)dB_t + b(X_t)dt$$

where one assumes that the functions σ and b satisfy the conditions

$$(\sigma(x) - \sigma(y))^2 \leq \rho(|x - y|), \quad |b(x) - b(y)| \leq K|x - y|,$$

for every $x, y \in \mathbb{R}$, with a constant $K < \infty$. Our goal is use local times to give a short proof of pathwise uniqueness for $E(\sigma, b)$ (this is slightly stronger than the result of Exercise 8.14).

1. Let Y be a continuous semimartingale such that, for every $t > 0$,

$$\int_0^t \frac{d\langle Y, Y \rangle_s}{\rho(|Y_s|)} < \infty, \quad \text{a.s.}$$

Prove that $L_t^0(Y) = 0$ for every $t \geq 0$, a.s.
2. Let X and X' be two solutions of $E(\sigma, b)$ on the same filtered probability space and with the same Brownian motion B. By applying question 1. to $Y = X - X'$, prove that $L_t^0(X - X') = 0$ for every $t \geq 0$, a.s., and therefore,

$$|X_t - X_t'| = |X_0 - X_0'| + \int_0^t (\sigma(X_s) - \sigma(X_s')) \operatorname{sgn}(X_s - X_s') \, dB_s$$

$$+ \int_0^t (b(X_s) - b(X_s')) \operatorname{sgn}(X_s - X_s') \, ds.$$

3. Using Gromwall's lemma, prove that if $X_0 = X_0'$ then $X_t = X_t'$ for every $t \geq 0$, a.s.

Notes and Comments

The local time of Brownian motion was first discussed by Lévy [54] under the name "mesure du voisinage". In 1958, Trotter [80] established the joint continuity of Brownian local times viewed as densities of the occupation measure. Tanaka [79] obtained the formulas of Proposition 9.2 in the Brownian case. The local time of semimartingales was discussed by Meyer [61], who derived Theorem 9.6 in this general setting (after the earlier work of Tanaka [79] in the Brownian setting). Yor [84] then developed the powerful approach that leads to Theorem 9.4. The upcrossing approximation of local time (Proposition 9.11) is due to Itô and McKean [42, Chapter 2] for Brownian motion, and was extended to semimartingales by El Karoui [22]. Other approximation results for the Brownian local time were obtained by Lévy [54] (see [42] and [70]). Theorem 9.14 is essentially due to Lévy [54], but our proof is based on an argument from Skorokhod [74] (see [23] for a related study of the so-called "reflection problem" in the semimartingale setting). For further properties of local times, the reader may consult Chapter V of [70] or Chapter 6 of [49], as well as the classical book [42] of Itô and McKean. Local times are also a key ingredient of excursion theory, which is treated in the general setting of Markov processes in Blumenthal's book [4]. Exercise 9.22 deals with the singular stochastic equation first studied by Harrison and Shepp [31], whose solution is the so-called skew Brownian motion. Exercise 9.23 gives the one-dimensional version of the Kallianpur–Robbins law, which can also be found in [48]. Exercise 9.25 is from [51].

Erratum to: Brownian Motion, Martingales, and Stochastic Calculus

Jean-François Le Gall

Erratum to:
J.-F. Le Gall, *Brownian Motion, Martingales, and Stochastic Calculus*,
Graduate Texts in Mathematics 274, DOI 10.1007/978-3-319-31089-3
© Springer International Publishing Switzerland 2016

The following information about the original edition was missing in the front-matter:

Translated from the French language edition:
'Mouvement brownien, martingales et calcul stochastique' by Jean-François Le Gall
Copyright © Springer-Verlag Berlin Heidelberg 2013
Springer International Publishing is part of Springer Science+Business Media
All Rights Reserved.

The online version of the updated book can be found under
DOI 10.1007/978-3-319-31089-3

Jean-François Le Gall
Département de Mathématiques
Université Paris-Sud
Orsay Cedex, France

© Springer International Publishing Switzerland 2016   E1
J.-F. Le Gall, *Brownian Motion, Martingales, and Stochastic Calculus*,
Graduate Texts in Mathematics 274, DOI 10.1007/978-3-319-31089-3_10

Appendix A1
The Monotone Class Lemma

The monotone class lemma is a tool of measure theory which is very useful in several arguments of probability theory. We give here the version of this lemma that is used in several places in this volume.

Let E be an arbitrary set, and let $\mathscr{P}(E)$ denote the set of all subsets of E. If $\mathscr{C} \subset \mathscr{P}(E)$, $\sigma(\mathscr{C})$ stands for the smallest σ-field on E containing $\mathscr{C}$ (it is also the intersection of all σ-fields containing $\mathscr{C}$).

Definition A subset $\mathscr{M}$ ōf $\mathscr{P}(E)$ is called a *monotone class* if the following properties hold:

  (i) $E \in \mathscr{M}$.
 (ii) If $A, B \in \mathscr{M}$ and $A \subset B$, then $B \backslash A \in \mathscr{M}$.
(iii) If $(A_n)_{n \geq 0}$ is an increasing sequence of subsets of E such that $A_n \in \mathscr{M}$ for
     every $n \geq 0$, then $\bigcup_{n \geq 0} A_n \in \mathscr{M}$.

A σ-field is a monotone class. As in the case of σ-fields, one immediately checks that the intersection of an arbitrary collection of monotone classes is again a monotone class. If $\mathscr{C}$ is an arbitrary subset of $\mathscr{P}(E)$, the monotone class generated by $\mathscr{C}$, which is denoted by $\mathscr{M}(\mathscr{C})$, is defined by setting

$$\mathscr{M}(\mathscr{C}) = \bigcap_{\mathscr{M} \text{ monotone class, } \mathscr{C} \subset \mathscr{M}} \mathscr{M}.$$

Monotone class lemma *If $\mathscr{C} \subset \mathscr{P}(E)$ is stable under finite intersections, then $\mathscr{M}(\mathscr{C}) = \sigma(\mathscr{C})$.*

Proof Since any σ-field is a monotone class, it is clear that $\mathscr{M}(\mathscr{C}) \subset \sigma(\mathscr{C})$. To prove the reverse inclusion, it is enough to verify that $\mathscr{M}(\mathscr{C})$ is a σ-field. However, a monotone class is a σ-field if and only if it is stable under finite intersections (indeed, considering the complementary sets shows that it is then stable under finite

© Springer International Publishing Switzerland 2016
J.-F. Le Gall, *Brownian Motion, Martingales, and Stochastic Calculus*,
Graduate Texts in Mathematics 274, DOI 10.1007/978-3-319-31089-3

unions, and via an increasing passage to the limit one gets that it is also stable under countable unions). Let us show that $\mathcal{M}(\mathscr{C})$ is stable under finite intersections.

For every $A \in \mathscr{P}(E)$, set

$$\mathcal{M}_A = \{B \in \mathcal{M}(\mathscr{C}) : A \cap B \in \mathcal{M}(\mathscr{C})\}.$$

Fix $A \in \mathscr{C}$. Since $\mathscr{C}$ is stable under finite intersections, we have $\mathscr{C} \subset \mathcal{M}_A$. Let us verify that $\mathcal{M}_A$ is a monotone class:

- $E \in \mathcal{M}_A$ is trivial.
- If $B, B' \in \mathcal{M}_A$ and $B \subset B'$, we have $A \cap (B' \backslash B) = (A \cap B') \backslash (A \cap B) \in \mathcal{M}(\mathscr{C})$ and thus $B' \backslash B \in \mathcal{M}_A$.
- If $B_n \in \mathcal{M}_A$ for every $n \geq 0$ and the sequence $(B_n)_{n \geq 0}$ is increasing, we have $A \cap (\cup B_n) = \cup(A \cap B_n) \in \mathcal{M}(\mathscr{C})$ and therefore $\cup B_n \in \mathcal{M}_A$.

Since $\mathcal{M}_A$ is a monotone class that contains $\mathscr{C}$, $\mathcal{M}_A$ also contains $\mathcal{M}(\mathscr{C})$. We have thus obtained that

$$\forall A \in \mathscr{C}, \ \forall B \in \mathcal{M}(\mathscr{C}), \ A \cap B \in \mathcal{M}(\mathscr{C}).$$

This is not yet the desired result, but we can use the same idea another time. Precisely, we now fix $A \in \mathcal{M}(\mathscr{C})$. According to the first part of the proof, $\mathscr{C} \subset \mathcal{M}_A$. By exactly the same arguments as in the first part of the proof, we get that $\mathcal{M}_A$ is a monotone class. It follows that $\mathcal{M}(\mathscr{C}) \subset \mathcal{M}_A$, which shows that $\mathcal{M}(\mathscr{C})$ is stable under finite intersections, and completes the proof.   □

Here are a few consequences of the monotone class lemma that are used above.

1. Let $\mathscr{A}$ be a σ-field on E, and let μ and ν be two probability measures on $(E, \mathscr{A})$. Assume that there exists a class $\mathscr{C} \subset \mathscr{A}$, which is stable under finite intersections, such that $\sigma(\mathscr{C}) = \mathscr{A}$ and $\mu(A) = \nu(A)$ for every $A \in \mathscr{C}$. Then $\mu = \nu$. (Use the fact that $\mathscr{G} := \{A \in \mathscr{A} : \mu(A) = \nu(A)\}$ is a monotone class.)
2. Let $(X_i)_{i \in I}$ be an arbitrary collection of random variables, and let $\mathscr{G}$ be a σ-field on the same probability space. In order to show that the σ-fields $\sigma(X_i, i \in I)$ and $\mathscr{G}$ are independent, it is enough to verify that $(X_{i_1}, \ldots, X_{i_p})$ is independent of $\mathscr{G}$, for any choice of the finite set $\{i_1, \ldots, i_p\} \subset I$. (Observe that the class of all events that depend on a finite number of the variables $X_i, i \in I$, is stable under finite intersections and generates $\sigma(X_i, i \in I)$.)
3. Let $(X_i)_{i \in I}$ be an arbitrary collection of random variables, and let Z be a bounded real variable. Let $i_0 \in I$. In order to verify that $E[Z \,|\, X_i, i \in I] = E[Z \,|\, X_{i_0}]$, it is enough to show that $E[Z \,|\, X_{i_0}, X_{i_1}, \ldots, X_{i_p}] = E[Z \,|\, X_{i_0}]$ for any choice of the finite collection $\{i_1, \ldots, i_p\} \subset I$. (Observe that the class of all events A such that $E[\mathbf{1}_A Z] = E[\mathbf{1}_A E[Z \,|\, Y_{i_0}]]$ is a monotone class.)

This last consequence of the monotone class lemma is useful in the theory of Markov processes.

Appendix A2
Discrete Martingales

In this appendix, we recall without proof the results about discrete time martingales and supermartingales that are used in Chap. 3. The proof of the subsequent statements can be found in Neveu's book [65], and in many other books dealing with discrete time martingales (see in particular Williams [82] and Chapter XII in Grimmett and Stirzaker [30]).

We use the notation $\mathbb{N} = \{0, 1, 2, \ldots\}$. Let us start by recalling the basing definitions. We consider a probability space $(\Omega, \mathscr{F}, P)$, and we fix a discrete filtration, that is, an increasing sequence $(\mathscr{G}_n)_{n \in \mathbb{N}}$ of sub-σ-fields of $\mathscr{F}$. We also let

$$\mathscr{G}_\infty = \bigvee_{n=0}^{\infty} \mathscr{G}_n$$

be the smallest σ-field that contains all the σ-fields $\mathscr{G}_n$.

Definition A sequence $(Y_n)_{n \in \mathbb{N}}$ of integrable random variables, such that Y_n is $\mathscr{G}_n$-measurable for every $n \in \mathbb{N}$, is called

- a *martingale* if, whenever $0 \le m < n$, $E[Y_n \mid \mathscr{G}_m] = Y_m$;
- a *supermartingale* if, whenever $0 \le m < n$, $E[Y_n \mid \mathscr{G}_m] \le Y_m$;
- a *submartingale* if, whenever $0 \le m < n$, $E[Y_n \mid \mathscr{G}_m] \ge Y_m$.

All these notions obviously depend on the choice of the filtration $(\mathscr{G}_n)_{n \in \mathbb{N}}$, which is fixed in what follows.

Maximal inequality *If $(Y_n)_{n \in \mathbb{N}}$ is a supermartingale, then, for every $\lambda > 0$ and every $k \in \mathbb{N}$,*

$$\lambda P\left(\sup_{n \le k} |Y_n| > \lambda \right) \le E[|Y_0|] + 2E[|Y_k|].$$

© Springer International Publishing Switzerland 2016
J.-F. Le Gall, *Brownian Motion, Martingales, and Stochastic Calculus*,
Graduate Texts in Mathematics 274, DOI 10.1007/978-3-319-31089-3

Doob's inequality in L^p *If $(Y_n)_{n \in \mathbb{N}}$ is a martingale, then, for every $k \in \mathbb{N}$ and every $p > 1$,*

$$E\left[\sup_{0 \leq n \leq k} |Y_n|^p \right] \leq \left(\frac{p}{p-1} \right)^p E[|Y_k|^p].$$

Remark This inequality is interesting only if $E[|Y_k|^p] < \infty$, since otherwise both sides are infinite.

If $y = (y_n)_{n \in \mathbb{N}}$ is a sequence of real numbers, and $a < b$, the upcrossing number of this sequence along $[a, b]$ before time n, denoted by $M_{ab}^y(n)$, is the largest integer k such that there exists a strictly increasing finite sequence

$$m_1 < n_1 < m_2 < n_2 < \cdots < m_k < n_k$$

of nonnegative integers smaller than or equal to n with the properties $y_{m_i} \leq a$ and $y_{n_i} \geq b$, for every $i \in \{1, \ldots, k\}$. In what follows we consider a sequence $Y = (Y_n)_{n \in \mathbb{N}}$ of real random variables, and the associated upcrossing number $M_{ab}^Y(n)$ is then an integer-valued random variable.

Doob's upcrossing inequality *If $(Y_n)_{n \in \mathbb{N}}$ is a supermartingale, then, for every $n \in \mathbb{N}$ and every $a < b$,*

$$E[M_{ab}^Y(n)] \leq \frac{1}{b-a} E[(Y_n - a)^-].$$

This inequality is a crucial tool for proving the convergence theorems for discrete-time martingales and supermartingales. Let us recall two important instances of these theorems.

Convergence theorem for discrete-time supermartingales *If $(Y_n)_{n \in \mathbb{N}}$ is a supermartingale, and if the sequence $(Y_n)_{n \in \mathbb{N}}$ is bounded in L^1, then there exists a random variable $Y_\infty \in L^1$ such that*

$$Y_n \xrightarrow[n \to \infty]{a.s.} Y_\infty.$$

Convergence theorem for uniformly integrable discrete-time martingales *Let $(Y_n)_{n \in \mathbb{N}}$ be a martingale. The following are equivalent:*

(i) *The martingale $(Y_n)_{n \in \mathbb{N}}$ is closed, in the sense that there exists a random variable $Z \in L^1(\Omega, \mathscr{F}, P)$ such that $Y_n = E[Z \mid \mathscr{G}_n]$ for every $n \in \mathbb{N}$.*
(ii) *The sequence $(Y_n)_{n \in \mathbb{N}}$ converges a.s. and in L^1.*
(iii) *The sequence $(Y_n)_{n \in \mathbb{N}}$ is uniformly integrable.*

If these properties hold, the a.s. limit of the sequence $(Y_n)_{n \in \mathbb{N}}$ is $Y_\infty = E[Z \mid \mathscr{G}_\infty]$.

We now recall two versions of the optional stopping theorem in discrete time. A (discrete) stopping time is a random variable T with values in $\mathbb{N} \cup \{\infty\}$, such that $\{T = n\} \in \mathscr{G}_n$ for every $n \in \mathbb{N}$. The σ-field of the past before T is then $\mathscr{G}_T = \{A \in \mathscr{G}_\infty : A \cap \{T = n\} \in \mathscr{G}_n, \text{ for every } n \in \mathbb{N}\}$.

Optional stopping theorem for uniformly integrable discrete-time martingales
Let $(Y_n)_{n\in\mathbb{N}}$ be a uniformly integrable martingale, and let Y_∞ be the a.s. limit of Y_n when $n \to \infty$. Then, for every choice of the stopping times S and T such that $S \leq T$, we have $Y_T \in L^1$ and

$$Y_S = E[Y_T \mid \mathscr{G}_S]$$

with the convention that $Y_T = Y_\infty$ on the event $\{T = \infty\}$, and similarly for Y_S.

Optional stopping theorem for discrete-time supermartingales (bounded case)
If $(Y_n)_{n\in\mathbb{N}}$ is a supermartingale, then, for every choice of the bounded stopping times S and T such that $S \leq T$, we have

$$Y_S \geq E[Y_T \mid \mathscr{G}_S].$$

We conclude with a variant of the convergence theorem for supermartingales in the backward case. We consider a backward filtration, that is, an increasing sequence of filtrations $(\mathscr{H}_n)_{n\in-\mathbb{N}}$ indexed by negative integers (in such a way that the σ-field $\mathscr{H}_n$ is "smaller and smaller" when $n \to -\infty$). A sequence $(Y_n)_{n\in-\mathbb{N}}$ of integrable random variables indexed by negative integers is called a backward supermartingale if, for every $n \in -\mathbb{N}$, Y_n is $\mathscr{H}_n$-measurable and, for every $m \leq n \leq 0$, $E[Y_n \mid \mathscr{H}_m] \leq Y_m$.

Convergence theorem for backward discrete-time supermartingales *If $(Y_n)_{n\in-\mathbb{N}}$ is a backward supermartingale, and if the sequence $(Y_n)_{n\in-\mathbb{N}}$ is bounded in L^1, then the sequence $(Y_n)_{n\in-\mathbb{N}}$ converges a.s. and in L^1 when $n \to -\infty$.*

It is crucial for the applications developed in Chap. 3 that the convergence also holds in L^1 in the backward case (compare with the analogous result in the "forward" case).

References

1. Adler, R.J.: An Introduction to Continuity, Extrema and Related Topics for General Gaussian Processes. Institute of Mathematical Statistics, Hayward (1990)
2. Bachelier, L.: Théorie de la spéculation. Ann. Sci. Ecole Norm. Sup. **17**, 21–86 (1900)
3. Bertoin, J.: Lévy Processes. Cambridge University Press, Cambridge (1996)
4. Blumenthal, R.M.: Excursions of Markov Processes. Birkhäuser, Boston (1992)
5. Blumenthal, R.M., Getoor, R.K.: Markov Processes and Potential Theory. Academic, New York (1968)
6. Burkholder, D.L.: Distribution function inequalities for martingales. Ann. Probab. **1**, 19–42 (1973)
7. Burkholder, D.L., Davis, B.J., Gundy, R.F.: Integral inequalities for convex functions of operators on martingales. In: Proceedings of Sixth Berkeley Symposium on Mathematical Statistics and Probability, vol. 2, pp. 223–240. University of Calfornia Press, Berkeley (1972)
8. Cameron, R.H., Martin, W.T.: Transformation of Wiener integrals under translations. Ann. Math. **45**, 386–396 (1944)
9. Chung, K.L.: Lectures from Markov Processes to Brownian Motion. Springer, Berlin (1982)
10. Chung, K.L., Williams, R.J.: Introduction to Stochastic Integration. Birkhäuser, Boston (1983)
11. Dambis, K.E.: On the decomposition of continuous martingales. Theor. Probab. Appl. **10**, 401–410 (1965)
12. Davis, B.J.: Brownian motion and analytic functions. Ann. Probab. **7**, 913–932 (1979)
13. Dellacherie, C., Meyer, P.-A.: Probabilités et potentiel, Chapitres I à IV. Hermann, Paris (1975)
14. Dellacherie, C., Meyer, P.-A.: Probabilités et potentiel, Chapitres V à VIII. Théorie des martingales. Hermann, Paris (1980)
15. Doob, J.L.: Stochastic Processes. Wiley, New York (1953)
16. Doob, J.L.: Classical Potential Theory and its Probabilistic Counterpart. Springer, Berlin (1984)
17. Dubins, L., Schwarz, G.: On continuous martingales. Proc. Nat. Acad. Sci. USA **53**, 913–916 (1965)
18. Durrett, R.: Brownian Motion and Martingales in Analysis. Wadsworth, Belmont (1984)
19. Dynkin, E.B.: Inhomogeneous strong Markov processes. Dokl. Akad. Nauk. SSSR **113**, 261–263 (1957)
20. Dynkin, E.B.: Theory of Markov Processes. Pergamon, Oxford (1960)
21. Dynkin, E.B.: Markov Processes (2 Volumes). Springer, Berlin (1965)
22. El Karoui, N.: Sur les montées des semimartingales. Astérisque **52–53**, 63–88 (1978)

© Springer International Publishing Switzerland 2016
J.-F. Le Gall, *Brownian Motion, Martingales, and Stochastic Calculus*,
Graduate Texts in Mathematics 274, DOI 10.1007/978-3-319-31089-3

23. El Karoui, N., Chaleyat-Maurel, M.: Un problème de réflexion et ses applications au remps local et aux équations différentielles stochastiques sur ℝ, cas continu. Astérisque **52–53**, 117–144 (1978)

24. Ethier, S.N., Kurtz, T.G.: Markov Processes: Characterization and Convergence. Wiley, New York (1986)

25. Fisk, D.L.: Quasimartingales. Trans. Am. Math. Soc. **120**, 369–389 (1965)

26. Friedman, A.: Stochastic Differential Equations and Applications, vol. 1. Academic Press, New York (1975)

27. Friedman, A.: Stochastic Differential Equations and Applications, vol. 2. Academic Press, New York (1976)

28. Getoor, R.K., Sharpe, M.J.: Conformal martingales. Invent. Math. **16**, 271–308 (1972)

29. Girsanov, I.V.: On transforming a certain class of stochastic processes by absolutely continuous substitution of measures. Theory Probab. Appl. **5**, 285–301 (1960)

30. Grimmett, G.R., Stirzaker, D.R.: Probability and Random Processes. Oxford University Press, Oxford (1992)

31. Harrison, J.M., Shepp, L.A.: On skew Brownian motion. Ann. Probab. **9**, 309–313 (1981)

32. Hunt, G.A.: Some theorems concerning Brownian motion. Trans. Am. Math. Soc. **81**, 294–319 (1956)

33. Hunt, G.A.: Markov processes and potentials. Ill. J. Math. **1**, 44–93 (1957)

34. Hunt, G.A.: Markov processes and potentials. Ill. J. Math.**2**, 151–213 (1957)

35. Ikeda, N., Watanabe, S.: Stochastic Differential Equations and Diffusion Processes. North-Holland, Amsterdam/New York (1981)

36. Itô, K.: Stochastic integral. Proc. Imp. Acad. Tokyo **20**, 518–524 (1944)

37. Itô, K.: On a stochastic integral equation. Proc. Imp. Acad. Tokyo **22**, 32–35 (1946)

38. Itô, K.: On stochastic differential equations. Mem. Am. Math. Soc. **4**, 1–51 (1951)

39. Itô, K.: Multiple Wiener integral. J. Math. Soc. Jpn. **3**, 157–169 (1951)

40. Itô, K.: On a formula concerning stochastic differentials. Nagoya Math. J. **3**, 55–65 (1951)

41. Itô, K.: Selected Papers. Edited and with an Introduction by S.R.S. Varadhan and D.W. Stroock. Springer, New York (1987)

42. Itô, K., McKean, H.P.: Diffusion Processes and their Sample Paths. Springer, Berlin (1965)

43. Itô, K., Watanabe, S.: Transformation of Markov processes by multiplicative functionals. Ann. Inst. Fourier **15**, 15–30 (1965)

44. Jacod, J., Shiryaev, A.: Limit Theorems for Stochastic Processes. Springer, Berlin (2003)

45. Kakutani, S.: On Brownian motion in n-space. Proc. Imp. Acad. Tokyo **20**, 648–652 (1944)

46. Kakutani, S.: Two-dimensional Brownian motion and harmoinc functions. Proc. Imp. Acad. Tokyo **20**, 706–714 (1944)

47. Kallenberg, O.: Foundations of Modern Probability. Springer, Berlin/Heidelberg/New York (2002)

48. Kallianpur, G., Robbins, H.: Ergodic property of Brownian motion process. Proc. Nat. Acad. Sci. USA **39**, 525–533 (1953)

49. Karatzas, I., Shreve, S.: Brownian Motion and Stochastic Calculus. Springer, Berlin (1987)

50. Kunita, H., Watanabe, S.: On square-integrable martingales. Nagoya J. Math. **36**, 1–26 (1967)

51. Le Gall, J.-F.: Applications du temps local aux équations différentielles stochastiques unidimensionnelles. In: Lectures Notes in Mathematics, vol. 986, pp. 15–31. Springer, Berlin (1983)

52. Le Gall, J.-F.: Some Properties of Planar Brownian Motion. Lecture Notes in Mathematics, vol. 1527, pp. 111–235. Springer, Berlin (1992)

53. Lévy, P.: Le mouvement brownien plan. Am. J. Math. **62**, 487–550 (1940)

54. Lévy, P.: Processus stochastiques et mouvement brownien. Gauthier-Villars, Paris (1948)

55. Lifshits, M.A.: Gaussian Random Functions. Kluwer Academic, Dordrecht (1995)

56. Marcus, M.B., Rosen, J.: Markov Processes, Gaussian Processes, and Local Times. Cambridge University Press, Cambridge (2006)

57. McKean, H.P.: Stochastic Integrals. Academic, New York (1969)

58. Meyer, P.-A.: A decomposition theorem for supermartingales. Ill. J. Math. **6**, 193–205 (1962)

59. Meyer, P.-A.: Processus de Markov. Lecture Notes in Mathamatics, vol. 26. Springer, Berlin (1967)

60. Meyer, P.-A.: Intégrales stochastiques. In: Lecture Notes Mathamatics, vol. 39, pp. 72–162. Springer, Berlin (1967)

61. Meyer, P.-A.: Un cours sur les intégrales stochastiques. In: Lecture Notes in Mathamatics, vol. 511, pp. 245–400. Springer, Berlin (1976)

62. Mörters, P., Peres, Y.: Brownian Motion. Cambridge University Press, Cambridge (2010)

63. Protter, P.: Stochastic Integration and Differential Equations. Springer, Berlin (2005)

64. Neveu, J.: Mathematical Foundations of the Calculus of Probability. Holden–Day, San Francisco (1965)

65. Neveu, J.: Discrete-Parameter Martingales. North-Holland, Amsterdam (1975)

66. Øksendal, B.: Stochastic Differential Equations. An Introduction with Applications. Springer, Berlin (2003)

67. Pitman, J.W., Yor, M.: A decomposition of Bessel bridges. Z. Wahrscheinlichkeitstheorie verw. Gebiete **59**, 425–457 (1982)

68. Pitman, J.W., Yor, M.: Asymptotic laws of planar Brownian motion. Ann. Probab. **14**, 733–779 (1986)

69. Port, S.C., Stone, C.J.: Brownian Motion and Classical Potential Theory. Academic Press, New York (1978)

70. Revuz, D., Yor, M.: Continuous Martingales and Brownian Motion. Springer, Berlin (1991)

71. Rogers, L.C.G., Williams, D.: Diffusions, Markov Processes and Martingales, vol. 1: Foundations. Wiley, New York (1994)

72. Rogers, L.C.G., Williams, D.: Diffusions, Markov Processes and Martingales, vol. 2: Itô calculus. Wiley, New York (1987)

73. Sharpe, M.J.: General Theory of Markov Processes. Academic, New York (1989)

74. Skorokhod, A.V.: Stochastic equations for diffusion processes in a bounded domain. Theory Probab. Appl. **6**, 264–274 (1961)

75. Spitzer, F.: Some theorems concerning two-dimensional Brownian motion. Trans. Am. Math. Soc. **87**, 187–197 (1958)

76. Stroock, D.W.: An Introduction to Markov Processes. Springer, Berlin (2005)

77. Stroock, D.W., Varadhan, S.R.S.: Multidimensional Diffusion Processes. Springer, Berlin (1979)

78. Talagrand, M.: The Generic Chaining. Springer, Berlin (2005)

79. Tanaka, H.: Note on continuous additive functionals of the 1-dimensional Brownian path. Z. Wahrscheinlichkeitstheorie verw. Gebiete **1**, 251–257 (1963)

80. Trotter, H.F.: A property of Brownian motion paths. Ill. J. Math. **2**, 425–433 (1958)

81. Wiener, N.: Differential space. J. Math. Phys. **2**, 132–174 (1923)

82. Williams, D.: Probability with Martingales. Cambridge University Press, Cambridge (1991)

83. Yamada, T., Watanabe, S.: On the uniqueness of solutions of stochastic differential equations. J. Math. Kyoto Univ. **11**, 155–167 (1971)

84. Yor, M.: Sur la continuité des temps locaux associés à certaines semimartingales. Astérisque **52–53**, 23–36 (1978)

Index

© Springer International Publishing Switzerland 2016
J.-F. Le Gall, *Brownian Motion, Martingales, and Stochastic Calculus*,
Graduate Texts in Mathematics 274, DOI 10.1007/978-3-319-31089-3

Printed in the United States
by Bookmasters

Printed in the United States
By Bookmasters